Advances in Water Resource Monitoring and Modelling: Water Quantity and Quality Issues

Advances in Water Resource Monitoring and Modelling: Water Quantity and Quality Issues

Editor

Diego Copetti

MDPI • Basel • Beijing • Wuhan • Barcelona • Belgrade • Manchester • Tokyo • Cluj • Tianjin

Editor
Diego Copetti
National Research Council of
Italy
Brugherio, Italy

Editorial Office
MDPI
St. Alban-Anlage 66
4052 Basel, Switzerland

This is a reprint of articles from the Special Issue published online in the open access journal *Resources* (ISSN 2079-9276) (available at: https://www.mdpi.com/journal/resources/special_issues/water_monitoring_modelling).

For citation purposes, cite each article independently as indicated on the article page online and as indicated below:

LastName, A.A.; LastName, B.B.; LastName, C.C. Article Title. *Journal Name* **Year**, *Volume Number*, Page Range.

ISBN 978-3-0365-8210-8 (Hbk)
ISBN 978-3-0365-8211-5 (PDF)

Cover image courtesy of Diego Copetti
The picture represents the Sainte-Croix Reservoir, Verdon region (France)

Contents

About the Editor

Diego Copetti

Diego Copetti is a researcher at the National Research Council of Italy, Water Research Institute (CNR-IRSA), whose research interests mainly include lake ecology. He is interested in how these environments respond to local (e.g., nutrient loads) and global (e.g., climate change) forcings. During his research career, he used both experimental and modelling approaches. He is interested in the water cycle in urban agglomerations as well as the impact that they have on surface aquatic ecosystems. He likes photography and the study of history and philosophy of science.

resources

Editorial

Integration of Water Quantity/Quality Needs with Socio-Economical Issues: A Focus on Monitoring and Modelling

Diego Copetti

Unit of Brugherio, National Research Council of Italy—Water Research Institute (CNR-IRSA), Via del Mulino, 19, 20861 Brugherio, Italy; diego.copetti@irsa.cnr.it; Tel.: +39-03921694215

1. Introduction

Freshwater is essential for a multitude of different uses, including drinking supply, irrigation, and energy production [1]. Healthy freshwater ecosystems are also necessary for the protection of biodiversity [2]. A global water crisis is currently affecting our planet [3], with impacts even in regions considered rich in water, such as the subalpine areas north and south of the Alps [4,5]. Climate change is also exerting direct and indirect impacts on both water availability and biogeochemical cycles [6,7].

The relationship between water quantity and quality is, thus, becoming increasingly stringent, particularly when high quality standards are required, such as for drinking and sanitary uses, with consistent economic consequences [8]. Therefore, our ability to monitor and model both the water availability and its quality assumes a fundamental strategic significance in the management of the water resource.

This Special Issue hosts twelve papers exploring advances in the monitoring and modelling of both water quality and quantity. Most of the contributions (10) are research papers that cover a wide range of topics including hydrology, environmental economy, and ecotoxicology. The remaining two contributions are review papers dealing with the development of circular economy approaches in the recycle of phosphorus and the monitoring of the water resource using remote sensing techniques.

2. Water Resource Monitoring and Modelling, Insights from the Publications Included in the Special Issue

Wurl et al. [9] studied the impact of seawater intrusion due to overexploitation of groundwater in the Los Planes aquifer in the Baja California Peninsula (Mexico). The study showed a general increase in mineralization from 2014 to 2016, except for the northwestern part of the bay affected by thermal water inflow, with high mineralization, through the El Sargento fault. This thermal water is further mixed with seawater, resulting in higher mineralization. The study highlighted the importance of considering hydrothermal activities when interpreting hydrochemical data to address management strategies.

Somorowska [10] analyzed changes in cold season temperature from 1951 to 2021 in the Liwiec River basin (Poland) and found that increasing temperature significantly impacted snowfall, rainfall, and baseflow metrics. The snowfall-to-precipitation ratio decreased by 16% and the baseflow index rose by 18%, resulting in a shift from snow-dominated to snow-affected river regime. The study provided evidence of a gradual temperature increase over the last 71 years. The results may be of interest for middle-latitude regions.

Graf and Vyshnevskyi [11] focused their attention on river-flow forecasts in Ukraine through direct multistep-ahead forecasting using an XGBoost (extreme gradient boosting) model, to estimate long-term changes and predict monthly flows of selected rivers. A single multi-output model was proposed to forecast short- and long-term scenarios. Three forecast stages were considered using measurements of monthly flows, precipitation, and

Citation: Copetti, D. Integration of Water Quantity/Quality Needs with Socio-Economical Issues: A Focus on Monitoring and Modelling. *Resources* **2023**, *12*, 60. https://doi.org/10.3390/resources12050060

Received: 4 May 2023
Accepted: 15 May 2023
Published: 18 May 2023

air temperature in the period 1961–2020. Responses of different hydrological systems showed varying degrees of sensitivity to changes in precipitation and air temperature with different projections for future time horizons. The results may be of interest in other catchments, irrespective of their geographic location.

Graf et al. [12] developed two machine learning models to predict ice phenomena in the Warta River (Poland) using observational data from eight river gauges. The models, MLPNN (perceptron neural network) and XGBoost, provided promising results for the forecasting of ice phenomena. Most important predictors were the nature of phenomena on the day before an observation, as well as water and air temperatures, while river flow and water level were less important. These findings could be useful for predicting ice phenomena in other regions characterized by similar boundary conditions.

Somorowska [13] evaluated the trends in terrestrial evaporation (ET) across Poland using data from 1980 to 2020. The study aimed to assess annual and monthly ET trends and revealed significant changes of this variable over different seasons and regions. The results indicated that rising temperatures and small increases in precipitation led to an increase in ET during the summer months. The study found that in the period 2007–2020, annual ET increased by 7% compared to the reference period 1980–2006. These findings have important implications for water resources management in Poland and potentially in other in Eastern European countries.

Smol and Koneczna [14] discussed the importance of water and wastewater management in the transition towards a circular economy in the European Union and suggested a set of economic indicators to assess progress in this sector. These indicators are grouped under actions such as reduction, reclamation, reuse, recycling, recovery, and landfilling, and can be used by water supply and sewage companies to assess their progress towards a circular economy. The proposed set of indicators is flexible and can be adapted to maintain effectiveness throughout the transition period.

Sakai Cid et al. [15] addressed antiretrovirals (ARVs) in coastal waters. These pharmaceuticals are present in aquatic ecosystems worldwide and can have ecotoxicological effects on marine aquatic organisms. This study predicted environmental concentration (PEC) of 13 ARVs in the Santos Bay (Brazil) and found that all exceeded the European Medicines Agency's guideline limits. Three ARVs were then selected for acute and chronic toxicity tests with sea urchins (*Echinometra lucunter*) and showed potential hazards for aquatic life in this bay. The study highlighted the need for both specific ARVs monitoring and stringent policies to reduce the introduction of ARVs into the aquatic environment.

Szatten et al. [16] discussed the impact of human induced hydrologic alteration of river systems. The research focused on the Lower Brda river cascade dams (Poland) and used hydrological and water quality data from 1984 to 2017 and the Indicators of Hydrologic Alteration method to show how changes in regime operation affect sediment and nutrient balance. The study found that sustainable management of sediments and nutrients in the altered catchment can help in achieving a good river ecological status, suggesting important management insights.

Boryczko et al. [17] discussed the importance of a proactive strategy to prevent failures on water supply. The article presented the results of a simulation (using the EPANET 2.0 software developed by the United States Environmental Protection Agency) aiming at estimating the consequences of failures in a water supply system in Poland. Simulation results were used to create a water supply risk map. The study showed that the highest risk was related to the failure of the main pipe. Recommendations include attention on the system modernization and maintenance. The results of the study may be of interest for Eastern European countries.

Tchórzewska-Cieślak et al. [18] discussed the importance of investing in a water supply system (WSS) to provide high-quality water to residents. The paper presented a methodology to analyze consumers' willingness to accept additional costs for improving the operational safety of the WSS. The study was based on a regional WSS located in the

Poland's Podkarpacie province. The assessment suggested that consumers are willing to pay additional costs for better water supply services.

Smol et al. [19] highlighted the importance of phosphorus raw materials in the Visegrad Group countries (Poland, Slovakia, Czech Republic, and Hungary). These countries lack mineral deposits of phosphate rock. The authors analyzed the structure of import and export of phosphorus raw materials and revised primary and secondary phosphorus sources, suggesting their use in agricultural systems, and showcased examples of good phosphorus recovery practices in the Visegrad countries. Finally, the authors indicated that the proposed approach could ensure the safety of food production in the region and contribute to faster independence from phosphorus raw material imports.

Water quality monitoring is important to manage aquatic ecosystems and cope with inland water degradation. Remote sensing, particularly using multispectral and hyperspectral sensors, is recognized as an essential technique for water quality monitoring. Bresciani et al. [20] carried out a review focused on the use of data gathered from currently orbiting hyperspectral sensors (i.e., PRISMA and DESIS) to retrieve water quality parameters in various aquatic ecosystems, including deep clear lakes and river dammed reservoirs. The examples were from case studies in northern, central, and southern Italy.

3. Conclusions

The demand for fresh water of good quality is continuously increasing worldwide for reasons linked to both the development of human society and the onset of climate change. Our hope is that the articles published in this Special Issue could make a substantial and innovative contribution to the monitoring and modelling of this precious non-renewable resource.

Acknowledgments: The Guest Editor thanks the editorial team of Resources.

Conflicts of Interest: The authors declare no conflict of interest.

References

1. UN Water (Ed.) *Water for a Sustainable World*; The United Nations World Water Development Report; UNESCO: Paris, France, 2015; ISBN 978-92-3-100071-3.
2. Tickner, D.; Opperman, J.J.; Abell, R.; Acreman, M.; Arthington, A.H.; Bunn, S.E.; Cooke, S.J.; Dalton, J.; Darwall, W.; Edwards, G.; et al. Bending the Curve of Global Freshwater Biodiversity Loss: An Emergency Recovery Plan. *BioScience* **2020**, *70*, 330–342. [CrossRef] [PubMed]
3. UN Water (Ed.) *Water and Climate Change*; The United Nations World Water Development Report; UNESCO: Paris, France, 2020; ISBN 978-92-3-100371-4.
4. Dresti, C.; Fenocchi, A.; Copetti, D. Modelling Physical and Ecological Processes in Medium-to-Large Deep European Perialpine Lakes: A Review. *J. Limnol.* **2021**, *80*. [CrossRef]
5. Brunner, M.I.; Björnsen Gurung, A.; Zappa, M.; Zekollari, H.; Farinotti, D.; Stähli, M. Present and Future Water Scarcity in Switzerland: Potential for Alleviation through Reservoirs and Lakes. *Sci. Total Environ.* **2019**, *666*, 1033–1047. [CrossRef]
6. Mosley, L.M. Drought Impacts on the Water Quality of Freshwater Systems; Review and Integration. *Earth-Sci. Rev.* **2015**, *140*, 203–214. [CrossRef]
7. Copetti, D.; Carniato, L.; Crise, A.; Guyennon, N.; Palmeri, L.; Pisacane, G.; Struglia, M.; Tartari, G. Impacts of Climate Change on Water Quality. In *Regional Assessment of Climate Change in the Mediterranean*; Navarra, A., Tubiana, L., Eds.; Advances in Global Change Research; Springer: Berlin, Germany, 2013; Volume 50, pp. 307–332, ISBN 978-94-007-5780-6.
8. Gosling, S.N.; Arnell, N.W. A Global Assessment of the Impact of Climate Change on Water Scarcity. *Clim. Change* **2016**, *134*, 371–385. [CrossRef]
9. Wurl, J.; Imaz-Lamadrid, M.A.; Mendez-Rodriguez, L.C.; Hernández-Morales, P. Hydrochemical Indicator Analysis of Seawater Intrusion into Coastal Aquifers of Semiarid Areas. *Resources* **2023**, *12*, 47. [CrossRef]
10. Somorowska, U. Warming Air Temperature Impacts Snowfall Patterns and Increases Cold-Season Baseflow in the Liwiec River Basin (Poland) of the Central European Lowland. *Resources* **2023**, *12*, 18. [CrossRef]
11. Graf, R.; Vyshnevskyi, V. Forecasting Monthly River Flows in Ukraine under Different Climatic Conditions. *Resources* **2022**, *11*, 111. [CrossRef]
12. Graf, R.; Kolerski, T.; Zhu, S. Predicting Ice Phenomena in a River Using the Artificial Neural Network and Extreme Gradient Boosting. *Resources* **2022**, *11*, 12. [CrossRef]

13. Somorowska, U. Changes in Terrestrial Evaporation across Poland over the Past Four Decades Dominated by Increases in Summer Months. *Resources* **2022**, *11*, 6. [CrossRef]
14. Smol, M.; Koneczna, R. Economic Indicators in Water and Wastewater Sector Contributing to a Circular Economy (CE). *Resources* **2021**, *10*, 129. [CrossRef]
15. Cid, R.S.; Roveri, V.; Vidal, D.G.; Dinis, M.A.P.; Cortez, F.S.; Salgueiro, F.R.; Toma, W.; Cesar, A.; Guimarães, L.L. Toxicity of Antiretrovirals on the Sea Urchin Echinometra Lucunter and Its Predicted Environmental Concentration in Seawater from Santos Bay (Brazilian Coastal Zone). *Resources* **2021**, *10*, 114. [CrossRef]
16. Szatten, D.; Habel, M.; Babiński, Z. Influence of Hydrologic Alteration on Sediment, Dissolved Load and Nutrient Downstream Transfer Continuity in a River: Example Lower Brda River Cascade Dams (Poland). *Resources* **2021**, *10*, 70. [CrossRef]
17. Boryczko, K.; Piegdoń, I.; Szpak, D.; Żywiec, J. Risk Assessment of Lack of Water Supply Using the Hydraulic Model of the Water Supply. *Resources* **2021**, *10*, 43. [CrossRef]
18. Tchórzewska-Cieślak, B.; Pietrucha-Urbanik, K.; Kuliczkowska, E. An Approach to Analysing Water Consumers' Acceptance of Risk-Reduction Costs. *Resources* **2020**, *9*, 132. [CrossRef]
19. Smol, M.; Marcinek, P.; Šimková, Z.; Bakalár, T.; Hemzal, M.; Klemeš, J.J.; Fan, Y.V.; Lorencz, K.; Koda, E.; Podlasek, A. Inventory of Good Practices of Sustainable and Circular Phosphorus Management in the Visegrad Group (V4). *Resources* **2023**, *12*, 2. [CrossRef]
20. Bresciani, M.; Giardino, C.; Fabbretto, A.; Pellegrino, A.; Mangano, S.; Free, G.; Pinardi, M. Application of New Hyperspectral Sensors in the Remote Sensing of Aquatic Ecosystem Health: Exploiting PRISMA and DESIS for Four Italian Lakes. *Resources* **2022**, *11*, 8. [CrossRef]

resources

MDPI

Article

Hydrochemical Indicator Analysis of Seawater Intrusion into Coastal Aquifers of Semiarid Areas

Jobst Wurl [1,*], Miguel Angel Imaz-Lamadrid [2], Lía Celina Mendez-Rodriguez [3] and Pablo Hernández-Morales [4]

1 Departamento Académico de Ciencias de la Tierra, Universidad Autónoma de Baja California Sur, Carretera al Sur km 5.5, La Paz 23080, Mexico
2 Departamento Académico de Ingeniería en Pesquerías, Universidad Autónoma de Baja California Sur, Carretera al Sur km 5.5, La Paz 23080, Mexico
3 Centro de Investigaciones Biológicas del Noroeste S.C (CIBNOR), Calle IPN 195, La Paz 23096, Mexico
4 Departamento Académico de Ciencias Marinas y Costeras, Universidad Autónoma de Baja California Sur, Carretera al Sur km 5.5, La Paz 23080, Mexico
* Correspondence: jwurl@uabcs.mx

Abstract: Saltwater intrusion into groundwater systems is a problem worldwide and is induced mainly by human activities, such as groundwater overexploitation and climate change. The coastal Los Planes aquifer in the southern part of the Baja California Peninsula (Mexico) is affected by seawater intrusion due to more than 40 years of groundwater overexploitation. A dataset of 55 samples was compiled, including 18 samples from our campaigns between 2014 and 2016. Several methods exist to define the impact of seawater in a coastal aquifer, such as the "seawater fraction", the "Chloro-Alkaline Indices", the "Hydrochemical Facies Evolution Diagram", and the "Saltwater Mixing Index". These methods provide reasonable results for most of the coastal zone of the Los Planes aquifer. A slight increase in mineralization was observed from 2014 to 2016 compared with the situation in 2003. However, in its northwestern part, samples from hydrothermal wells were not recognized by these methods. Here, the aquifer is affected mainly by thermal water with elevated mineralization, introduced through the El Sargento fault, a main fault, which cuts through the study area in the north–south direction. By considering known hydrothermal manifestations in the interpretation, samples could be classified as a combination of four end-members: fresh groundwater, seawater, and the composition of two types of thermal water. One thermal endmember with very low mineralization coincides with the thermal water described from the Los Cabos Block, where meteoric water represents the source (found in the Sierra la Laguna). The second endmember is comparable to coastal thermal manifestations where seawater represents the main source. Therefore, the higher mineralization in the northwestern zone is the result of the mobilization of thermal groundwater and direct mixing with seawater, which is introduced locally at the coast due to overextraction. This finding is important for future management strategies of the aquifer.

Keywords: hydrogeochemical characterization; seawater intrusion processes; geothermal water; mixing processes; cation exchange; Los Cabos block

Citation: Wurl, J.; Imaz-Lamadrid, M.A.; Mendez-Rodriguez, L.C.; Hernández-Morales, P. Hydrochemical Indicator Analysis of Seawater Intrusion into Coastal Aquifers of Semiarid Areas. *Resources* **2023**, *12*, 47. https://doi.org/10.3390/resources12040047

Academic Editor: Diego Copetti

Received: 24 January 2023
Revised: 26 March 2023
Accepted: 27 March 2023
Published: 7 April 2023

1. Introduction

Coastal aquifers are an important source of drinking water and water for agriculture and industry, especially in arid regions; however, these resources are significantly threatened by salinization due to seawater intrusion [1]. Seawater intrusion is one of the most severe environmental problems in coastal aquifers worldwide, exacerbated by the overexploitation of freshwater in coastal zones, and is susceptible to the influences of changing climates [2–4].

Mexico has experienced an important growth in population and agriculture surfaces in coastal areas over the last few years, which has led to the overexploitation of several aquifer systems [5]. In 2018, 105 aquifers were classified as overexploited, and 50 had salinization problems [6]. In 2020, the number of overexploited aquifers increased to 175 [7]. The problems of seawater intrusion are most notable in northwest Mexico [5] because of overexploitation and reduced annual precipitation, with an average of 200 mm in Baja California Sur (BCS) [6].

In the southern portion of the Baja California Peninsula, the salinization of aquifers has been related to agriculture, pollution, and seawater intrusion [8–12] without considering other phenomena such as hydrothermalism as a source of salinization. After 50 years of geothermal resource exploration in Mexico, it can be concluded that the peninsula has several regions with high potential [13]. Seven geothermal areas have been identified in the southern portion of the Baja California Peninsula. Two of these areas are related to recent volcanic activity, two are associated with the spreading tectonic zone in the middle of the Gulf of California, and the remaining three are due to movement through deep faults; the Los Planes aquifer (Figure 1) belongs to the last type [13]. In the Los Cabos region, four wells (SL1, SL2, SL4, SL5) were originally bored to obtain saltwater for a desalination plant project, but they resulted in the production of thermal water with an estimated deep reservoir temperature of 200 °C [13].

Our study area, the coastal Los Planes aquifer in the southern part of the Baja California Peninsula (Figure 1), is affected by seawater intrusion as a result of more than 40 years of groundwater overexploitation, as stated in former investigations (see Table 1) [14]. This is reasonable for most of the coastal zone, but it cannot be the explanation for the northwestern part of the aquifer because the water table and resulting flow lines indicate that this zone is still under an effluent flow regimen [14,15]. According to Del Rosal-Pardo, Busch et al., and Coyan et al. [16–18], at least four main faults with normal displacement and a north–south strike direction are affecting the Los Planes aquifer. These faults are related to hydrothermalism and may represent an additional source of salinity in the aquifer (Figure 1).

Table 1. Water balance in the aquifer between 2003 and 2020, after CONAGUA [7].

Year	Extraction Volume Million m^3/Year	Recharge Million m^3/Year	Deficit Million m^3/Year
2003	12.29	9.89	−2.40
2007	12.29	8.40	−3.89
2013	12.29	8.26	−4.03
2015	12.29	8.40	−3.89
2020	13.10	8.40	−4.70

The purpose of this research is to assess the impact of seawater intrusion on the groundwater quality of a coastal aquifer in 2003 and in 2016 and to evaluate the extent to which hydrothermalism contributes to salinization. Different methods of analyzing groundwater affected by seawater intrusion and/or hydrothermal water are tested to determine how well they can distinguish between the two sources of salinization. Additionally, different types of hydrothermal water can be identified, and their corresponding mixing proportions with seawater can be calculated.

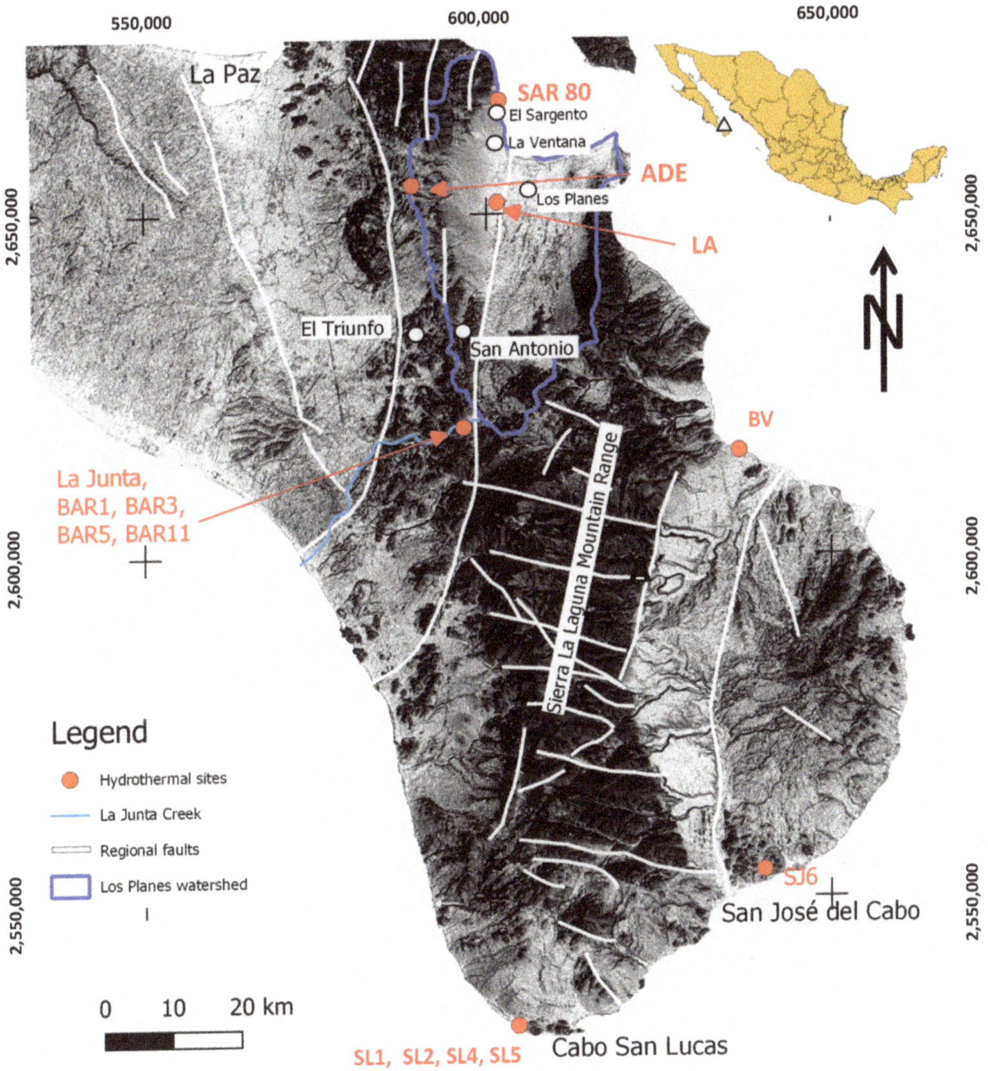

Figure 1. The southern tip of the Baja California Peninsula with main faults and important thermal manifestations, as described in the article. Surface relief was obtained from a 30 m DEM from NASA [19].

2. Regional Setting

The study was conducted in the southwestern portion of the State of BCS (Figure 1), within the Municipality of La Paz, in an area of 230 km^2 (Figure 2). The area is located at a distance of 50 km southeast of the city of La Paz. The Sierra de la Laguna is the main high mountain range in this region, with a maximum altitude of 2200 m. The mountains are formed by a massif crystalline basement, called the Los Cabos block [20], which is limited to the east by the San José del Cabo fault with a topographic escarpment above 1000 m [21,22]. The study area includes the San Juan de Los Planes basin (SJPB).

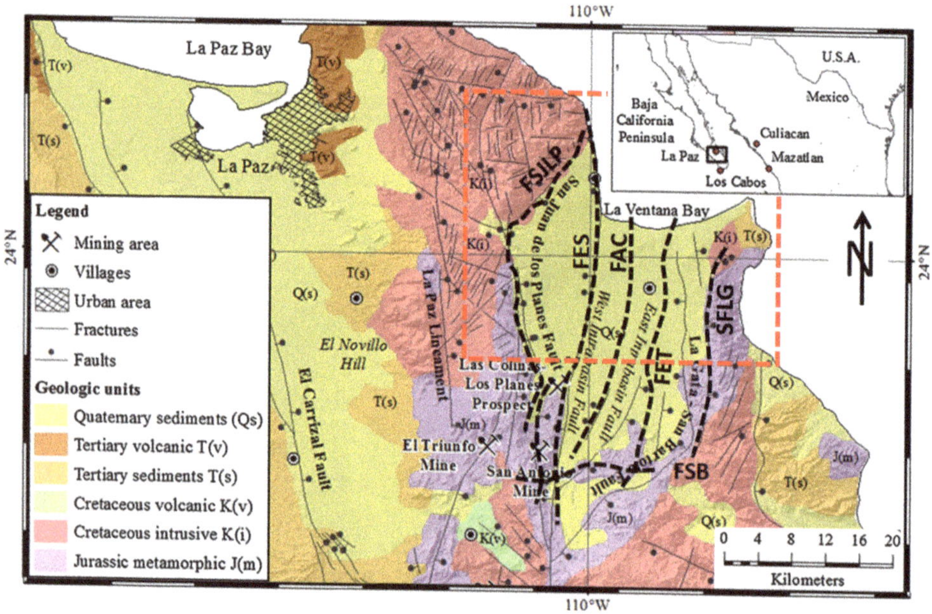

Figure 2. Geological map of the study area (dotted red line) with the main faults in the Los Planes basin: San Juan de Los Planes fault (FSJLP); El Sargento fault (FES); Agua Caliente fault (FAC); Tecuán fault (FET); La Gata fault (SFLG); and San Bartolo fault (FSB) [16–18].

2.1. Climate

CONAGUA [23] describes the climate as follows. According to the climate classification criteria proposed by Köppen, the prevailing climate in Los Planes is very dry (type BW) presenting most rain in summer and less in winter. The average annual temperature is between 22 °C and 24 °C. The highest monthly average temperatures occur in the summer, from June to September, varying from 27.3 °C to 29.8 °C; the absolute maximum temperatures recorded are from 41 °C to 45.5 °C. The coldest temperatures occur from December to February, with monthly mean values of 16.7 °C to 18.2 °C. The average annual rainfall for the basin is 281 mm and is affected by tropical cyclones during the rainy season (from May to November) [24,25]. The average rainfall varies from 175 mm/year in the coastal plain to 450 mm/year in the mountain range. The potential evaporation reaches up to 2000 mm/year. Tropical cyclones occur frequently during the rainy season (most likely between July and October).

2.2. Geological Setting

The Los Cabos Block, a batholithic massif of granitic and granodioritic rocks from the Cretaceous, appears as a mountainous complex that intrudes into Jurassic heterogeneous meta-sediments [22]. The block is partly covered by Miocene volcanic and volcaniclastic rocks from the Comondú Formation, which predates the opening of the Gulf of California [22]. The eastern part of the study area is formed by the Los Planes (SJLP) basin, a wide tectonic depression (pit) of elongated shape, with a preferential direction N-S that covers an area of 790 km^2. The basin represents an active graben structure, limited to the west by the San Juan de Los Planes normal fault (FSJLP) and to the east by the San Bartolo–La Gata normal fault (FSLG) (Figure 2). Additional north–south faulting is affecting the SJPB: El Sargento fault (FES); Agua Caliente fault (FAC); and Tecuán fault (FET) [16–18,26]. Busch et al. [17] verified the existence of the west-dipping intra-basin faults based on gravity sections and estimated that the FSJLP and FES reach depths of more than 3 km. Another fault, the so-called San Bartolo fault (FSB), with a preferential E-W direction, is in the southern part of the Los Planes basin.

The San Juan de Los Planes (SJLP) basin is limited to the south by the Sierra de San Antonio, which is formed by granodiorites, granites, and metasedimentary rocks (tonalite, quartz diorite, quartz monzonite, gabbro, aplite andesite, and rhyolite dams). The Sierra La Gata forms the eastern limit of the Los Planes basin with a north–south orientation and is composed predominantly of intrusive rocks (granites, granodiorites, diorites, and tonalities) and metasedimentary rocks (originating from shale and sandstone and, to a lesser extent, graywacke and limestone) with a high degree of metamorphism [27]. The filling material is constituted by unconsolidated sediments of the Tertiary and Quaternary ages, consisting of fluvial sediments and alluvial deposits, mainly composed of medium- to coarse-grained sands, and dunes; the depth of the porous medium is variable within 30 to 250 m [28].

2.3. Hydrogeological Framework

The SJLP basin forms a semi-confined regional aquifer. It consists of unconsolidated granular sediments that cover an impermeable granitic basement [14]. The SJLP basin, which represents the lower part of the study area, is filled with sediments that consist of Quaternary alluvial deposits, mainly composed of medium- to coarse-grained sands (Figures 2 and 3). The hydraulic conductivity varies from 5 to 50 m/day. The aquifer is only occasionally recharged by the infiltration of rainwater and stream contributions [14]. The streams have an intermittent regime and infiltrate the alluvial plain; on occasion, the runoff reaches the Gulf of California.

The position of the aquifer basement and main structural elements are known mainly from geoelectrical resistivity and gravity sections [14,17]. Near the village of San Juan de Los Planes, a depth of more than 103 m is verified by four drillings; the base of the sandy aquifer consists of a 20 m layer of sandy silt [29] (Figure 3). CNA [14] reported an average hydraulic conductivity of 4.1 m/day, obtained from nineteen pumping tests (maximum value of 11 m/day and minimum value of 0.6 m/day); storage coefficients of 0.11 and 2×10^{-3} were obtained at two sites, indicating unconfined to semi-confined conditions.

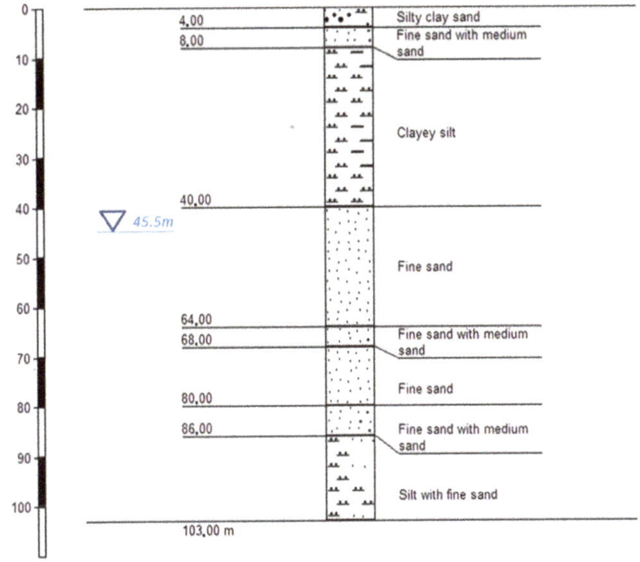

Figure 3. Typical sediments of the Los Planes aquifer in the center of the Los Planes aquifer (near Los Planes village; see Figure 4A), after TMI [29].

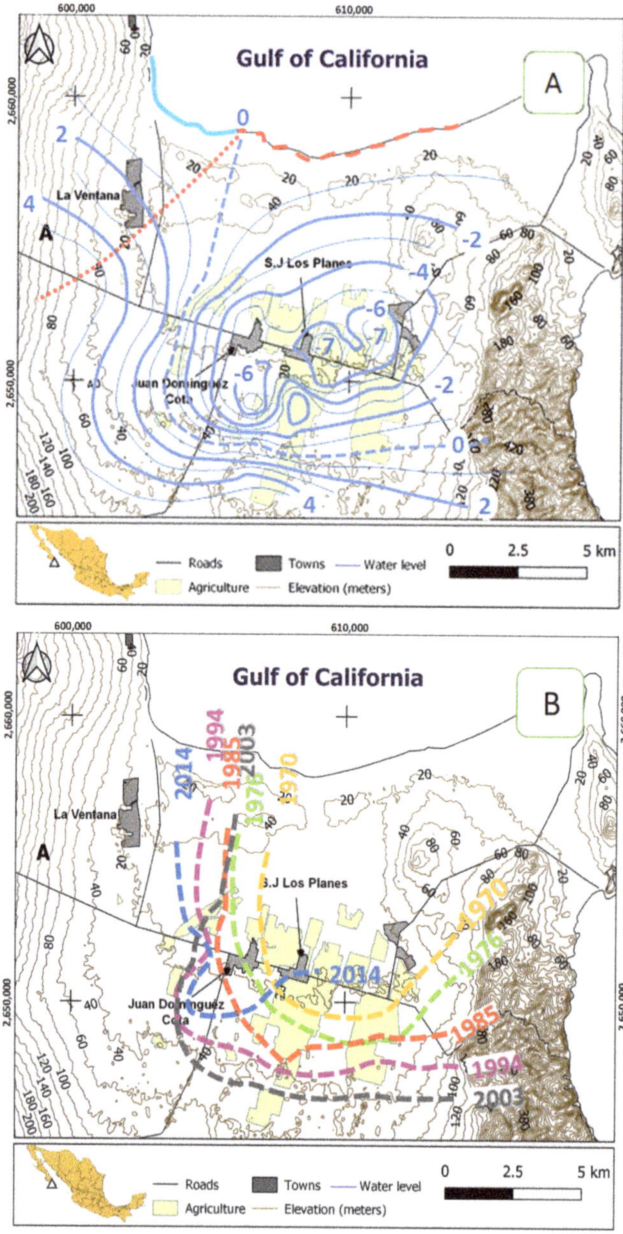

Figure 4. (**A**) The water table of the San Juan de Los Planes aquifer, after [14]. The dotted red line indicates the limit between the northeastern zone with groundwater flow toward the sea and the long dashed red line indicates the coastline with seawater intrusion (the red star indicates the position of a lithological column in Figure 3). (**B**) Variation of the 0 masl contour line of the water table between 1970 and 2014 [14,15].

The overexploitation of the aquifer started in the 1970s and caused the water table to lower by more than seven meters below sea level in the center of farmland near the village of San Juan de Los Planes (Figure 4; [14]). This changed the hydraulic gradient near the coast, causing the intrusion of seawater at a length of 7 to 12 km inland [14] (Figure 4). All official balance calculations, elaborated by CONAGUA over the last 20 years, resulted in negative volumes per year due to overexploitation (Table 1).

Due to elevated groundwater extractions, a cone of depression formed (documented first in 1970), which deepened in the following years. In 2003, the lowest water levels were observed in the agricultural areas near the villages of Los Planes and Juan Cota Domínguez, reaching a minimum value of −7 m with respect to the sea level, consequently causing seawater intrusion. The 0 m contour line moved southward, with a maximum distance of 11 km, to the coastline in 2003 (Figure 4).

As is visible in Figure 4, an ongoing process of saltwater intrusion and water table lowering is notable from 1970 to 2003, followed by a phase of increasing water levels between 2003 and 2014; meanwhile, the northwestern part (near La Ventana) presented decreasing water levels between from 1970 to 2014.

2.4. Geothermal Framework

The hydrothermal activity on the peninsula is linked to regional normal striking faults that allow for the deep penetration of water in areas of high heat flow; these faults are related to the extensional tectonics of the Gulf of California in the Tertiary [13,30,31]. Since the installation of the geothermal power plant Tres Virgenes in the northern part of the peninsula in 1997, many new sites with geothermal manifestations on the peninsula of Baja California have been discovered. A compilation of the main sites was presented by Arango-Galván et al. [13], who defined 14 geothermal areas and described their characteristics.

In the western part of the Los Planes basin, several geothermal wells and springs are situated along the coastline in a north–south direction. These thermal manifestations follow the north–south striking El Sargento fault (FES). About one kilometer north of the village of El Sargento, a geothermal anomaly is situated at the beach called "Agua Caliente". A water temperature of 82 °C was recorded in April 2016 at this beach at a 70 cm depth (Figure 5). Four wells are situated along the El Sargento fault (FES), and one is close to the Los Planes fault (Figure 5).

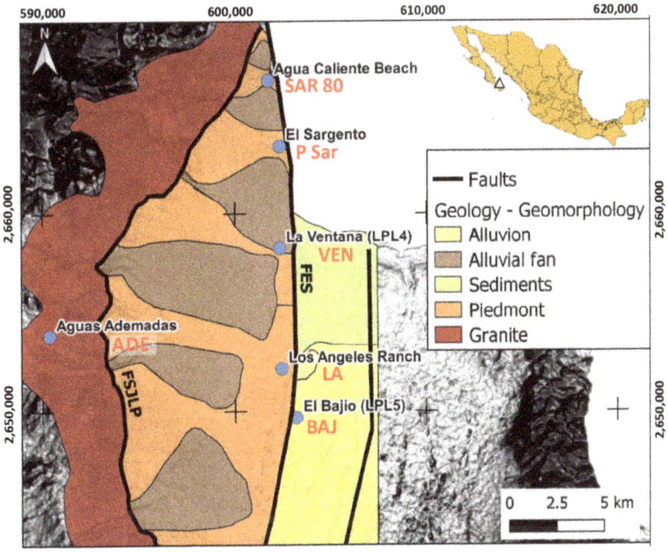

Figure 5. Geologic map of the northwestern part of the Los Planes aquifer with the main faults (El Sargento fault (FES) and San Juan de Los Planes fault (FSJPL)) and related geothermal manifestations (blue points).

2.5. Agriculture and Industrial Activities

Agriculture represents the main economic activity in the Los Planes watershed, with a total irrigated surface of 820 ha [14]. Chile represents the most important crop, taking up 61% of the irrigated surface [14]. Only 16.5 percent of the irrigated surface is used during the summer season, mostly for corn production. The total water volume for irrigation is 6.5 million m^3, with a resulting irrigation return flow of 16.6% (Table 2).

Table 2. List of the main crop types; corresponding irrigation systems, irrigated surfaces, and water volumes; and the resulting return flows [14].

Season	Crop	Irrigation System	Irrigated Surface (ha)	Net Volume Used (m^3)	Irrigation Return (m^3/Year)
Winter	Chile	Drip	160	1,200,000	48,000
	Chile	Flood gates	340	2,550,000	484,500
	Tomato	Drip	60	462,000	18,480
	Corn	Irrigation channel	65	286,000	111,540
	Cotton	Flood gates	30	210,000	39,900
	Cucumber	Drip	50	200,000	8000
	Beans	Aspersion irrigation	50	190,000	26,600
	Subtotal		*755*	*5,098,000*	*737,020*
Summer	Corn	Irrigation channel	135	594,000	231,660
	Subtotal		*135*	*594,000*	*231,660*
Perennial	Alfalfa	Aspersion irrigation	50	600,000	84,000
	Fruit trees	Aspersion irrigation	15	180,000	25,200
	Subtotal		*65*	*780,000*	*109,200*
	Total		**955**	**6,472,000**	**1,077,880**

At present, there is no industrial activity of importance going on, but the western part of the Los Planes aquifer is affected by historical gold mining. This activity took place in the region between 1878 and 1911, including the area around the village of San Antonio. Carrillo [32] described three types of ore deposits in the San Antonio–El Triunfo mining district, where epithermal veins represent the main type, containing high sulfide concentrations associated with gold and silver (gold associated with arsenopyrite). The ore processing techniques employed included cyanide-based milling, gold ore roasting, and amalgamation, among others [33]. An estimated 800,000 to 1,000,000 tons of mine waste materials were scattered in an area of approximately 350–400 km^2, which contained different byproducts resulting from arsenopyrite oxidation, mainly arsenolite (As_2O_3) [34]. Therefore, the region shows contamination problems with high concentrations of As, Cd, Pb, and Zn in the mine waste [35,36]. The contamination was widely distributed, mainly by winds and runoff from tropical storms, so the contamination was distributed in arroyo sediments at distances of up to 18 km.

3. Materials and Methods

3.1. Water Samples Documented in Former Studies (for Water Quality and Seawater Intrusion Estimation)

A compilation of own and published data was elaborated to describe the seawater intrusion status in 2003 and from 2014 to 2016 (documented in the Supplementary Materials File).

The CNA (2003) [14] water sampling campaign included 45 sampling sites, 19 of which were coastal wells, included in this study. The in situ measurements included pH, temperature, dissolved oxygen, electrical conductivity, and alkalinity. The water samples were collected in HDPE bottles and filtered, preserved, and sent to specialized labs for the determination of major ions and cations, as well as metals [14].

The dataset for 2014–2016 included seven wells (9, 10, 14, 16–19) located in the central part of the Los Planes basin and documented by Briseño-Arellano [15]. Another 16 samples from coastal wells were taken in 2016, including 12 sites where data from 2003 were available and 4 at known hydrothermal anomalies (3 wells and one spring (Agua Caliente); see Figure 6). In all samples, the field measurements included pH, temperature, dissolved oxygen, and electrical conductivity. All samples were filtered through 0.45 µm membrane filters in the field and collected in polyethylene bottles, previously washed, and samples were stored in ice boxes at <4 °C to minimize bacterial activity. The samples collected for cation analysis were acidified with concentrated Suprapur HCl or concentrated Suprapur HNO_3 to obtain pH < 2. The samples were analyzed in specialized laboratories using ICP-OES (metals) and ion chromatography (major ions) following QA/QC protocols for the analysis of water [37,38]). Blanks were included during sampling trips [39]. Values for pH, electrical conductivity, carbonates, chlorides, and sulfates were determined in the field [37,40,41].

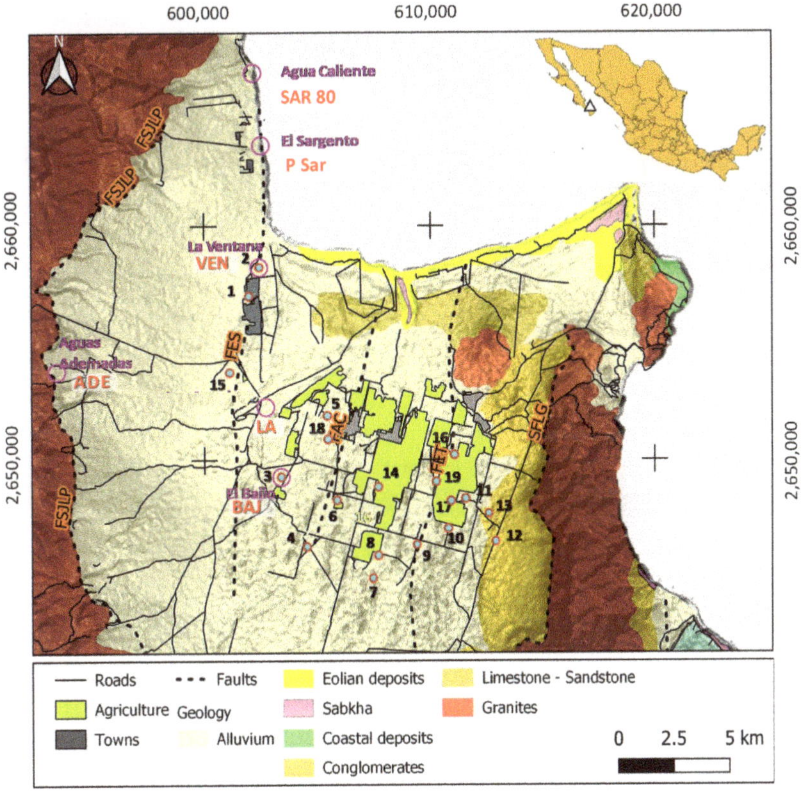

Figure 6. Position of 19 wells with hydrochemical data from 2003 and 2014 to 2016 (green dots) and wells with elevated temperatures (red dots).

3.2. *Recognizing Seawater Intrusion by Different Methods*

Seawater intrusion is a dynamic process that results from periodic changes in the recharge–discharge balance of the aquifer, where any direct or indirect influence on the aquifer's water balance affects the position and movement of the seawater interface [42,43]. The variations in groundwater chemistry then result from differences in recharge-source chemistries, varied aquifer materials, and fluctuations in groundwater flow [44]. To recognize the effects of seawater intrusion on the hydrochemical composition of groundwater, several parameters can be considered, such as the higher total mineralization or the chloride concentration of the groundwater [45]. Electrical conductivity (EC) is another basic indicator used to determine whether an aquifer is contaminated and also indicates if this process is increasing [45]. Commonly, an EC of less than 1000 μS/cm indicates that the groundwater is under normal conditions [46]. There are several indicators and graphical representation methods to understand the hydrochemical characteristics and facies of freshwater–saltwater systems. However, groundwater may also acquire salinity through contamination from agriculture, industries, improper sewage disposal, domestic wastewater, the mobilization of deeper groundwater or thermal water, etc. The following methods were applied to recognize the seawater fraction in the groundwater, ion exchange processes, and evaporation processes and obtain information on anthropogenic contamination and geothermal influence. A flowchart (Figure 7) explains the sequence in which these methods were applied.

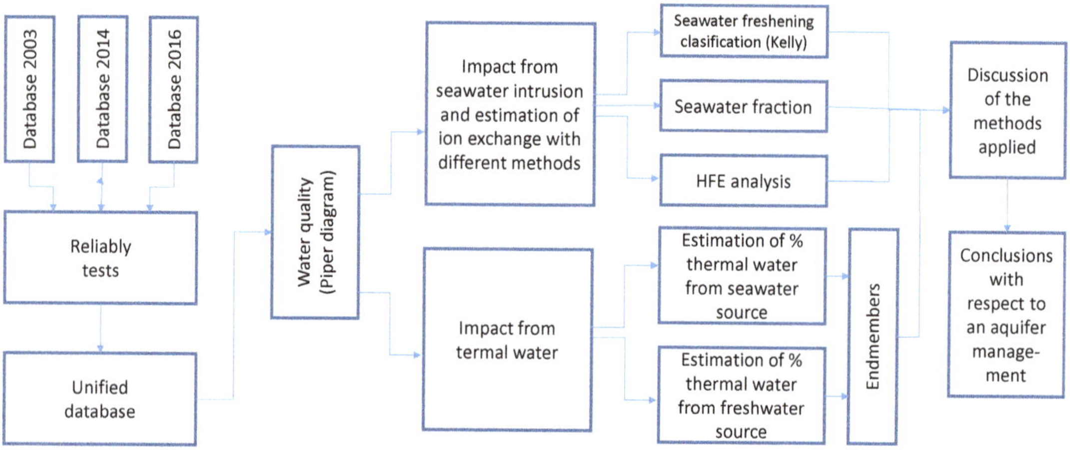

Figure 7. The flowchart of the hydrochemical data interpretation.

3.2.1. Classification of Seawater Freshening

Kelly [47] defined seven different criteria ranging from fresh groundwater to seawater to interpret the effects of saline water intrusion in a Piper diagram. A "mixture" field between the seawater and freshwater field indicates conservative mixing without ion exchange. As the water is mixed in the presence of aquifer materials, ion exchange may occur, modifying the chemical composition of the groundwater. This will cause the sample position in the diagram to migrate from the conservative mixing line to the upper part of the diamond field during the intrusion phase (after Equation (1)) and downward during the refreshment phase (Equation (2)):

$$2Na^{+} + CaX_2 \rightarrow Ca^{++} + 2NaX \quad (1)$$

$$Ca^{++} + 2NaX \rightarrow 2Na^{+} + CaX_2 \quad (2)$$

3.2.2. Calculation of the Seawater Fraction and Mixing Index

Seawater Fraction

The mixing reactions that occur in the transition zone between fresh groundwater and seawater can be recognized by comparing the measured water composition with the one calculated for the conservative mixing of freshwater and seawater [48]. The freshwater–seawater mixing proportion in each sample was calculated using the Appelo and Postma formula [41] and assuming that there were no other contributions to salinity other than freshwater and seawater:

$$\text{Seawater fraction}: \quad f_{5ea} = \frac{m_{sample} - m_{fresh}}{m_{sea} - m_{fresh}} \qquad (3)$$

where m is the molar concentration of the anion (Cl^- or Br^-) in the sample in freshwater (fresh) or seawater (sea).

Saltwater Mixing Index (SMI)

In open seawater, the conservative elements are found in constant proportions to one another and to salinity, although salinity varies. All the major ions in seawater, except for bicarbonate, and some trace elements are included in this group [49]. Therefore, element ratios can be used as an indicator of seawater intrusion. The "Saltwater Mixing Index" (SMI) proposed by Park and Aral [50] can be used to estimate the relative degree of saltwater/brackish water mixing with freshwater, taking four major ions into account; it is calculated as follows:

$$SMI = a * \frac{C\, Na^+}{T\, Na^+} \quad b * \frac{C\, Mg^{2+}}{T\, Mg^{2+}} + c * \frac{C\, Cl^-}{T\, Cl^-} + d * \frac{C\, SO_4{}^{2-}}{T\, SO_4{}^{2-}} \qquad (4)$$

where letters a, b, c, and d represent the relative concentration proportion of ions Na^+, Mg^{2+}, Cl^-, and $SO_4{}^{2-}$ in seawater ($a = 0.31$, $b = 0.04$, $c = 0.57$, $d = 0.08$); C is the measured concentration (in mg/L) of the ions in the groundwater [51]. The letter T represents the estimated regional threshold values, which were obtained, for example, from probability curves [51,52]. A probability plot was elaborated to calculate the threshold values for the natural concentrations of the major ions in the fresh groundwater of the study area. Based on a probability plot of a certain ion, the group of samples with to the lowest concentration was separated, and the mean value and standard deviation were obtained. Finally, the threshold values were calculated from the mean concentrations plus 2 standard deviations. The following threshold values were obtained: Na^+ (85.4 mg/L), Mg^{2+} (9.5 mg/L), Cl^- (60.5 mg/L), $SO_4{}^{2-}$ (35 mg/L), $HCO_3{}^-$ (160.7 mg/L), Ca^{2+} (22.3 mg/L), K^+ (1.6 mg/L).

3.2.3. Hydrochemical Facies Evolution Diagram

The hydrochemical facies evolution diagram (HFE-D), proposed by Giménez-Forcada [42], allows for the identification of the origin of groundwater salinization; it also indicates different freshening or salinization trends in a coastal aquifer [53,54]. In the HFE-diagram, the abscissas represent the evolution of % Na^+ + K^+ and % Ca^{2+} (or % Mg^{2+}), and the ordinates identify the evolution of % Cl^- and % $HCO_3{}^-$ + $CO_3{}^{2-}$ (or % $SO_4{}^{2-}$). The percentages are estimated based on total cation and anion concentrations, including those not explicitly delineated (usually Mg^{2+} or $SO_4{}^{2-}$); based on a percentage of ions of more than 50%, the following four main facies can be distinguished: Na-HCO_3 (salinized water where direct cation exchange reactions occurred), Na-Cl (seawater), Ca-HCO_3 (freshwater), and $CaCl_2$ (salinized water where reverse cation exchange reactions took place). If a percentage of an ion is less than 50%, then the prefix "Mix" is added in case the corresponding freshwater corresponds to the $CaHCO_3$ type [42,53].

3.3. Thermal Water

Scatter diagrams and a diagram after Giggenbach and Goguel [55] were used to recognize two hydrothermal water endmembers on the Baja California Peninsula and their mixing relationships with seawater. The diagram is based on the relationship between Mg/(Mg + Ca) and K/(K + Na). A curved line indicates the existence of water–rock equilibrium at different temperatures.

The sampling sites influenced by geothermal activity are shown in Figure 5. Four wells and one spring are situated close to the El Sargento fault, and another well is located close to the Los Planes fault. At these sites, a mixture of groundwater and geothermal water can be expected. Analyses from another 11 known hydrothermal sites of the region were also included in the interpretation: 4 from thermal manifestations in the Sierra La Laguna mountains [39], 5 described by López-Sánchez et al. [56] (4 wells are located near Cabo San Lucas and 1 is located near San Jose del Cabo), and 1 well at Buenavista reported by Hernández-Morales and Wurl [57].

4. Results and Discussion

4.1. Water Quality

Groundwater presented a wide range of total dissolved solids (TDS) for the whole dataset from freshwater to brackish water. In 2003, the TDS ranged from 309 to 2007 mg/L with a median value of 907 mg/L. For 2014–2016, salinity ranged from 340 mg/L to 4145 mg/L, with a median value of 1190 mg/L. The pH ranged from 6.9 to 8.2 in 2003 and 7.2 to 8.5 for 2014–2016 (Table 3). Changes in water quality were notable with an increase in the total mineralization (TDS +31.2%) and in Na^+ (+33.7%), Mg^{2+} (46.1%), Cl^- (44.6%), $SO_4{}^{2-}$ (20.7%), and $HCO_3{}^-$ (23.0%) as ions. A slight reduction was observed for Ca^{2+} (−1.5%).

Table 3. The results of the physicochemical parameters from the 2003 and 2014–2016 campaigns.

Campaign		Temp	pH	Ca^{2+}	Mg^{2+}	Na^+	K^+	$HCO_3{}^-$	$SO_4{}^{2-}$	Cl^-	TDS
		°C		mg/L	mg/L	mg/L	mg/L	mg/L	mg/L	mg/L	mg/L
2003 *	Minimum	38.6	6.9	13.30	4.90	66.00	2.30	91.12	23.00	50.30	309
2003 *	Maximum	32.0	8.2	172.40	78.80	550.00	19.60	257.28	375.00	893.40	2007
2003 *	Average.	30.0		77.81	31.90	205.52	6.22	152.02	137.47	309.66	907
2003 *	St. Dev.	1.8		48.84	24.68	145.79	4.31	43.86	105.85	264.69	580
2014–2016 **	Minimum	30.5	7.2	9.37	8.38	63.01	0.86	122.00	21.78	45.65	340
2014–2016 **	Maximum	37.8	8.5	213.62	176.93	1239.5	43.25	316.07	454.61	1917.4	4145
2014–2016 **	Average.	26.9		76.66	46.62	274.74	7.02	186.94	158.14	447.71	1190
2014–2016 **	St. Dev.	2.35		57.51	43.65	269.87	9.52	47.68	150.93	456.44	916

* Taken from [14]; ** data is from this study and from [15]. St. Dev. = standard deviation.

Regarding water quality for human use, 31.6% and 47.4% of the samples from 2003 did not comply with the Mexican drinking water standard (NOM-127-SSA1-1994) for Na^+ and Cl^-, respectively. For 2014–2016, 36.8%, 47.4%, and 10.5% of the samples did not comply with the standard for Na^+, Cl^-, and $SO_4{}^-$, respectively.

4.2: Groundwater Types

The relationship between major ions can be observed in the Piper diagram for 2003 and 2014–2016 (Figure 8A,B). In both diagrams, sodium is the most dominant cation, while chloride and bicarbonates are the most abundant anions. Interpreting the Piper diagram after the Kelly classification [47], it can be observed that in 2003 and 2014–2016. A total of 13 samples corresponded to the "intrusion" or "conservative mixing" fields. A general trend of increasing salinization between 2003 and 2014 to 2016 was observed; however, in some wells, freshening occurred. In 2003, three samples fell in the "freshening" and "slight freshening" fields, while for 2014–2016, there were four samples in these fields (Table 4).

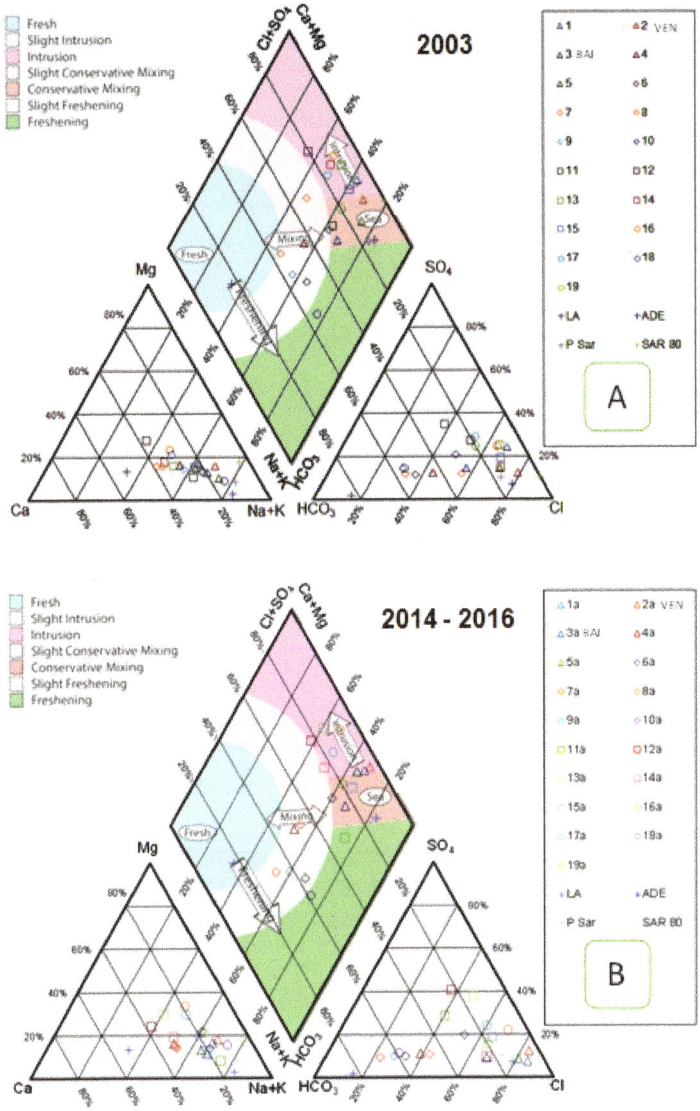

Figure 8. Piper diagram with seawater-freshening classification, after Kelly [47]: **(A)** 2003; **(B)** 2014–2016.

Table 4. Classification of 19 water samples, after Kelly [47].

Classification	2003	2014–2016
Intrusion	7	8
Conservative mixing	6	5
Slight conservative mixing	3	2
Slight freshening	2	2
Freshening	1	2
SUMA	19	19

4.3. Calculation of the Seawater Fraction Based on Ion Ratios

In open seawater, the conservative elements are found in constant proportions to one another and to salinity, although salinity varies. All the major ions in seawater, except for bicarbonate, and some trace elements are included in this group [49]. Therefore, element ratios can be used as an indicator of seawater intrusion. The calculated seawater fractions (%) are provided in Table 5).

Table 5. Calculated seawater fractions (%) for 2003 and 2014–2016 and their differences (%).

Ref.	*fsea %*	*fsea %*	*fsea %*
	2003	**2014–2016**	**Difference**
1	3.69	4.11	0.42
2 VEN	4.21	9.39	5.18
3 BAJ	*0.49*	*0.97*	0.48
4	*0.09*	*0.09*	−0.00
5	1.11	3.04	1.93
6	*−0.03*	*−0.03*	−0.00
7	*0.01*	*−0.08*	−0.09
8	*0.47*	*0.24*	−0.23
9	*−0.03*	*−0.01*	0.02
10	*0.43*	*0.67*	0.23
11	*0.29*	*0.26*	−0.03
12	1.67	1.23	−0.44
13	1.76	1.61	−0.15
14	1.05	*0.80*	−0.24
15	3.64	2.63	−1.01
16	1.73	4.17	2.44
17	1.27	1.87	0.60
18	*−0.05*	*−0.02*	0.03
19	*0.49*	*0.87*	0.38
*LA **		*3.1*	
*SAR **		*88.5*	
*P Sar **		*3.6*	
*ADE **		*−0.2*	

* At this site there was no data available from 2003.

Although an average increase of 0.5% in the seawater fraction was obtained for all samples from 2014 to 2016 compared with the situation in 2003, most wells showed a low seawater fraction (below 1% in 53% of the samples from 2003 and 58% from 2014 to 2016). According to Mahlknecht et al. [5], and regarding human uses, mixing only 2% seawater into a freshwater aquifer exceeds organoleptic objectives for the upper limit of chlorides, while mixing in 4% will make water mostly unusable, and 6% will make it only suitable for cooling and flushing purposes. In 2003, three samples surpassed the 2% limit, and in 2014–2016, five samples did so (Table 6). In 2003, one sample also surpassed the 4% limit, and three surpassed the limit for 2014–2016, of which one sample even surpassed the 6% limit.

Changes in the calculated seawater fraction (of less than +/−0.1%) were observed at six wells; eight showed an increase (red color) of up to 5.2%, and five showed a decrease with a maximum of 1%. However, the salinity of freshwater can vary depending on other factors such as the infiltration of irrigation return flows. Stigter et al. [58] defined an increase in salinity from a factor of three in a rural aquifer in Portugal with heavy pumping rates, and [5] used a factor of five for the more urban-influenced La Paz aquifer in northwestern Mexico. Under both definitions, only well 2 showed a clear increase in salinity.

4.4. Salinity Origin after the Hydrochemical Facies Evolution Diagram

In shallow sedimentary aquifers, groundwater evaporation, cation exchange, ion effects, and salt effects are important geochemical factors that occur widely. Cation exchange involves the replacement of the bivalent cations Ca^{++} and Mg^{++} in the aquifer matrix with the monovalent cations Na^+ and K^+ in groundwater see Equations (1) and (2).

The hydrochemical facies evolution diagram (HFE-D) was applied to the samples for 2003 and 2014–2016 to understand the state of the Los Planes aquifer and discriminate between the two principal phases (intrusion or freshening). Both phases are separated in the HFE-D by the conservative mixing line (blue line in Figure 9).

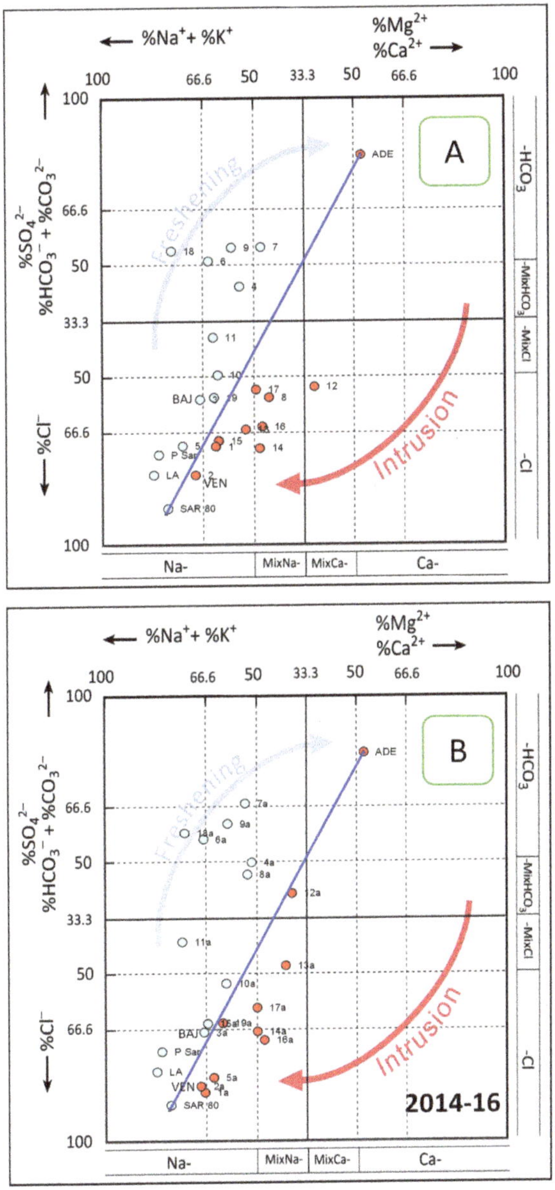

Figure 9. HFE-diagram for 2003 (**A**) and 2014–2016 (**B**).

In 2003, 10 samples can be identified as the freshening phase and 9 as the salinization phase. For 2014–2016, only 9 samples can still be defined as the freshening phase, but 14 can be identified as the salinization phase, an indication that seawater intrusion was advancing (Figures 9 and 10). The diagram allows us to identify five substages for each phase: f1, f2, f3, and f4 for freshening and FW for freshwater; i1, i2, i3, and i4 for the salinization and SW for saltwater (Figure 9). Four samples (LPL 5, 6, 7, and 16) show a shift from freshwater toward seawater intrusion (Table 6).

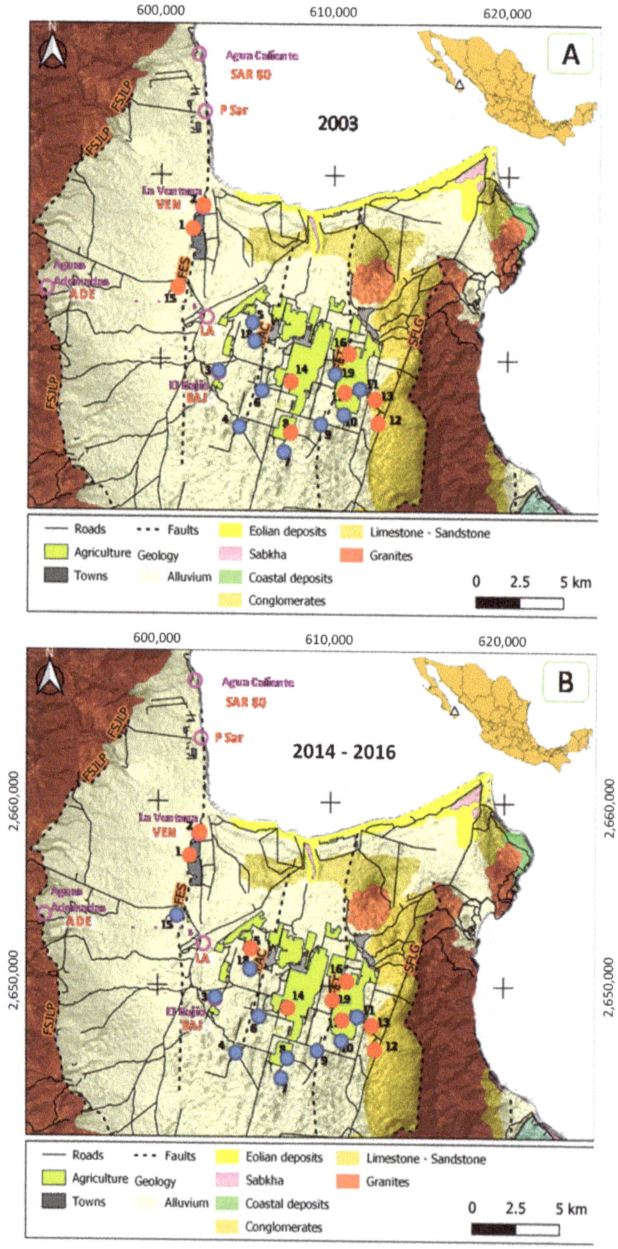

Figure 10. Changes in seawater intrusion between (**A**) 2003 and (**B**) 2014–2016, as defined by the HFE diagram.

Table 6. Facies evolution between 2003 and 2014–2016.

Site	2003			2014–2016		
No.	Phase	Facies		Phase	Facies	
1	**Intrus.**	Na	Cl	**Intrus.**	Na	Cl
2 VEN	**Intrus.**	Na	Cl	**Intrus.**	Na	Cl
3 BAJ	**Fresh.**	Na	Cl	**Fresh.**	Na	Cl
4	**Fresh.**	Na	MixHCO_3	**Fresh.**	Na	HCO_3
5	**Fresh.**	Na	Cl	**Intrus.**	Na	Cl
6	**Fresh.**	Na	HCO_3	**Fresh.**	Na	HCO_3
7	**Fresh.**	MixNa	HCO_3	**Fresh.**	Na	HCO_3
8	**Intrus.**	MixNa	Cl	**Fresh.**	Na	MixHCO_3
9	**Fresh.**	Na	HCO_3	**Fresh.**	Na	HCO_3
10	**Fresh.**	Na	MixCl	**Fresh.**	Na	Cl
11	**Fresh.**	Na	MixCl	**Fresh.**	Na	MixCl
12	**Intrus.**	MixCa	Cl	**Intrus.**	MixNa	MixSO_4
13	**Intrus.**	Na	Cl	**Intrus.**	MixNa	MixCl
14	**Intrus.**	MixNa	Cl	**Intrus.**	MixNa	Cl
15	**Intrus.**	Na	Cl	**Fresh.**	Na	Cl
16	**Intrus.**	MixNa	Cl	**Intrus.**	MixNa	Cl
17	**Intrus.**	MixNa	Cl	**Intrus.**	MixNa	Cl
18	**Fresh.**	Na	HCO_3	**Fresh.**	Na	HCO_3
19	**Fresh.**	Na	Cl	**Intrus.**	Na	Cl
LA				**Fresh.**	Na	Cl
ADE				**Intrus.**	Ca	HCO_3
P Sar				**Fresh.**	Na	Cl
SAR 80				**Fresh.**	Na	Cl

In the HFE diagram for 2003, two of the four main facies can be identified: seven wells of Na-Cl facies (seawater dominant) and three that correspond to the Na-HCO_3 facies (salinized water with direct exchange). With respect to the subfacies, four samples correspond to MixNa-Cl; two to Na-MixCl; and one to each of the subfacies, Na-MixHCO_3, Mix Na –HCO_3, and MixCa-Cl.

In the HFE diagram, for 2014–2016, two of the four main facies can be identified: seven wells of Na-Cl facies (seawater dominant) and five that correspond to the Na-HCO_3 facies (salinized water with direct exchange). With respect to the subfacies, three samples correspond to MixNa-Cl and one to each of the four subfacies, Na-MixCl, Na-MixHCO_3, Mix Na–MixSO_4, and MixCa-MixCl.

The number of wells with seawater intrusion was constant between 2003 and 2014–2016 (10 samples correspond to freshwater and 9 to the intrusion facies), but wells 5 and 19 changed from freshwater in 2003 to intrusion in 2014–2016, and wells 15 and 8 refreshed in 2014–2016. In the freshwater group for 2014–2016, an average seawater fraction of 0.28 was obtained, as compared with 2.16 in the intrusion group. The four geothermal-influenced samples are included in the HFE diagram, but their position is not well represented. All three samples along the El Sargento fault are defined as freshwater, although their seawater fractions elevated from 3.1 to 88.5%. On the other hand, the least mineralized sample, ADE, which had a negative seawater fraction calculated, was classified in the HFE diagram as intrusion.

5. Geothermal Water

The results of field measurements (in situ) are reported in Table 7. The low redox values (maximum 20 mV, minimum –410 mV) exclude the presence of oxygen in the water, and, therefore, the measured oxygen concentrations are not reported. The concentrations of major cations and anions for all water samples included in Figure 11 are contained in Tables 8 and 9, respectively. The Charge–Balance Error (CBE) results were within an acceptable range ($\leq \pm 7\%$) for all water samples.

Table 7. The results of the physicochemical parameters, measured in the field.

Site	Key	pH	Temperature (°C)	Redox (mV)	Electrical Conductivity (μS/cm)
Los Angeles	LA	7.37	48.7	−157.5	5500
Agua Ademadas	ADE	6.59	26.8	−1163.1	120
La Ventana	VEN	7.09	32.3	–	2480
El Bajio	BAJ	7.69	38.6	−1139.8	960
Pozo Sargento	P Sar	7.91	29.0	–	5990
Sargento 80 °C	SAR 80	7.8	80.0	–	–
La Junta Creek *	BAR 1	9.49	25.6	−370.0	393
La Junta Creek *	BAR 3	9.33	26.4	−287.7	325
La Junta Creek *	BAR 5	9.61	26.1	−270.0	632
La Junta Creek *	BAR 11	9.44	28.5	−410.0	390
Buenavista **	BV	8.04	41.3	36.1	773
C. San Lucas 1 ***	SL1	6.9	25	–	1700
C. San Lucas 2 ***	SL2	6.4	22	–	3530
C. San Lucas 4 ***	SL4	5.6	42	–	28,200
C. San Lucas 5 ***	SL5	5.7	72	–	49,600
San Jose del Cabo. 6 ***	SJ6	7.3	36	–	6750
Seawater M7 ****	M7	7.8	25	–	–

* Wurl et al., 2014 [39]; ** Hernandez Morales and Wurl 2016 [57]; *** Lopez Sanchez et al. 2006 [56]; **** Prol-Ledesma et al., 2004 [17].

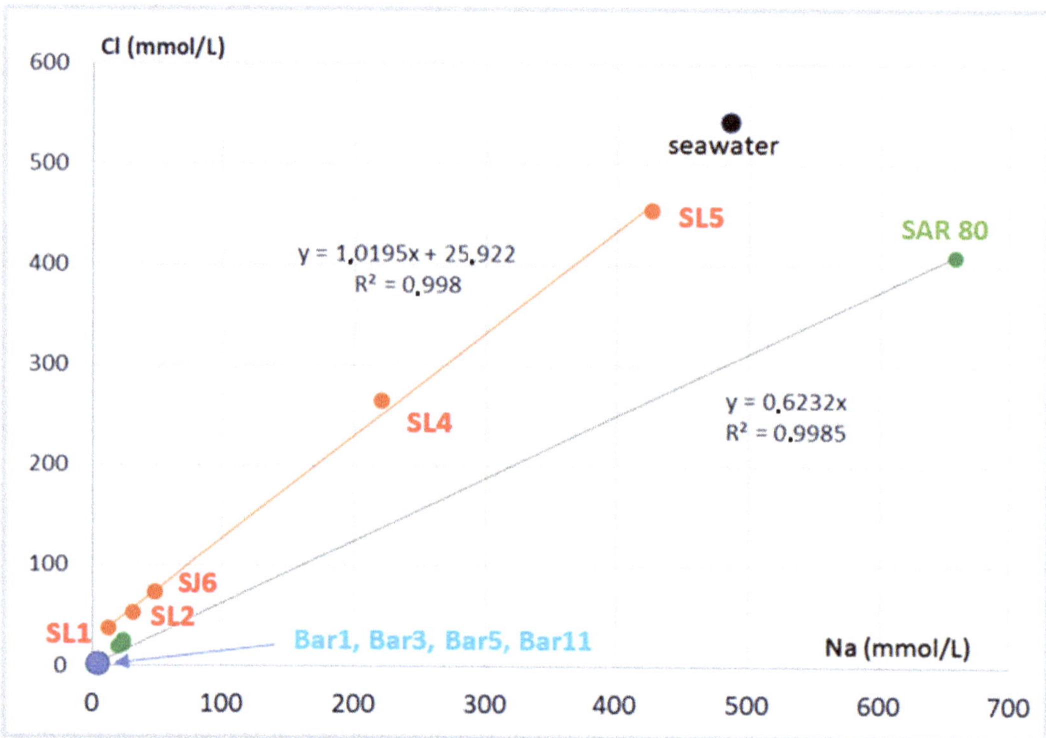

Figure 11. Scatter diagram of the relationship between Na and Cl in samples in Tables 7–9, visualizing the mixing between seawater and geothermal water.

Table 8. Concentration of main cations (mg/L).

Sample	Na	K	Ca	Mg	B
LA	460.6	15.0	68.6	8.9	1.1
ADE	29.9	0.5	40.0	6.2	0.1
VEN	550.0	19.6	105.6	67.6	–
BAJ	146.0	4.7	36.7	15.1	–
P Sar	554.5	9.2	60.8	30.7	1.0
SAR 80	15,136.9	509.0	589.1	569.4	8.2
Bar 1 *	90.2	1.5	2.0	0.1	9.5
Bar 3 *	97.5	1.2	2.6	0.1	6.0
Bar 5 *	87.9	1.4	1.9	0.1	0.8
Bar 11 *	85.2	0.5	4.8	0.2	2.0
BV **	152	4.31	5.2	0.63	0.3
SL1 ***	290.0	14.1	27.0	47.1	–
SL2 ***	722.0	35.0	32.0	59.0	–
SL4 ***	5070.0	283.0	1210.0	75.0	–
SL5 ***	9820.0	631.0	2430.0	69.6	–
SJ6 ***	1090.0	33.0	190.0	83.7	–
M7 ****	11,176.0	487.0	392.0	1400.0	–

* Wurl et al., 2014 [39]; ** Hernandez Morales and Wurl 2016 [57]; *** Lopez Sanchez et al. 2006 [56]; **** Prol-Ledesma et al., 2004 [17].

Table 9. Concentration of main anions (mg/L).

Sample	Cl	F	SO_4	HCO_3
LA	674.0	0.1	85.4	200
ADE	21.0	0.1	3.4	195
VEN	893.4	0.1	187.5	205.7
BAJ	158.1	0.1	55.0	170.0
P Sar	769.0	0.1	149.0	300
SAR 80	14,141.9	3.9	1670.6	100
Bar 1 *	58.3	1.8	56.3	70.3
Bar 3 *	88.6	3.2	33.6	75.8
Bar 5 *	53.0	0.6	48.6	237
Bar 11 *	36.5	2.3	40.8	59.7
BV **	121.5	2.22	59	7.2
SL1 ***	1279	0.1	105	371.9
SL2 ***	1826	0.1	200	278.9
SL4 ***	9132	1.7	625	93.0
SL5 ***	15,708	6.9	650	93.0
SJ6 ***	2557	2.9	500	325.4
M7 ****	18,744	0.1	2554	97.6

* Wurl et al., 2014 [39]; ** Hernandez Morales and Wurl 2016 [57]; *** Lopez Sanchez et al. 2006 [56]; **** Prol-Ledesma et al., 2004 [17].

Scatter diagrams were created to analyze the ratio of mixing between seawater and geothermal water samples from Los Planes. The relationship between Na and Cl was plotted for two types of thermal fluid: samples close to the El Sargento fault (green points in Figure 11) and samples described by Lopez-Sanchez et al. [56] (orange points in Figures 11 and 12). This comparison revealed a clear linear trend (R^2 = 0.9985) for the samples near the El Sargento fault, indicating a conservative mixing process between both fluids.

Two different main compositions of thermal water have been described for the Baja California Peninsula. The manifestations with the highest temperatures (SL5, SAR 80) were found near the coast [30,56,59], and in the case of SL5, seawater was identified as the main supply source, which is modified by geothermal processes [30,60,61]. This thermal water is enriched in Ca, As, Hg, Mn, Ba, HCO_3, Li, Sr, B, I, Cs, Fe, and Si relative to seawater [56,61]. The calculated temperatures for the deep reservoir reach 200 °C [56].

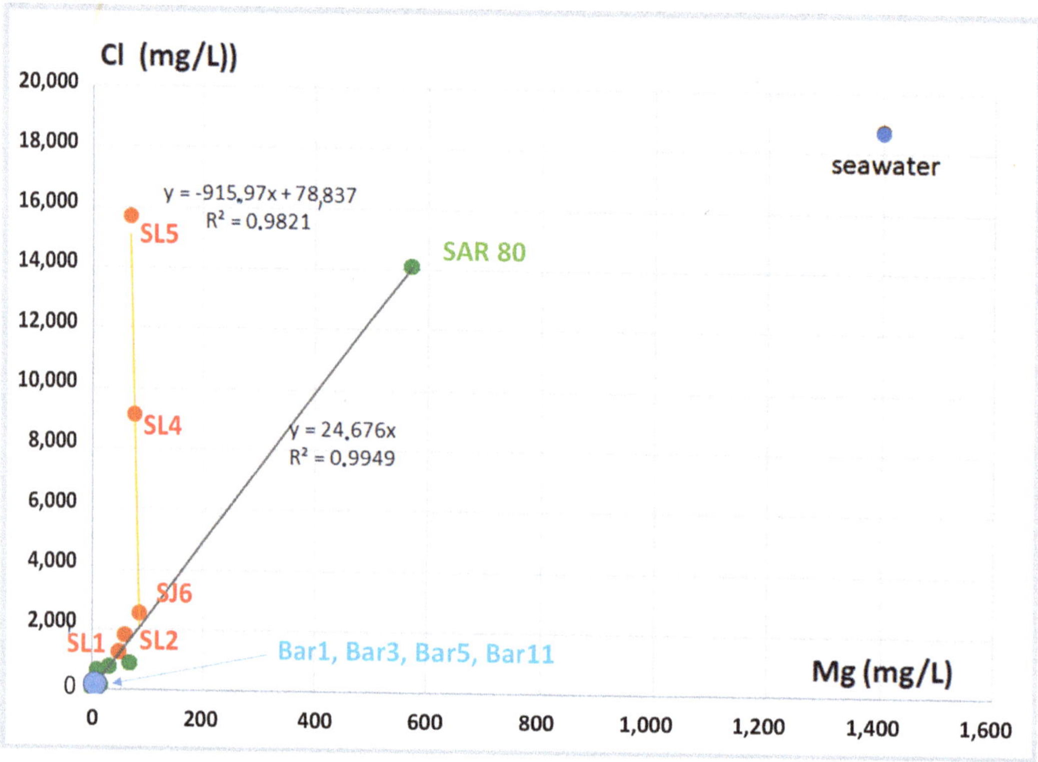

Figure 12. Scatter diagram of the relationship between Mg and Cl in samples in Tables 7–9, visualizing the mixing between seawater and geothermal water.

The relationship between Mg and Cl is presented in Figure 12. In this graph, three groups of water are identified. The first group corresponds to springs and wells in the La Junta area (Bars 1, 3, 5, and 11; blue dots) with the lowest Mg and Cl concentrations. The samples from Los Planes (green dots) and Cabo San Lucas (red dots) indicate a mixing process between seawater and thermal fluid. The thermal influence differs in both groups and can be recognized by the lower Mg content in the sample than in the seawater.

In contrast, Wurl et al. [39] and Hernández-Morales and Wurl [57], describe a different water composition for the thermal springs from the Los Cabos Block (Bar 1, Bar 3, Bar 5, and Bar 11), where meteoric water represents the main supply source. Here, very low mineralization and high pH values (up to 9.6) were observed. The calculated temperatures for the deep reservoir reach values between 86 °C and 115 °C (calculated with a quartz geothermometer) [62]. The highest temperature, measured at the surface, is less than 50 °C [57].

Mixtures between hydrothermal water and groundwater can be recognized based on elevated B/Cl ratios [63]. The following B/Cl ratios (for mg/L concentrations) were obtained: an average ratio of 0.075 for low mineralized thermal water, recharged by meteoric water (Bars 1,3, 5, and 11); higher mineralized thermal water with about 3% recharge from seawater 0.0016 (Los Angeles ranch); average seawater (0.00024); or freshwater (0.003 at LPL21). In the following section, the water samples are discussed based on scatter diagrams, as per the diagram proposed by Giggenbach and Goguel [55] (Figure 13).

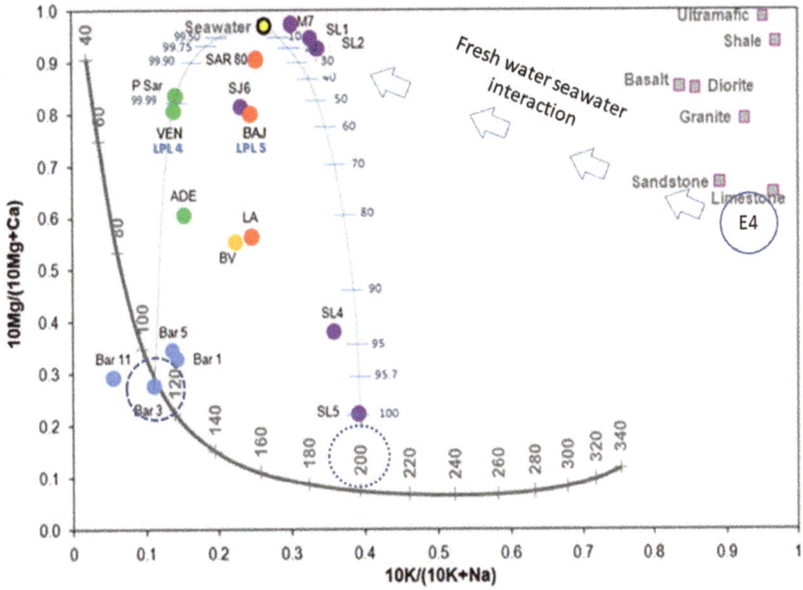

Figure 13. Diagram proposed by Giggenbach and Goguel [55] with all samples included and the percentage of mixing relationships between seawater and sample SL5 and Bar 3. Data were obtained from [11,30,39,56]. In the diagram taken from Giggenbach and Goguel [55], four water samples taken at thermal manifestations located near the La Junta creek in the Sierra la Laguna Mountains (Bar 1, Bar 3, Bar 5, and Bar 11 [39]) are closed to equilibrium conditions at a reservoir temperature between 100 and 110 °C. In the case of four wells drilled at the beach near Cabo San Lucas, only sample SL5 reaches the near-equilibrium temperature at 200 °C; the other three wells show mixtures with seawater in relationships of 7% at SL4, 70% at LS 2, and 80% at SL1. The seawater composition for the Gulf of California in Bahia Concepcion (M7), as described by Prol Ledesma [30], shows a slightly better fit with wells SL1 and SL2 than the composition of global seawater in Nordstrom et al. [64] (yellow circle in this figure).

Sample SL5 is representative of the thermal groundwater with seawater as the main source, and Bar 3 is typical for thermal water with meteoric water as the source, and these were recognized as endmembers. The resulting equilibrium temperatures of 200 °C and 110 °C for the deep reservoir coincide with the reported ones, obtained with different geo-thermometers by López-Sánchez et al. [56] and Wurl et al. [39]. Because the diagram permits the recognition of mixing processes between the different types of water [65], the mixing between both endmembers and seawater was calculated, and the mixing relation of different percentages was included in the diagram. The samples from the manifestation at Los Angeles and the thermal spring at the beach north of El Sargento coincide with those of Buenavista and San Jose del Cabo, indicating an intermediary position between both mixing lines, amounting to about 140–160 °C under equilibrium conditions and mixing with seawater. The shallow wells in the villages of El Sargento and La Ventana indicate lower equilibrium temperatures of 100–120 °C and mixing with less than 1% seawater. The fourth endmember (E 4 in Figure 13) is represented by the water–rock interaction with the sediment and was obtained from batch analyses conducted on sediments from the upper part of the aquifer (obtained from drillings in Briseño Arellano [15]; see Table 10). It represents the typical endmember for the pure reaction of rainwater with sediment.

Table 10. Typical endmember for the pure reaction of rainwater with sediment (obtained from batch analyses on sediments from the upper part of the aquifer), obtained from Briseño Arellano [15].

Sample	pH	Na	K	Ca	Mg	F	Cl	SO_4	HCO_3
07 PLRC 41	5.5	5.17	7.31	36.13	4.18	0.17	2.01	56.56	28.72
8 PLRC 103	5.5	3.96	6.32	40.45	5.28	0.23	1.67	99.3	27.28
10 PLRC 202	5.5	4.15	5.06	17.57	2.36	0.34	2.76	33.62	35.9
11 VWP3A	5.5	4.71	6.68	29.56	5.47	0	1.76	28.41	40.21
Mean value	**5.5**	**4.27**	**6.02**	**29.19**	**4.37**	**0.19**	**2.06**	**53.78**	**34.46**

6. Conclusions

Seawater intrusion contributes significantly to salinization, causing a decrease in groundwater quality and numerous environmental problems, such as high salinity levels, which may extend several kilometers inland in coastal aquifers. A negative effect on agriculture, water supply systems, and human health may result [41]. The existing methods cannot identify salinization due to mixing with mineralized thermal water. To recognize this additional impact, the differences in temperature and some dissolved solids allow us to distinguish between thermal water and typical surface water and groundwater. However, so far, only a few studies have focused on mixtures between thermal water and groundwater/surface water related to seawater intrusion.

Thus, the impact of higher mineralization caused by mixing with thermal water will not be recognized in most cases. In the case of the Los Planes aquifer, the elevated mineralization was reported correctly in former studies, but the influx of thermal water was confused with direct seawater introduced as an effect of ongoing aquifer overexploitation, which led to higher mineralization at the coastline; because of this, the water from most wells is not recommended for drinking. Our study indicates that the ongoing overextraction led to an average increase of 0.5% in the calculated seawater fraction since 2003. However, the Los Planes aquifer represents a complex system where thermal water affects groundwater quality in its western part. The geothermal manifestations along the faults of San Juan de Los Planes and El Sargento introduce thermal water with an important percentage of heated seawater into the aquifer. In a well near the Los Angeles ranch, we measured an increase in mineralization through thermal water, equivalent to 3.1 percent seawater. However, the water table indicates that this elevated mineralization cannot result from direct seawater inflow; rather, it is related to hydrothermal water with a mixture of seawater and freshwater as sources. This well is located on the El Sargento fault, where a gradual diminution of concentrations can be observed southward to the Sierra la Laguna Mountains, where meteoric water forms the source of the hydrothermal system. This low mineralized thermal water is associated with a rapid flow of meteoric water into deeper zones (reaching temperatures of around 100 °C) and a subsequent return to the surface [56]. The temperatures estimated for the deeper reservoir of the anomalies in the Los Planes aquifer at San Jose (SJ 6), El Bajio (BAJ), Los Angeles (LA), and El Sargento (SAR) show a tendency toward reservoir temperatures between 140 °C and 160 °C, providing several options for uses of this geothermal energy, for example, water desalination [66]. Future management strategies should be elaborated, taking into account our new findings.

Supplementary Materials: The datafile can be downloaded at: https://www.mdpi.com/article/10.3390/resources12040047/s1; Table S1: Groundwater; Table S2: Thermal water.

Author Contributions: Conceptualization, J.W. and M.A.I.-L.; methodology, J.W., M.A.I.-L. and L.C.M.-R. writing—original draft preparation, J.W., M.A.I.-L. and P.H.-M.; writing—review and editing, J.W., P.H.-M. and L.C.M.-R. visualization, M.A.I.-L., J.W. and P.H.-M. All authors have read and agreed to the published version of the manuscript.

Funding: This work was supported by the Mexican Secretariat of the Environment and Natural Resources [Grant No. SEMARNAT-2014-1-249423] within the project entitled "Effect of rainfall pattern on the concentration of trace elements in groundwater in arid areas".

Informed Consent Statement: Not applicable.

Data Availability Statement: The following external datasets were used in this study: CNA, 2003: http://defiendelasierra.org/wp-content/uploads/Estudio-CNA-Intrusion-Salina-y-Metales-Pesados-Los-Planes.pdf; Briseño Arellano, 2014 (accessed on 16 January 2023): http://132.248.9.195/ptd2014/agosto/0718072/0718072.pdf (accessed on 16 January 2023).

Conflicts of Interest: The authors declare no conflict of interest.

References

1. Polemio, M.; Walraevens, K. Recent research results on groundwater resources and saltwater intrusion in a changing environment. *Water* **2019**, *11*, 1118. [CrossRef]
2. Ferguson, G.; Gleeson, T. Vulnerability of coastal aquifers to groundwater use and climate change. *Nat. Clim. Chang.* **2012**, *2*, 342–345. [CrossRef]
3. Werner, A.D.; Bakker, M.; Post, V.E.A.; Vandenbohede, A.; Lu, C.; Ataie-Ashtiani, B.; Simmons, C.T.; Barry, D.A. Seawater intrusion processes, investigation and management: Recent advances and futures challenges. *Adv. Water Resour.* **2013**, *51*, 3–26. [CrossRef]
4. Tran, D.A.; Tsujimura, M.; Pham, H.V.; Nguyen, T.V.; Ho, L.H.; Vo, P.L.; Ha, K.Q.; Dang, T.D.; Vinh, D.V.; Doan, Q.-V. Intensified salinity intrusion in coastal aquifers due to groundwater overextraction: A case study in the Mekong Delta, Vietnam. *Environ. Sci. Pollut. Res.* **2022**, *29*, 8996–9010. [CrossRef]
5. Mahlknecht, J.; Merchán, D.; Rosner, M.; Meixner, A.; Ledesma-Ruis, R. Assessing seawater intrusion in an arid coastal aquifer under high anthropogenic influence using major constituents, Sr and B isotopes in groundwater. *Sci. Total Environ.* **2017**, *587*, 282–295. [CrossRef]
6. Comisión Nacional de Agua. Estadísticas del Agua en México. 2018. Available online: http://sina.conagua.gob.mx/publicaciones/EAM_2018.pdf (accessed on 16 January 2023).
7. Sistema Nacional de Información del Agua (SNIA-CONAGUA). Tablas de Sobreexplotación y Salinización de Acuíferos en México para el Año 2021. 2021. Available online: http://sina.conagua.gob.mx/sina/tema.php?tema=acuiferos (accessed on 9 January 2023).
8. Fichera, M.M. Numerical Modeling and Hydrochemical Analysis of the Current and Future State of Seawater Intrusion in the Todos Santos Aquifer, Mexico. Ph.D. Thesis, Colorado State University, Fort Collins, CO, USA, 2019. Available online: https://mountainscholar.org/bitstream/handle/10217/195327/Fichera_colostate_0053N_15368.pdf (accessed on 16 January 2023).
9. Chávez-López, S.; Brito-Castillo, L. Calidad del agua subterránea en la Reserva de La Biosfera de El Vizcaino BCS, México. *Rev. Latinoam. Recur. Nat.* **2010**, *6*, 7–20.
10. González-Abraham, A.; Fagundo-Castillo, J.R.; Carillo-Rivera, J.J.; Rodríguez-Estrella, R. Geoquímica de los sistemas de flujo de agua subterránea en rocas sedimentarias y vulcanogénicas de Loreto, BCS, México. *Bol. Soc. Geol. Mex.* **2012**, *6*, 319–333. [CrossRef]
11. Cardona, A.; Carrillo-Rivera, J.J.; Huizar-Alvarez, R.; Graniel-Castro, E. Salinization in coastal aquifers of arid zones: An example from Santo Domingo, Baja California Sur, Mexico. *Environ. Geol.* **2004**, *45*, 350–366. [CrossRef]
12. Wurl, J.; Gámez, A.E.; Ivanova-Boncheva, A.; Imaz-Lamadrid, M.A.; Hernández-Morales, P. Socio-hydrological resilience of an arid aquifer system, subject to changing climate and inadequate agricultural management: A case study from the Valle of Santo Domingo, México. *J. Hydrol.* **2018**, *559*, 486–498. [CrossRef]
13. Arango-Galván, C.; Prol-Ledesma, R.M.; Torres-Vera, M.A. Geothermal prospects in the Baja California Peninsula. *Geothermics* **2015**, *55*, 39–57. [CrossRef]
14. Comisión Nacional de Agua (CNA). Determinación de la Disponibilidad de Agua en el Acuífero (0323) Los Planes, Estado de Baja California Sur. 2003. Available online: http://defiendelasierra.org/wp-content/uploads/Estudio-CNA-Intrusion-Salina-y-Metales-Pesados-Los-Planes.pdf (accessed on 16 January 2023).
15. Briseño Arellano, 2014 Evaluación de la Evolución Hidrogeoquímica del Agua Subterránea en La Cuenca San Juan de Los Planes, Baja California Sur 2014 Tesis Lic. UNAM. Available online: https://repositorio.unam.mx/covers/133713/0/300.png (accessed on 16 January 2023).

16. Del Rosal Pardo, A. Recarga Natural en la Cuenca de San Juan de los Planes, Baja California Sur, México. Bachelor's Thesis, Universidad Autónoma de Baja California Sur, La Paz, México, 2003.
17. Busch, M.M.; Arrowsmith, J.R.; Umhoefer, P.J.; Coyan, J.A.; Maloney, S.J.; Gutiérrez, G.M. Geometry and evolution of rift-margin, normal-fault–bounded basins from gravity and geology, La Paz–Los Cabos region, Baja California Sur, Mexico. *Lithosphere* **2011**, *3*, 110–127. [CrossRef]
18. Coyan, M.M.; Arrowsmith, J.R.; Umhoefer, P.; Coyan, J.; Kent, G.; Driscoll, N.; Gutíerrez, G.M. Geometry and Quaternary slip behavior of the San Juan de los Planes and Saltito fault zones, Baja California Sur, Mexico: Characterization of rift-margin normal faults. *Geosphere* **2013**, *9*, 426–443. [CrossRef]
19. NASA. NASA Shuttle Radar Topography Mission (SRTM) Version 3.0 (SRTM Plus) Product Release. Land Process Distributed Active Archive Center, National Aeronautics and Space Administration. 2013. Available online: https://lpdaac.usgs.gov/news/nasa-shuttle-radar-topography-mission-srtm-version-30-srtm-plus-product-release/ (accessed on 28 October 2021).
20. Aranda-Gómez, J.J.; Pérez-Venzor, J.A. Estratigrafía del complejo cristalino de la región de Todos Santos, estado de Baja California Sur. *Revista* **1989**, *8*, 149–170.
21. Fletcher, J.M.; Kohn, B.P.; Foster, D.A.; Gleadow, A.J.W. Heterogeneous Neogene cooling and exhumation of the Los Cabos block, southern Baja California: Evidence from fission-track thermochronology. *Geology* **2000**, *28*, 107–110. [CrossRef]
22. Schaaf, P.; Böhnel, H.; Pérez-Venzor, J.A. Pre-Miocene palaeogeography of the Los Cabos block, Baja California Sur: Geochronological and palaeomagnetic constraints. *Tectonophysics* **2000**, *318*, 53–69. [CrossRef]
23. CONAGUA Comisión Nacional de Agua. Determinación de la Disponibilidad de Agua en el Acuífero (0323) Los Planes, Estado de Baja California Sur. 2020. Available online: https://sigagis.conagua.gob.mx/gas1/Edos_Acuiferos_18/BajaCaliforniaSur/DR_0323.pdf (accessed on 16 January 2023).
24. Wurl, J.; Martínez Gutiérrez, G. El efecto de ciclones tropicales sobre el clima en la cuenca de Santiago, Baja California Sur, México. In Proceedings of the III Simposio Internacional en Ingeniería y Ciencias para la Sustentabilidad Ambiental y Semana del Ambiente, Mexico City, Mexico, 5–6 June 2006.
25. Dominguez, C.; Magaña, V. The Role of Tropical Cyclones in Precipitation over the Tropical and Subtropical North America. *Front. Earth Sci.* **2018**, *6*, 19. [CrossRef]
26. Gutiérrez Caminero, L. Isótopos de Pb como Trazadores de Fuentes de Metales y Metaloides en el Distrito Minero San Antonio-El Triunfo, Baja California Sur. Master's Thesis, Centro de Investigación Científica y de Educación Superior de Ensenada, Ensenada, Mexico, 2013.
27. Aranda-Gómez, J.J.; Pérez-Venzor, J.A. *Excursión Geológica al Complejo Migmatítico de la Sierra de La Gata, BCS III Reunión Internacional Sobre la Geología de la Península de Baja California*; Sociedad Geológica Peninsular en la Universidad Autónoma de Baja California Sur: La Paz, Mexico, 1995; pp. 17–27.
28. CNA (Comisión Nacional del Agua), Gerencia de Aguas Subterránea. 1997; Estudio Hidrogeológico y Simulación Hidrodinámica del Acuífero del Valle de San Juan de los Planes, B.C.S. Final report; Dirección Local en Baja California Sur, unpublished.
29. Técnicas Modernas de la Ingeniería (TMI), S.A. 1977; Informe final del Estudio Geohidrológico del valle de San Juan de los Planes, en el estado de Baja California Sur, para la Subdirección de Geohidrología y de Zonas Áridas SARH, unpublished.
30. Prol-Ledesma, R.M.; Canet, C.; Torres-Vera, M.A.; Forrest, M.J.; Armienta, M.A. Vent fluid chemistry in Bahía Concepción coastal submarine hydrothermal system. *J. Volcanol. Geotherm. Res.* **2004**, *137*, 311–328. [CrossRef]
31. Prol-Ledesma, R.M.; Dando, P.R.; De Ronde, C.E.J. Special issue on shallow-water hydrothermal venting. *Chem. Geol.* **2005**, *224*, 1–4. [CrossRef]
32. Carrillo, A. Environmental Geochemistry of the San Antonio-El Triunfo Mining Area, Southernmost Baja California Peninsula Mexico. Ph.D. Thesis, University of Wyoming, Laramie, WY, USA, 1996.
33. COREMI (Consejo de Recursos Minerales). *Monografía Geológico-Minera del Estado de Baja California Sur*; Secretaría de Comercio y Fomento Industrial: Ciudad Juárez, México, 1999.
34. Carrillo-Chavez, A.; Drever, J.I.; Martinez, M. Arsenic content and groundwater geochemistry of the San Antonio-El Triunfo, Carrizal and Los Planes aquifers in southernmost Baja California, Mexico. *Environ. Geol.* **2000**, *39*, 1295–1303. [CrossRef]
35. Naranjo-Pulido, A.; Romero-Schmidt, H.; Mendez-Rodriguez, L.; Acosta-Vargas, B.; Ortega-Rubio, A. Soil arsenic contamination in the Cape Region, BCS, Mexico. *J. Environ. Biol.* **2002**, *23*, 347–352. [PubMed]
36. Carrillo, A.; Drever, J.I. Environmental assessment of the potential for arsenic leaching into groundwater from mine wastes in Baja California Sur, Mexico. *Geofísica Int.* **1998**, *37*, 35–39. [CrossRef]
37. Volke-Sepúlveda, T.; Solórzano-Ochoa, G.; Rosas-Domínguez, A.; Izumikawa, I.; Encarnación-Aguilar, G.; Velasco-Trejo, J.A.; Flores Martínez, S. Informe Final, Remediacion de Sitios Contaminados por Metales Provenientes de Jales Mineros en Los Distritos de El Triunfo-San Antonio y Santa Rosalía, Baja California Sur, Mexico, 2003; 37p. Available online: http://defiendelasierra.org/wp-content/uploads/Estudio-de-Remediaci%C3%B3n-El-Triunfo-y-San-Antonio-2003.pdf (accessed on 16 January 2023).
38. DVWK-Regeln Zur Wasserwirtschaft, H. Entnahme und Untersuchungsumfang von Grundwasserproben. In *DK 556.32.001.5 Grundwasseruntersuchung, DK 543.3.053 Probenahme. DVWK-Regeln zur Wasserwirtschaft 128, 36*; Deutscher Verband für Wasserwirtschaft und Kulturbau e.V.: Hamburg/Berlin, Germany, 1992. (In German)
39. Wurl, J.; Mendez-Rodriguez, L.; Acosta-Vargas, B. Arsenic content in groundwater from the southern part of the San Antonio-El Triunfo mining district, Baja California Sur, Mexico. *J. Hydrol.* **2014**, *518*, 447–459. [CrossRef]

40. U.S. Environmental Protection Agency. *Method 6020A—Inductively Coupled Plasma Mass Spectrometry*; Revision 1; U.S. Environmental Protection Agency: Washington, DC, USA, 2007.
41. Appelo, C.A.J.; Postma, D. *Geochemistry, Groundwater and Pollution*; A.A. Balkema: Rotterdam, The Netherlands, 1995.
42. Giménez-Forcada, E. Dynamic of Seawater Interface using Hydrochemical Facies Evolution Diagram (HFE-D). *Groundwater* **2010**, *48*, 212–216. [CrossRef]
43. Giménez-Forcada, E. Use of the Hydrochemical Facies Diagram (HFE-D) for the evaluation of salinization by seawater intrusion in the coastal Oropesa Plain: Comparative analysis with the coastal Vinaroz Plain, Spain. *HydroResearch* **2019**, *2*, 76–84. [CrossRef]
44. Ibrahim Hussein, H.A.; Ricka, A.; Kuchovsky, T.; El Osta, M.M. Groundwater hydrochemistry and origin in the south-eastern part of Wadi El Natrun, Egypt. *Arab. J. Geosci.* **2017**, *10*, 170. [CrossRef]
45. Klassen, J.; Allen, D.M.; Kirste, D. *Chemical Indicators of Salt Water Intrusion for the Gulf Islands, British Columbia*; Department of Earth Sciences, Simon Fraser University: Burnaby, BC, Canada, 2014; 43p.
46. Carreira, P.M.; Marques, J.M.; Nunes, D. Source of groundwater salinity in coastline aquifers based on environmental isotopes (Portugal): Natural vs. human interference. A review and reinterpretation. *Appl. Geochem.* **2014**, *41*, 163–175. [CrossRef]
47. Kelly, F. *Seawater Intrusion Topic Paper*; Island Country Health Department: Coupeville, WA, USA, 2005.
48. Anderson, W.P., Jr. Aquifer salinization from storm overwash. *J. Coast. Res.* **2002**, *18*, 413–420.
49. White, W.M. *Geochemistry*; John Wiley & Sons: Hoboken, NJ, USA, 2020.
50. Park, C.H.; Aral, M.M. Saltwater intrusion hydrodynamics in a tidal aquifer. *J. Hydrol. Eng.* **2008**, *13*, 863–872. [CrossRef]
51. Gupta, S.K.; Gupta, I.C. *Drinking Water Quality Assessment and Management*; Scientific Publishers: Jodhpur, India, 2020.
52. Edet, A. Hydrogeology and groundwater evaluation of a shallow coastal aquifer, southern AkwaIbom State (Nigeria). *Appl. Water Sci.* **2017**, *7*, 2397–2412. [CrossRef]
53. Van Le, T.T.; Lertsirivorakul, R.; Bui, T.V.; Schulmeister, M.K. An application of HFE-D for evaluating seawater intrusion in coastal aquifers of Southern Vietnam. *Groundwater* **2020**, *58*, 1012–1022. [CrossRef]
54. Sajil Kumar, P.J. Deciphering the groundwater–saline water interaction in a complex coastal aquifer in South India using statistical and hydrochemical mixing models. *Model. Earth Syst. Environ.* **2016**, *2*, 1–11. [CrossRef]
55. Giggenbach, W.F.; Goguel, R.L. *Collection and Analysis of Geothermal and Volcanic Water and Gas Discharges*; Report No. CD 2401; Department of Scientific and Industrial Research, Chemistry Division: Petone, New Zealand, 1989.
56. López-Sánchez, A.; Báncora-Alsina, C.; Prol-Ledesma, R.M.; Hiriart, G. A new geotermal resource in Los Cabos, Baja California Sur, Mexico. In Proceedings of the 28th New Zealand Geothermal Workshop, Auckland, New Zealand, 15–17 November 2006.
57. Hernández-Morales, P.; Wurl, J. Hydrogeochemical characterization of the thermal springs in northeastern of Los Cabos Block, Baja California Sur, Mexico. *Environ. Sci. Pollut. Res.* **2016**, *24*, 13184–13202. [CrossRef]
58. Stigter, T.Y.; Carvalho Dill, A.M.M.; Ribeiro, L.; Reis, E. Impact of the shift from groundwater to surface water irrigation on aquifer dynamics and hydrochemistry in a semi-arid region in the south of Portugal. *Agric. Water Manag.* **2006**, *85*, 121–132. [CrossRef]
59. Portugal, E.; Birkle, P.; Barragán, R.M.; Arellano, V.M.; Tello, E.; Tello, M. Hydrochemical-isotopic and hydrogeological conceptual model of the Las Tres Vírgenes geothermal field, Baja California Sur, México. *J. Vulcanol. Geotherm. Res.* **2000**, *101*, 223–244. [CrossRef]
60. Canet, C.; Prol-Ledesma, R.M.; Proenza, J.A.; Rubio-Ramos, M.A.; Forrest, M.J.; Torres-Vera, M.A.; Rodríguez-Díaz, A.A. Mn–Ba–Hg mineralization at shallow submarine hydrothermal vents in Bahía Concepción, Baja California Sur, Mexico. *Chem. Geol.* **2005**, *224*, 96–112. [CrossRef]
61. Villanueva-Estrada, R.E.; Prol-Ledesma, R.M.; Torres-Alvarado, I.S.; Canet, C. Geochemical modeling of a Shallow Submarine Hydrothermal System at Bahía Concepción, Baja California Sur, México. In Proceedings of the World Geothermal Congress, Antalya, Turkey, 24–29 April 2005.
62. Verma, S.P.; Santoyo, E. New improved equations for Na/K, Na/Li and SiO_2 geothermometers by outlier detection and rejection. *J. Volcanol. Geotherm. Res.* **1997**, *79*, 9–24. [CrossRef]
63. Vengosh, A.; Gill, J.; Davisson, M.L.; Hudson, G.B. A multi-isotope (B, Sr, O, H, and C) and age dating (^{3}H–^{3}He and ^{14}C) study of groundwater from Salinas Valley, California: Hydrochemistry, dynamics, and contamination processes. *Water Resour. Res.* **2002**, *38*, 9-1-9-17. [CrossRef]
64. Nordstrom, D.K.; Plummer, L.N.; Wigley, T.M.L.; Wolery, T.J.; Ball, J.W.; Jenne, E.A.; Bassett, R.L.; Crerar, D.A.; Florence, T.M.; Fritz, B. A Comparison of Computerized Chemical Models for Equilibrium Calculations in Aqueous Systems. In *Chemical Modeling in Aqueous Systems*; American Chemical Society: Washington, DC, USA, 1979; pp. 857–892. [CrossRef]
65. Powell, T.; Cumming, W. Spreadsheets for geothermal water and gas geochemistry. In Proceedings of the Thirty-Fifth Workshop on Geothermal Reservoir Engineering, Stanford, CA, USA, 1–3 February 2010. 10p.
66. Goosen, M.; Mahmoudi, H.; Ghaffour, N. Water desalination using geothermal energy. *Energies* **2010**, *3*, 1423–1442. [CrossRef]

resources

MDPI

Article

Warming Air Temperature Impacts Snowfall Patterns and Increases Cold-Season Baseflow in the Liwiec River Basin (Poland) of the Central European Lowland

Urszula Somorowska

Department of Hydrology, Faculty of Geography and Regional Studies, University of Warsaw, Krakowskie Przedmieście 30, 00-927 Warsaw, Poland; usomorow@uw.edu.pl

Abstract: The rapidly changing climate affects vulnerable water resources, which makes it important to evaluate multi-year trends in hydroclimatic characteristics. In this study, the changes in cold-season temperature (November–April) were analyzed in the period of 1951–2021 to reveal their impacts on precipitation and streamflow components in the Liwiec River basin (Poland). The temperature threshold approach was applied to reconstruct the snowfall/rainfall patterns. The Wittenberg filter method was applied to the hydrograph separation. The Mann–Kendall test and Sen's slope were applied to estimate the significance and magnitude of the trends. An assessment of the similarity between trends in temperature and hydroclimatic variables was conducted using the Spearman rank-order correlation. The shift-type changes in river regime were assessed via the Kruskal–Wallis test. The results revealed that temporal changes in both snowfall, rainfall, and baseflow metrics were significantly associated with increasing temperature. Over 71 years, the temperature rose by ~2.70 °C, the snowfall-to-precipitation ratio decreased by ~16%, the baseflow increased with a depth of ~17 mm, and the baseflow index rose by ~18%. The river regime shifted from the snow-dominated to the snow-affected type. Overall, this study provides evidence of a gradual temperature increase over the last seven decades that is affecting the precipitation phase and streamflow component partitioning in the middle-latitude region.

Keywords: cold season; 1951–2021; trends; snowfall-to-precipitation ratio; baseflow index; river regime shift; lowland river basin; middle-latitude region

Citation: Somorowska, U. Warming Air Temperature Impacts Snowfall Patterns and Increases Cold-Season Baseflow in the Liwiec River Basin (Poland) of the Central European Lowland. *Resources* **2023**, *12*, 18. https://doi.org/10.3390/resources12020018

Academic Editor: Diego Copetti

Received: 29 October 2022
Revised: 12 January 2023
Accepted: 13 January 2023
Published: 17 January 2023

1. Introduction

The warming climate, which is considered one of the most important factors that affects stream-flow regimes in many regions of the world, has environmental and socioeconomic implications, particularly with respect to the vulnerability of water resources [1–3]. With the increasing air temperature, altered precipitation patterns change the water quantities, thereby contributing to runoff components as quick flow and baseflow [4,5]. In addition to the amount of precipitation, the precipitation phase (snowfall or rainfall) plays a critical role in runoff generation [6,7]. Snowfall, if persistent, stores cold-season precipitation into the spring months and keeps the hydrological system dormant, while rainfall infiltrates soils, recharges aquifers, and feeds streams and rivers. The warming air temperature might reduce snowfall and amplify snow melt, thereby resulting in a decline in water storage in snowpacks, earlier snow, and soil thawing, and, as a consequence, causing a shift in the hydrological regime [8–10]. Since the warming climate modifies the ratio of snow to precipitation (S/P ratio) in many parts of the world [11–13], the detailed characterization of temperature and precipitation changes is a high priority in ongoing research [14–16]. However, significant uncertainties remain regarding the current and future trends in the hydrologic implications of warming due to the high sensitivity of hydrological processes to climate variability and change [17].

Divergent trends in snow characteristics occur in different regions of the world. When considering snowfall, climate warming might increase humidity, which enhances extreme snowfall, whereas the rising temperature reduces the likelihood of snowfall [18]. For example, the authors of [19] found that the frequency of daily snowfall events tended to decrease across much of the Northern Hemisphere except at the highest latitudes such as in northern Canada, northern Siberia, and Greenland. Divergent snowfall patterns were also uncovered in [18]; it was found that while higher-latitude regions experience increasing extreme snowfall percentiles, decreasing extreme snowfall percentiles are characteristic of lower-latitude regions in Western Europe. Observational evidence from the pan-European in situ data was provided in [20]; it was revealed that over the period of 1951–2017, the mean snow depth decreased more than the extreme snow depth, and widespread decreases in the maximum and mean snow depth were found over Europe except in the coldest climates. It is worth noting that interannual variability in the extent of snow is high, and new extremes in maximum snow metrics over Eurasia have occurred in recent years [16,21]. Generally, the increases in winter temperatures have resulted in a decrease in the S/P ratio and an increase in winter snowmelt for most of the Northern Hemisphere; this was particularly significant in the middle-latitude regions [11,22].

Divergent assessments also concern the hydrologic implications of rising temperature reported in global and regional studies. In the coldest river basins, the response to warming is manifested by an increase in the spring streamflow peak, whereas for the transitional basins, the spring runoff decreases [1]. This is because transitional river basins face large increases in winter streamflow. While many studies have generally focused on mountainous and arctic regions [23–26], relatively fewer studies have reported on snow hydrology changes across low-relief topography regions [27,28]. These are regions in which the river runoff is also sensitive to the effects of the changing patterns of snow accumulation and melt. In [1], regions with snowmelt-dominated runoff were selected using the ratio of accumulated annual snowfall divided by annual runoff (S/QT ratio or snowfall-to-runoff ratio), and the criterion of S/QT > 0.5 over the global land regions was applied. According to this criterion, vast areas of the Central European Lowland located within the borders of Poland are not classified as snowmelt-dominated areas. However, this may be due to the short observation period, which covered the years 1980–1999, and the coarse spatial resolution of the gridded data ($0.5° \times 0.5°$ latitude/longitude) used in the analysis. Therefore, it seems reasonable to evaluate the S/QT ratio using ground-based data with a higher spatial resolution for a longer period of time. Such an analysis might unmask multi-year changes in the S/QT ratio over time. In the east and northeast parts of Poland, the hydrological regime of lowland rivers is influenced by relatively persistent seasonal snowpack in winter, with the highest streamflow occurring in spring when the snow mass melts and feeds the rivers. The strong influence of the snowmelt-dominated river regime is manifested by the relatively high Pardé coefficient in the spring months, which reaches or exceeds 180%. Therefore, it is unclear whether this region also did not have a snowmelt-dominated character before the 1980s. While some previous studies investigated selected aspects of decreases in snow-cover depth driven by the rising winter temperature in Poland in the years 1966/1967–2019/20 [29], no studies have considered the trends in wintertime runoff components and their dependence on rising temperatures and changing snowfall patterns. To fill this gap, this paper focused on quantifying the changes in snowfall, rainfall, baseflow, and quick flow, as well as their association with temperature during the cold season. The multi-year trends were examined in an exemplary typical mesoscale river basin of the Central European Lowland. This basin is situated in the middle eastern region of Poland, where the most intensive temperature increase was recorded at selected weather stations followed by a decrease in snow cover depth [29]. Decreasing tendencies were found in the snow metrics; however, the variability in the snow characteristics within both the winter season and the multi-year period was high [30].

The key scientific questions that needed to be answered in this study were as follows:

(1) What are the signals of warming climate revealed by the changes in seasonal air temperature at the river basin scale?
(2) What is the contribution of snowfall to the total precipitation? How does this change over a multi-year period, and what are the magnitude of the changes?
(3) How does the temperature change impact the streamflow and the river regime?

It was hypothesized that the snowfall has remarkably decreased in the last seven decades, thereby leading to a decrease in S/P ratio. It was also hypothesized that the increasing trends in cold-season temperature are responsible for the changes in streamflow, with a higher component of baseflow being recharged by the increased fraction of rainfall to total precipitation.

2. Materials and Methods

2.1. Study Area

This study concerned the Liwiec River basin, a left tributary of the Bug River, which is situated in the Mazovian Lowland in central eastern Poland (Figure 1). The study area is located between 52°00′ N–52°30′ N and 21°30′ E–22°0′ E and belongs to the Central Poland Lowland, which is a part of the Central European Lowland [31]. The stream gauge is situated at the Łochów cross-section and closes the river basin at 2471 km^2. The elevation ranges from 83 m a.s.l. near the outlet to 227 m a.s.l. in the southern part of the basin. The Liwiec River is about 142 km in length. The basin is influenced by a snowy, humid climate (Dfb) with a warm summer (see the updated Köppen–Geiger classification [32]). Monthly precipitation and air temperature are evenly distributed across the catchment and differentiate seasonally in four meteorological seasons: spring (March–May), summer (June–August), autumn (September–November), and winter (December–February). The annual precipitation is about 550 mm, of which 250–300 mm falls in the winter half-year (November–April). The average annual air temperature is 7.5 °C. The lowest temperatures are recorded in January; the average monthly temperature for this month is −4 °C. The highest values are recorded in July (an average monthly value of 19 °C). The studied area is characterized by a greater amplitude of air temperature compared to central and western Poland due to moderate continental impacts. The river regime is considered to be nival, with the highest streamflow rates usually occurring in March; the lowest streamflow rates usually occur in July and August and sometimes continue throughout September–October. Low flows might also occur in winter when snow cover blocks the groundwater recharge and the groundwater resource gradually depletes. When snow melts, a pronounced peak flow occurs. The aquifers are found in the Quaternary formations. They are directly recharged by precipitation, and the Liwiec River and its tributaries constitute a zone of natural drainage. The depth of the first groundwater table varies depending on the lithology of the surface formations, but in most of the area it does not exceed 5 m. In the highlands, aquifer sands are often found under less permeable loams. In the river valleys and in the flat areas in the north and south of the basin area, the unconfined aquifers form the subsurface. The land use is dominated by agricultural land (63%), which includes arable land (48%) and meadows (15%). The forest type is the second-most dominant (33%), while the rest of the territory is covered by artificial surfaces (3%) and other categories (1%).

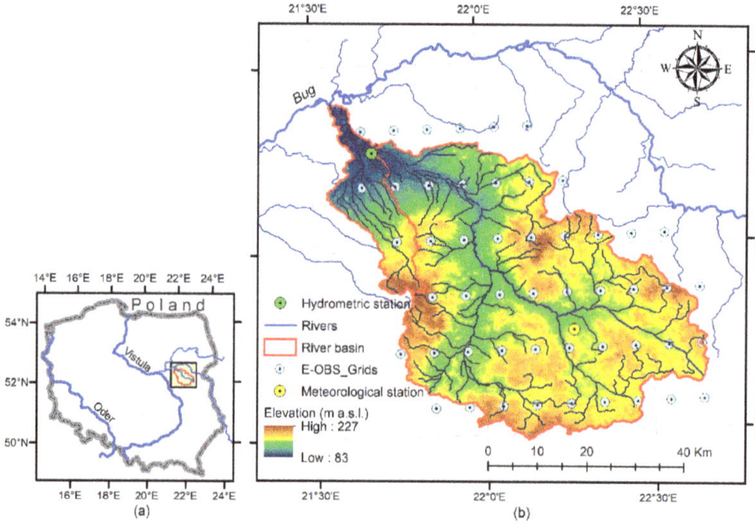

Figure 1. (**a**) Geographic location of the study area. (**b**) Elevation map of the Liwiec River basin according to EU-DEM v1.1 (acquired from https://land.copernicus.eu; accessed on 13 October 2022), and distribution of the E-OBS gridded temperature and precipitation points (acquired from https://www.ecad.eu; accessed on 19 June 2022).

2.2. Datasets

For the analysis of climate conditions, the air temperature and precipitation dataset was employed for 1950–2021 from version 25.0e of the station-based E-OBS gridded dataset (https://www.ecad.eu; accessed on 19 June 2022) available from the European Climate Assessment and Dataset Project [33]. It comprised daily precipitation (P) and air temperature (T) values acquired from a regular latitude/longitude grid of 0.1° × 0.1° (Figure 1b). For the Liwiec River basin, the data subset was extracted as a 3D-gridded dataset of 6 × 12 × 25,933 dimensions, which corresponded to 6 latitude grid cells, 12 longitude grid cells, and 25,933 daily solutions. Using the daily gridded values, the basin scale, monthly air temperature, and precipitation estimates were calculated, and then the average values for the months November–April were calculated to represent the cold-season conditions. This approach followed the convention of the water year, which was designated in Poland as a 12-month period (November–October) that consists of the "winter half-year" (cold season) and "summer half-year" (warm season) [34,35]. It is worth noting that the cold season is considered to be the 6-month period starting on November 1 of a particular year and lasting until the end of April in the next calendar year. For example, the values representing the cold season of the year 1951 were calculated using the monthly values starting in November 1950 and ending in April 1951. Using these monthly values, the cold-season precipitation and air-temperature time series were calculated and tested for the presence of a trend as explained in Section 2.3.2.

For the analysis of snowfall patterns, the air-temperature threshold approach was applied to partition the precipitation into rain and snow fractions [36]. Since the E-OBS data did not differentiate between rain and snow, it was assumed that all precipitation fell as snow below a threshold of 1.20 °C and as rain above this threshold. This threshold is near the upper bound of the transition temperature across the Liwiec River basin as revealed in [36]. The threshold temperature across the basin is evenly distributed in a narrow range of 1.19 to 1.22 °C. The threshold data covering the territory of Poland were prepared at a resolution of 0.5° × 0.625° (latitude × longitude) based on 89 meteorological stations distributed across the country. The spatial resolution of the threshold temperature data was coarser than the E-OBS dataset; thus, resampling to a common resolution of

$0.1^\circ \times 0.1^\circ$ was required to keep the information from the finest layers. Using the 3D-gridded air-temperature dataset of $6 \times 12 \times 25{,}933$ dimensions, a binary mask was prepared; grids with a temperature above the threshold were set to zero while the remaining grids received a value of 1. This binary representation of snowfall occurrence was used to differentiate between snow and rain in daily precipitation data stored in the 3D matrix of $6 \times 12 \times 25{,}933$ dimensions. Moreover, the daily snowfall and rainfall values were accumulated into monthly values and averaged for the entire river basin. For the 6-month cold season, the air temperature (T), snowfall (S), and rainfall (R) time series consisted of 71 values covering the period of 1951–2021.

To analyze the possible impacts of the warming climate on the river regime, the daily streamflow data for gauging station no. 152210120 (Liwiec River, Łochów cross-section) were acquired for the water years of 1951–2021 from https://danepubliczne.imgw.pl/data (accessed on 30 August 2022). The data were prepared and verified by the Institute of Meteorology and Water Management—National Research Institute in Warsaw, Poland (IMGW-PIB). The procedure used to separate the hydrograph into baseflow and quick flow is described in Section 2.3.1. For the cold season, the streamflow (QT), baseflow (QB), and quick flow (QQ) time series were tested for the presence of trends as explained in Section 2.3.2.

2.3. Methods

2.3.1. Hydrograph Separation into Quick Flow and Baseflow Components

The daily streamflow (QT) time series were analyzed for the water years of 1951–2021, and the two components of baseflow (QB) and quick flow (QQ) were separated from the total flow (QT). The HYDRORECESSION toolbox was used for the streamflow recession analysis [37]. Aksoy and Wittenberg's method [38] was applied to the extraction of recession segments, in which negative dQT/dt values were considered to represent the recession of the hydrograph composed of the baseflow. Here, the minimum duration of the recession segment was set to 10 days, and the filter criterion (removed days) was assumed to be 5 days. The Wittenberg filter method was applied to the daily streamflow for baseflow separation, which assumed that the baseflow recession segment satisfied the nonlinear relationship of $S = aQ^b$. The optimal model parameters were determined via linear regression (least squares). The cold-season and monthly values of QB and QQ were extracted and examined. Then, the baseflow index was calculated as BFI = QB/QT, and the quick flow index was determined as QFI = QQ/QT.

2.3.2. Trend Detection

A non-parametric Mann–Kendall (MK) test [39,40] was applied to detect trends in the precipitation, air temperature, and streamflow data time series. Moreover, the components of precipitation (snowfall and rainfall) and streamflow (baseflow and quick flow) were also tested for the presence of temporal trends. The MK test was used to test the null hypothesis of no trend (H_0) against the alternative hypothesis (H_1) that there was an increasing or decreasing monotonic trend. The trends were tested at a significance level of alpha = 0.05. The magnitude (slope) of an existing trend (as change per year) was calculated using the directional coefficient expressed by the Theil–Sen estimator [41,42]. A positive slope value suggested an increasing trend, and a negative slope value indicated a decreasing trend. If change was not statistically significant but showed an inclination, it was called a tendency. The Climate Data Toolbox (CDT) for MATLAB [43] was used to calculate the MK standardized test statistic (Z) and the p-value.

2.3.3. Spearman Rank-Order Correlation for Similarity Assessment between Trends

As a measure of the strength of the link between trends in the monthly time series of temperature and other variables (including snowfall, rainfall, baseflow, and quick flow), a non-parametric Spearman's rank-order correlation coefficient (R_S) was calculated and evaluated at a significance level of alpha = 0.05. This assessed the relationship between two variables without making any assumptions about the frequency distribution of the variables. The Spearman rank-order correlation was equal to the Pearson correlation between the rank values of the two variables and ranged between −1 and 1. For each variable, the monthly trend rates were ranked and the strength of the link between trends in air temperature and other variables was assessed. The tested hypothesis was that with the rising temperature, snowfall would decrease, rainfall would increase, and significantly more baseflow would occur. Statistically significant results were given a p-value < 0.05. The pairs with positive-correlation coefficients tended to increase or decrease together, while negative-correlation coefficients indicated a relationship between two variables in which an increase in one variable was associated with a decrease in the other. The MATLAB Statistics and Machine Learning Toolbox (Release 2021) was used for all statistical analyses.

2.3.4. Shift-Type Changes in the River Regime

In this study, the ratio of the accumulated annual snowfall to the annual total runoff (S/QT ratio) was used to examine the role of snowmelt in the seasonal streamflow patterns. This metric was introduced in [1] to determine whether or not runoff was snowmelt-dominated using the criterion of S/QT > 0.5. The shift-type changes were analyzed, and the time series of S/QT was partitioned into two subseries. The separation into two subperiods was achieved by minimizing the sum of the residual (squared) error of each subset from its local mean and finally returning the index, which in this case was the year in which the change occurred. The change point was identified using the MATLAB function "findchangepts". A complete 71-element S/QT time series was divided into subperiods of differing lengths that consisted of 20 and 51 records covering the years of 1951–1970 and 1971–2021, respectively. Finally, a Kruskal–Wallis test was applied to test for statistically significant differences between the subseries in the two selected subperiods [44,45]. This was a non-parametric test that compared the mean ranks (i.e., medians). For this test, the null hypothesis was that the subseries medians would be equal (versus the alternative of a difference between them).

3. Results

3.1. *Cold-Season Trends in Hydroclimatic Variables over the Period of 1951–2021*

Figure 2 shows the course of the cold season's hydrometeorological variables over the years of 1951-2021, including the precipitation ($P_{11\text{-}04}$), air temperature ($T_{11\text{-}04}$), snowfall ($S_{11\text{-}04}$), snowfall-to-precipitation ratio ($RSP_{11\text{-}04} = S_{11\text{-}04}/P_{11\text{-}04}$), rainfall ($R_{11\text{-}04}$), rainfall-to-precipitation ratio ($RRP_{11\text{-}04} = R_{11\text{-}04}/P_{11\text{-}04}$), streamflow ($QT_{11\text{-}04}$), baseflow ($QB_{11\text{-}04}$), quick flow ($QQ_{11\text{-}04}$), and baseflow index ($BFI_{11\text{-}04}$). Statistically significant changes occurred in the time series of air temperature (increasing trend; Figure 2b), snowfall-to-precipitation ratio (decreasing trend; Figure 2d), rainfall-to-precipitation ratio (increasing trend; Figure 2f), baseflow (increasing trend; Figure 2h), quick flow (decreasing trend; Figure 2i), and baseflow index (increasing trend; Figure 2j). The $T_{11\text{-}04}$ showed a trend rate of ~0.38 °C/decade, while a trend rate of ~−2.27%/decade was detected in $RSP_{11\text{-}04}$. Consequently, $RRP_{11\text{-}04}$ showed an increase of ~2.27%/decade. Thus, the gradual temperature increase was accompanied by a decrease in the snow-to-precipitation ratio and an increase in the proportion of the liquid phase in the precipitation (RRP). The warmest cold season occurred in 2020 with a $T_{11\text{-}04}$ of 4.4 °C, while the coldest occurred in 1963 with a $T_{11\text{-}04}$ of −3.2 °C (Figure 2b). Generally, the lowest values of $T_{11\text{-}04}$ were recorded in the first half of the analyzed period, while the highest were recorded in the last two decades. In response to the slightly increasing tendency in $P_{11\text{-}04}$ and the significant increase in $T_{11\text{-}04}$, $QT_{11\text{-}04}$ did show a slight decreasing tendency of ~0.048 m^3s^{-1}/decade (Figure 2g), which was equivalent to 0.06 mm/decade, while $QB_{11\text{-}04}$ increased with a trend rate of ~0.39 m^3s^{-1}/decade (Figure 2h), which was equivalent to 2.46 mm/decade. Thus, the changes in $QB_{11\text{-}04}$ reached ~17.44 mm over 71 years.

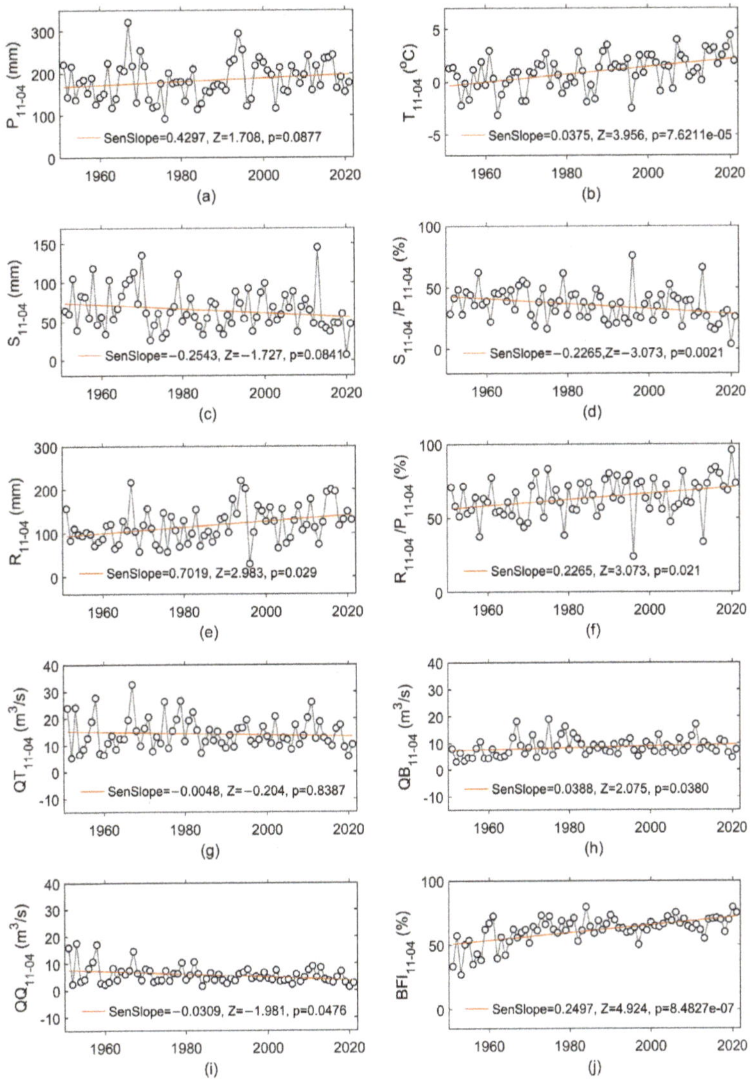

Figure 2. Changes in cold-season (**a**) precipitation ($P_{11\text{-}04}$), (**b**) air temperature ($T_{11\text{-}04}$), (**c**) snowfall ($S_{11\text{-}04}$), (**d**) snowfall-to-precipitation ratio ($RSP_{11\text{-}04} = S_{11\text{-}04}/P_{11\text{-}04}$), (**e**) rainfall ($R_{11\text{-}04}$), (**f**) rainfall-to-precipitation ratio ($RRP_{11\text{-}04} = R_{11\text{-}04}/P_{11\text{-}04}$), (**g**) streamflow ($QT_{11\text{-}04}$), (**h**) baseflow ($QB_{11\text{-}04}$), (**i**) quick flow ($QQ_{11\text{-}04}$), and (**j**) baseflow index ($BFI_{11\text{-}04}$). The MK test statistic is denoted as Z. The presence of a trend was determined at a significance level of alpha = 0.05; in cases with a *p*-value > 0.05, the changes were not statistically significant. Sen's slope is expressed in the variable unit per year.

3.2. *Trends in Monthly Hydroclimatic Variables over the Period of 1951–2021*

Temperature is a prime factor that determines the occurrence of precipitation as rain or snow [46]. Therefore, this study primarily examined the air-temperature trends over the multi-year period of 1951−2021 as an area-weighted average across the entire river basin. Figure 3 shows the course of monthly air temperature (T) in the cold season that lasted from November to April. The multi-year rate of change in the monthly T was equal to 0.24, 0.27, 0.32, 0.49, 0.53, and 0.31 °C/decade for consecutive months of the cold-season period, respectively.

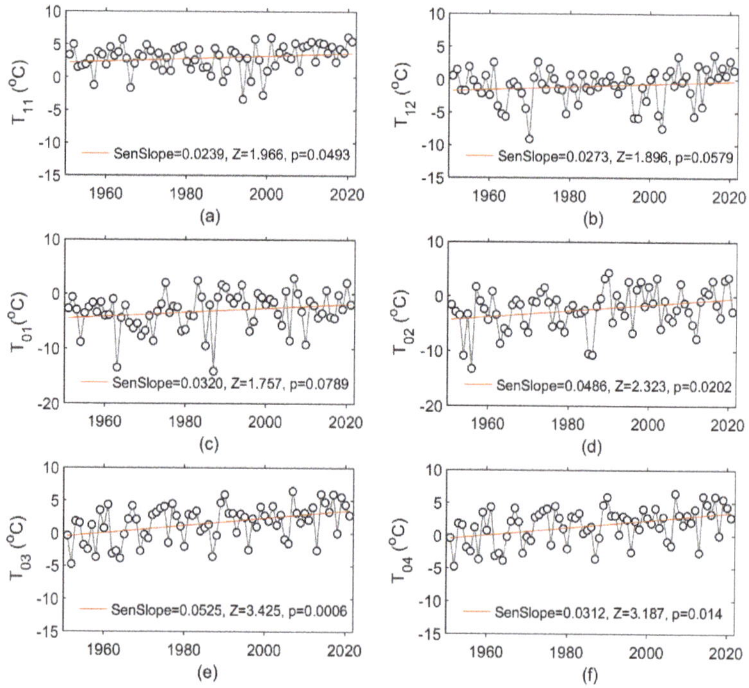

Figure 3. Changes in monthly air temperature in (**a**) November (T_{11}), (**b**) December (T_{12}), (**c**) January (T_{01}), (**d**) February (T_{02}), (**e**) March (T_{03}), and (**f**) April (T_{04}). The MK test statistic is denoted as Z. The straight red line in each figure represents the linear trend (Sen's slope) of the monthly average temperature (expressed in °C/y).

For the months of November, February, March, and April, these changes were statistically significant; the highest rates occurred in February and March. In December and January, the change also showed an increase.

Examining changes in the monthly snowfall and snowfall-to-precipitation ratio yielded substantially consistent results (Figure 4). With an increasing temperature, in all six months of the cold season, a decreasing tendency was seen regarding the snowfall amount, which was statistically significant in February. In January, February, and March, the snowfall-to-precipitation ratio decreased ~4%/decade in each month (Figure 4f,h,j); therefore, in 71 years, it decreased by ~28%. When considering the change in the six-month cold season, $S_{11\text{-}04}$ decreased at a rate of ~2.5mm/decade (Figure 2c) for a total decrease of ~18 mm in 71 years, and $RSP_{11\text{-}04}$ decreased by 2.3%/decade (Figure 2d), which accounted for 16% in 71 years. Consequently, the monthly rainfall increased in five of the six months of the cold season (from December to April, with the highest increase in March) by ~2.56 mm/decade (Figure 5i), which represented a rate of ~18 mm in 71 years. The highest increase in RRP occurred in the months of December, January, February, and March within a range of ~3.2–4.4%/decade (Figure 5d,f,h,j), which represented a range of 23–31% in 71 years.

Changes in the monthly baseflow and BFI are shown in Figure 6, while the changes in the quick flow and QFI are presented in Figure 7. A statistically significant increase in QB occurred in January, February, and March; while in November, December, and April, growing tendencies were registered. In all six months, BFI showed an increase that was statistically significant in March and April (Figure 6j,l). Inverse changes were noted for QFI, which showed statistically significant decreasing changes in March and April (Figure 7j,l).

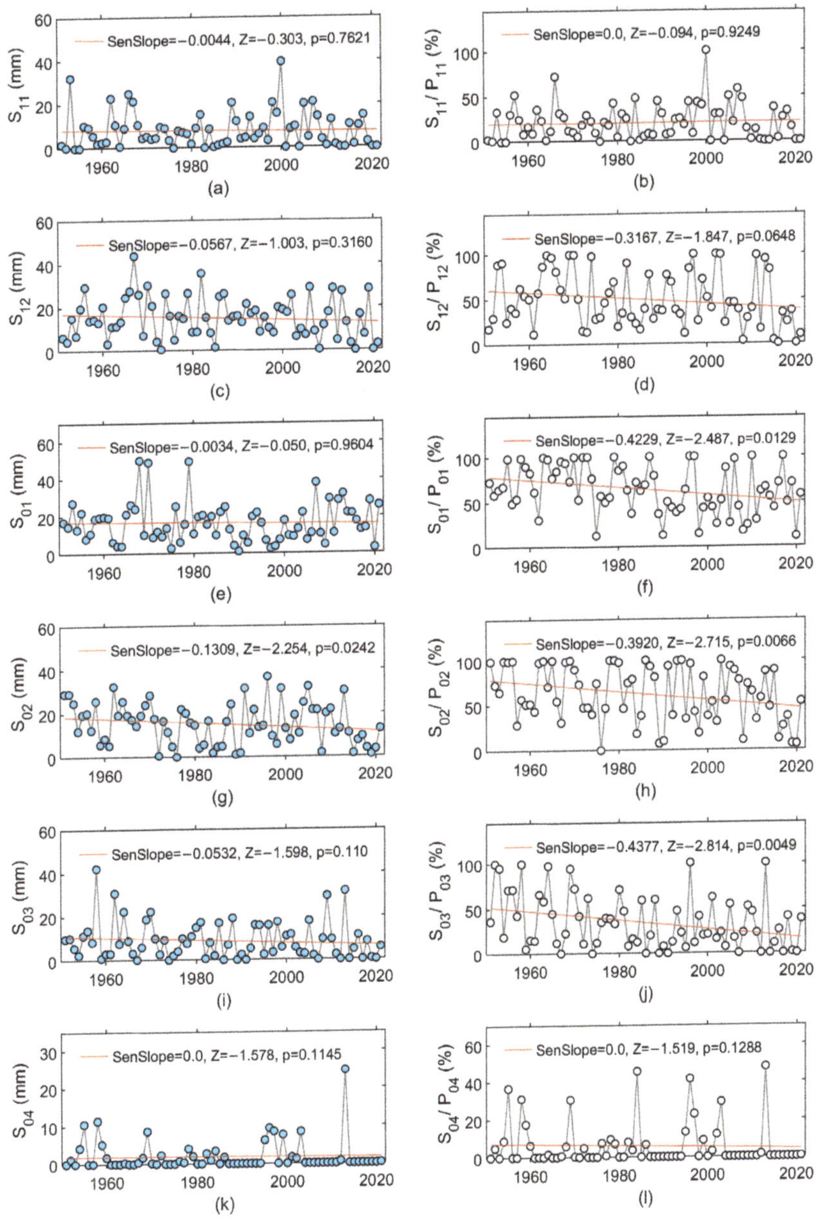

Figure 4. Changes in monthly snowfall in (**a**) November (S_{11}), (**c**) December (S_{12}), (**e**) January (S_{01}), (**g**) February (S_{02}), (**i**) March (S_{03}), and (**k**) April (S_{04}); and changes in the monthly snowfall-to-precipitation ratio in (**b**) November (S_{11}/P_{11}), (**d**) December (S_{12}/P_{12}), (**f**) January (S_{01}/P_{01}), (**h**) February (S_{02}/P_{02}), (**j**) March (S_{03}/P_{03}), and (**l**) April (S_{04}/P_{04})). The MK test statistic is denoted as Z. The straight red line in each figure represents the linear trend (Sen's slope) of monthly snowfall (expressed in mm/y) or the monthly snowfall-to-precipitation ratio (expressed in %/y).

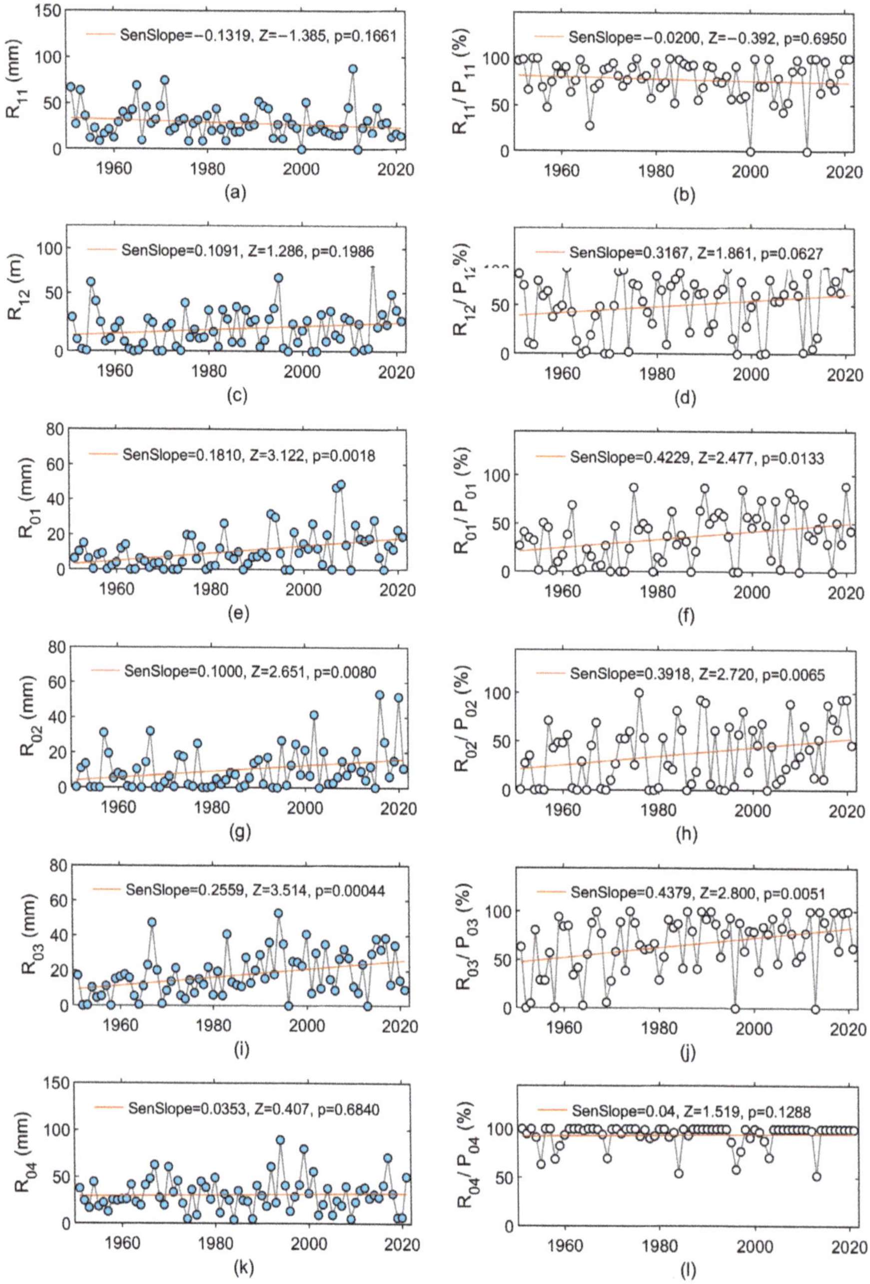

Figure 5. Changes in monthly rainfall in (**a**) November (R_{11}), (**c**) December (R_{12}), (**e**) January (R_{01}), (**g**) February (R_{02}), (**i**) March (R_{03}), and (**k**) April (R_{04}), and changes in monthly rainfall-to-precipitation ratio in (**b**) November (R_{11}/P_{11}), (**d**) December (R_{12}/P_{12}), (**f**) January (R_{01}/P_{01}), (**h**) February (R_{02}/P_{02}), (**j**) March (R_{03}/P_{03}), and (**l**) April (R_{04}/P_{04}). The MK test statistic is denoted as Z. The straight red line in each figure represents the linear trend (Sen's slope) of monthly rainfall (expressed in mm/y) or monthly rainfall-to-precipitation ratio (expressed in %/y).

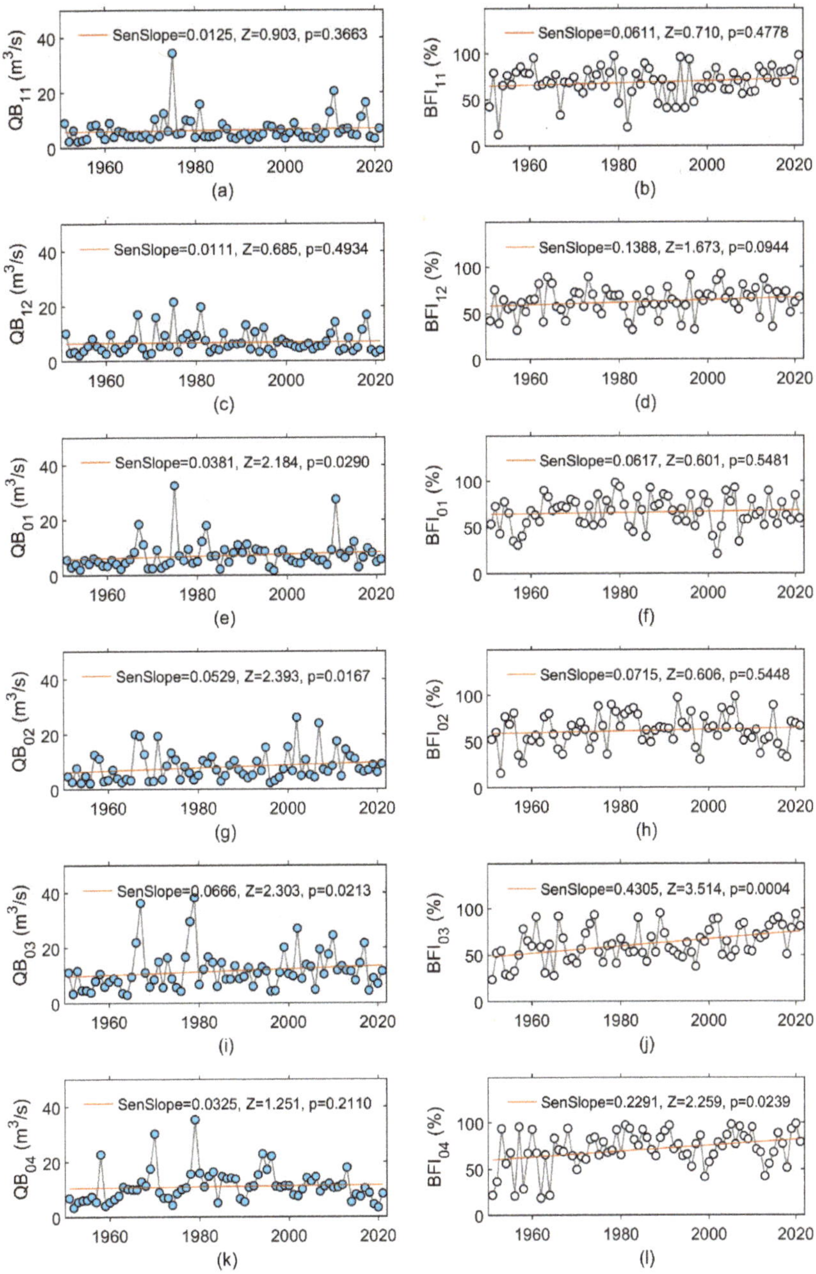

Figure 6. Changes in monthly baseflow in (**a**) November (QB_{11}), (**c**) December (QB_{12}), (**e**) January (QB_{01}), (**g**) February (QB_{02}), (**i**) March (QB_{03}), and (**k**) April (QB_{04}), and changes in monthly baseflow index in (**b**) November (BFI_{11}), (**d**) December (BFI_{12}), (**f**) January (BFI_{01}), (**h**) February (BFI_{02}), (**j**) March (BFI_{03}), and (**l**) April (BFI_{04}). The MK test statistic is denoted as Z. The straight red line in each figure represents the linear trend (Sen's slope) of the monthly baseflow (expressed in m^3s^{-1}/y) or monthly baseflow index (expressed in %/y).

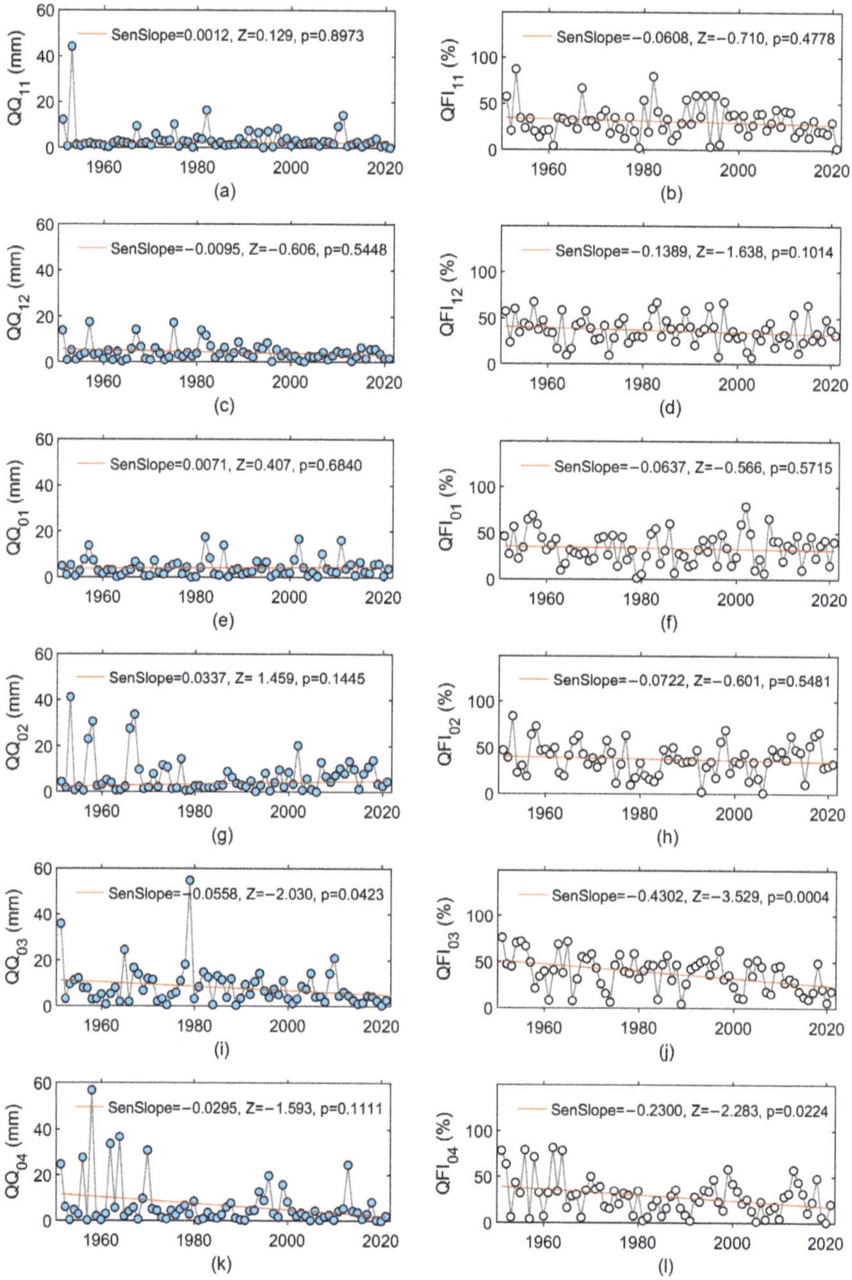

Figure 7. Changes in monthly quick flow in (**a**) November (QQ_{11}), (**c**) December (QQ_{12}), (**e**) January (QQ_{01}), (**g**) February (QQ_{02}), (**i**) March (QQ_{03}), and (**k**) April (QQ_{04}); and changes in quick flow index in (**b**) November (QFI_{11}), (**d**) December (QFI_{12}), (**f**) January (QFI_{01}), (**h**) February (QFI_{02}), (**j**) March (QFI_{03}), and (**l**) April (QFI_{04}). The MK test statistic is denoted as Z. The straight red line in each figure represents the linear trend (Sen's slope) of the monthly quick flow (expressed in m^3s^{-1}/y) or the monthly quick flow index (expressed in %/y).

3.3. Similarity between Trends over the Period of 1951–2021

The Spearman rank-order correlation coefficients (R_S) expressed the strength of a link between trends in the monthly temperature time series and other variables, including snowfall, rainfall, baseflow, and quick flow (Table 1). With the rising temperature, the tendency toward increases in the rainfall, rainfall-to-precipitation ratio, baseflow, and baseflow index was confirmed by the positive values of R_S. Negative values of R_S manifested that rising temperature where the snowfall, snowfall-to-precipitation ratio, quick flow and quick flow index had a tendency to decrease. Strong statistically significant results were revealed for the snowfall-to-precipitation ratio, rainfall-to-precipitation ratio, and baseflow.

Table 1. Spearman rank-order correlations (R_s) between multi-year trends (Sen's slopes) in the monthly temperature (T) and hydroclimatic variables of snow (S), snow-to-precipitation ratio (S/P), rainfall (R), rainfall-to-precipitation ratio (R/P), baseflow (QB), baseflow index (BFI), quick flow (QQ), and quick flow index (QFI).

Variable	S	S/P	R	R/P	QB	BFI	QQ	QFI
T	−0.2571	−0.8407 [1]	0.7142	0.8857 [1]	0.9429 [1]	0.5429	−0.0857	−0.5429

[1] Correlation was statistically significant at a significance level of alpha = 0.05.

3.4. Shift in River Regime

A shift-type change in the snow-to-runoff ratio (S/QT) occurred in 1971 as detected by the change point analysis (Figure 8). The Kruskal–Wallis test confirmed the significance of the difference between the median values of S/QT in the two subperiods. The mean values decreased from 0.61 in the subperiod of 1951–1970 to 0.44 in the subperiod of 1971–2021. In the first snowy period (1951–1970), the highest values of S/QT reached 0.95, while the lowest reached 0.35. In 1971–2021, the highest S/QT did not exceed 0.82 and very often dropped below 0.35; the lowest value of 0.09 occurred in 2020. When considering the mean S/QT in the two subperiods and the threshold criterion of S/QT = 0.5 applied in [1], it was concluded that the river regime shifted from the snow-dominated to the snow-affected type with a mixed recharge by both the snow and rainfall precipitation phases.

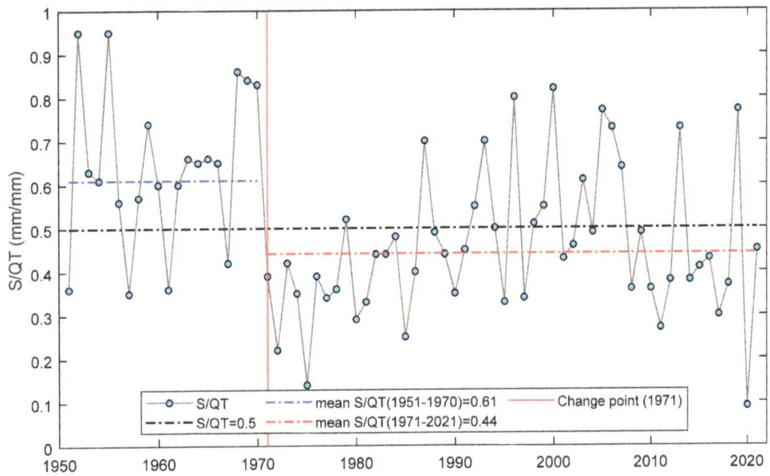

Figure 8. Snow-to-runoff ratio (S/QT) in two subperiods (1951–1970 and 1971–2021) separated by a change point in 1971.

It is worth noting that the term "snow-dominated" refers to the S/QT ratio. As shown in Figure 2d, the snowfall ($S_{11\text{-}04}$) rarely exceeded rainfall ($R_{11\text{-}04}$). In the subperiod of 1951–1970, the $S_{11\text{-}04}/P_{11\text{-}04}$ ratio exceeded 50% in 1958 (63%), 1968 (53%), 1969 (56%), and 1970 (53%); the multi-year mean value was 43%. In the subperiod of 1971–2021, the mean of $S_{11\text{-}04}/P_{11\text{-}04}$ dropped to 33%. The $S_{11\text{-}04}/P_{11\text{-}04}$ had maximum values in 1979 (62%), 1996 (76%), 2005 (53%), and 2013 (66%) and exhibited strong interannual variations. On average, rainfall ($R_{11\text{-}04}$) exceeds snowfall ($S_{11\text{-}04}$) (Figure 2f), and the $S_{11\text{-}04}/P_{11\text{-}04}$ ratio gradually decreased (Figure 2d).

4. Discussion

This study demonstrated that the air temperature remarkably increased in the last seven decades at both seasonal and monthly time scales over the cold season of 1951–2021. The results of the analyses supported the study's hypotheses. Warming winters have gradually reduced the snowfall amount and snowfall fraction of total precipitation. However, this did not exclude the occurrence of extremely snowy winters, an example of which was the snowfall in 2012/2013, which was the highest in the entire 70 years in this river basin (Figure 2c). However, the 2019/2020 season was marked by extremely low snowfall and the lowest share of snow in the precipitation, which was caused by the exceptionally high temperatures that occurred throughout the cold season. This was the mildest winter on record across Europe, particularly in the north and east [47]. Overall, in the Liwiec River basin, the cold-season temperature rose by ~2.70 °C over 71 years, and the snowfall-to-precipitation ratio (S/P ratio) decreased by ~16% over 71 years. The warming winter temperatures across Poland were previously confirmed; it was found that at the majority of weather stations in Poland, the snow-cover depth significantly decreased in recent decades [29,30]. It is worth mentioning that atmospheric thaws alternately occur with cool and frosty periods and are characteristic features of the climate of Poland. These are caused by the variable weather conditions in winter seasons [48]. In the analyzed river basin, in the years of 1960/1961–2009/2010, the mean number of days with atmospheric thaw in December–February was in the range of 10–12 days [48]. However, the extreme thaw-start and thaw-end dates could differ by more than three months due to the high interannual variability. Thus, with increasing temperatures, it might be expected that the frequency of atmospheric thaws would increase, thereby accelerating snowmelt and making the snow cover less persistent. An increased activity of the hydrological system is expected to manifest with the amplified infiltration process occurring on large flat surfaces, which recharges the groundwater from which an increased baseflow is generated.

The follow-up hypothesis was also confirmed; it was found that the relationship between the monthly trends in air temperature and baseflow was strong and statistically significant. Hence, the increasing cold-season temperature trend contributed to the changes in streamflow marked by an increase in baseflow and baseflow index (BFI). The baseflow component of streamflow increased to a depth of ~17 mm over 71 years, and the baseflow index rose by ~18% in 71 years. The BFI increased from 0.51 in the subperiod of 1951–1970 to 0.66 in the subperiod of 1971–2021. The opposite results were found regarding the quick flow metrics (QFI); the time series of the quick flow and quick flow index showed decreasing trends. This may have been due to more frequent snow- and soil-thawing periods and the lack of a sudden amount of melting snow quickly entering the river in the form of overland flow. The results of this study seemed to be consistent with findings in [49], in which it was proved that runoff along the Vistula River in the winter season has become more uniform and shows decreasing maxima and increasing minima of daily flows and a stable mean runoff volume. The study in [4] also made significant contributions to the knowledge of global trends by showing that changes in both the baseflow and BFI were significantly region-dependent. The rivers in the eastern part of Poland were not examined, but the western part (covering the Oder River basin) showed a decreasing baseflow and BFI for the winter season (December–January–February) when evaluated for the period of 1970–2016. Such contrasting results were, as proved in [4], region-dependent; the eastern part of Poland where the Liwiec River basin is situated has a much cooler and humid climate than the western part.

5. Conclusions

The main conclusions of this work can be summarized as follows:

(1) The lowland, mesoscale river basin in the humid middle-latitude climate rapidly warmed by ~0.38 °C/decade ($p < 0.05$) in the cold season of the water years (November–April) of 1951–2021, and the strongest warming was observed in recent decades. The highest level of warming occurred in February and March, reaching 0.49 and 0.53°C/decade, respectively.

(2) The warming climate has directly affected the hydrological system, thereby leading to decreases in the snowfall and snowfall fraction ($p < 0.05$), increases in the rainfall and rainfall fraction ($p < 0.05$), increases in the river baseflow and baseflow index ($p < 0.05$), and a slight decrease in the total runoff ($p > 0.05$).

(3) A warmer cold season of the water year and related changes in snowfall/rainfall patterns threaten the river regime by shifting it from the snow-dominated to snow-affected type as defined by the snowfall-to-runoff ratio. A change in river regime occurred in the 1970s, thereby transforming the river system into a less snowy and more rain-dependent system.

Overall, this study provided evidence of the link between trends in temperature and other hydroclimatic characteristics. However, the obtained results only applied to the analyzed river basin and require further verification for a broader range of geographic and climatic conditions in the middle-latitude region. Thus, future studies should consider other river basins. In Poland, the existing historical climate and hydrometric data enable further analysis of this area. Comparative studies that examine other river basins could explain the regional similarities or differences related to snow-hydrology and river-regime transitions. Such research could provide an answer regarding which river basins are still characterized by a snow regime and which have already undergone this transformation.

The changes in the hydrological regime analyzed in this paper only concerned the amount of snowfall related to the total annual runoff as expressed by the snowfall/runoff ratio. When considering the river regime changes, shifts in the time of the maximum winter/spring flows and their impact on the occurrence of low summer flows are also important. Such an analysis would be particularly valuable from the perspective of seasonal water resource availability.

The obtained results suggested that diminishing snowfall is expected to alter the groundwater recharge and streamflow dynamics if warming trends continue. Thus, the analysis presented here could be followed by a consideration of climate change scenarios and their impacts on snowfall and streamflow seasonality. Although this analysis was limited to showing similarities in the studied hydroclimatic variable trends, it seemed to provide valuable insights into the drivers and causes of changes in the river regime in the middle-latitude lowland region.

Funding: This research was funded by the Faculty of Geography and Regional Studies, University of Warsaw, Poland, under grant nos. SWIB66/2022 and SWIB85/2022.

Data Availability Statement: All of the data used during this study are available at the locations cited in the Acknowledgements section.

Acknowledgments: The author thanks the editors and reviewers for their insightful comments and suggestions. The author acknowledges the E-OBS dataset (version 23.1e) from the EU-FP6 project ENSEMBLES (http://ensembles-eu.metoffice.com) and the data providers in the ECA&D project (http://www.ecad.eu); version 25.0e of the dataset was downloaded from the Copernicus Climate Change Service (https://surfobs.climate.copernicus.eu/dataaccess/access_eobs.php) (accessed on 19 June 2022). The author also acknowledges the gridded Northern Hemisphere 50% rain–snow Ts threshold product (a formatted version of the observational dataset) that was downloaded from the DRYAD: spatial variations in the rain–snow temperature threshold across the Northern Hemisphere (https://doi.org/10.5061/dryad.c9h35) (accessed on 31 January 2022). The author also acknowledges the streamflow dataset that was prepared and verified by the Institute of Meteorology and Water Management—National Research Institute in Warsaw, Poland (IMGW-PIB) and downloaded from https://danepubliczne.imgw.pl/data (accessed on 30 August 2022).

Conflicts of Interest: The author declares no conflict of interest. The funder had no role in the design of the study; in the collection, analyses, or interpretation of data; in the writing of the manuscript; or in the decision to publish the results.

References

1. Barnett, T.P.; Adam, J.C.; Lettenmaier, D.P. Potential impacts of a warming climate on water availability in snow-dominated regions. *Nature* **2005**, *438*, 303–309. [CrossRef] [PubMed]
2. Nijssen, B.; O'Donnell, G.M.; Hamlet, A.F.; Lettenmaier, D.P. Hydrologic Sensitivity of Global Rivers to Climate Change. *Clim. Chang.* **2001**, *50*, 143–175. [CrossRef]
3. Wieder, W.R.; Kennedy, D.; Lehner, F.; Musselman, K.N.; Rodgers, K.B.; Rosenbloom, N.; Simpson, I.R.; Yamaguchi, R. Pervasive alterations to snow-dominated ecosystem functions under climate change. *Proc. Natl. Acad. Sci. USA* **2022**, *119*, e2202393119. [CrossRef]
4. Tan, X.; Liu, B.; Tan, X. Global Changes in Baseflow Under the Impacts of Changing Climate and Vegetation. *Water Resour. Res.* **2020**, *56*, e2020WR027349. [CrossRef]
5. Lilhare, R.; Déry, S.J.; Stadnyk, T.A.; Pokorny, S.; Koenig, K.A. Warming soil temperature and increasing baseflow in response to recent and potential future climate change across northern Manitoba, Canada. *Hydrol. Process.* **2022**, *36*, e14748. [CrossRef]
6. Dierauer, J.R.; Allen, D.M.; Whitfield, P.H. Snow Drought Risk and Susceptibility in the Western United States and Southwestern Canada. *Water Resour. Res.* **2019**, *55*, 3076–3091. [CrossRef]
7. Moraga, J.S.; Peleg, N.; Fatichi, S.; Molnar, P.; Burlando, P. Revealing the impacts of climate change on mountainous catchments through high-resolution modelling. *J. Hydrol.* **2021**, *603*, 126806. [CrossRef]
8. Kang, D.H.; Gao, H.; Shi, X.; Islam, S.U.; Déry, S.J. Impacts of a Rapidly Declining Mountain Snowpack on Streamflow Timing in Canada's Fraser River Basin. *Sci. Rep.* **2016**, *6*, 19299. [CrossRef]
9. Dierauer, J.R.; Allen, D.M.; Whitfield, P.H. Climate change impacts on snow and streamflow drought regimes in four ecoregions of British Columbia. *Can. Water Resour. J. Rev. Can. Ressour. Hydr.* **2021**, *46*, 168–193. [CrossRef]
10. Musselman, K.N.; Addor, N.; Vano, J.A.; Molotch, N.P. Winter melt trends portend widespread declines in snow water resources. *Nat. Clim. Chang.* **2021**, *11*, 418–424. [CrossRef]
11. Aygün, O.; Kinnard, C.; Campeau, S. Impacts of climate change on the hydrology of northern midlatitude cold regions. *Prog. Phys. Geogr. Earth Environ.* **2019**, *44*, 338–375. [CrossRef]
12. Dong, W.; Ming, Y. Seasonality and Variability of Snowfall to Total Precipitation Ratio over High Mountain Asia Simulated by the GFDL High-Resolution AM4. *J. Clim.* **2022**, *35*, 5573–5589. [CrossRef]
13. Yao, J.; Chen, Y.; Guan, X.; Zhao, Y.; Chen, J.; Mao, W. Recent climate and hydrological changes in a mountain–basin system in Xinjiang, China. *Earth-Sci. Rev.* **2022**, *226*, 103957. [CrossRef]
14. Sippel, S.; Fischer, E.M.; Scherrer, S.C.; Meinshausen, N.; Knutti, R. Late 1980s abrupt cold season temperature change in Europe consistent with circulation variability and long-term warming. *Environ. Res. Lett.* **2020**, *15*, 94056. [CrossRef]
15. Mudryk, L.R.; Kushner, P.J.; Derksen, C.; Thackeray, C. Snow cover response to temperature in observational and climate model ensembles. *Geophys. Res. Lett.* **2017**, *44*, 919–926. [CrossRef]
16. Mudryk, L.; Santolaria-Ot\'\in, M.; Krinner, G.; Ménégoz, M.; Derksen, C.; Brutel-Vuilmet, C.; Brady, M.; Essery, R. Historical Northern Hemisphere snow cover trends and projected changes in the CMIP6 multi-model ensemble. *Cryosphere* **2020**, *14*, 2495–2514. [CrossRef]
17. Hattermann, F.F.; Vetter, T.; Breuer, L.; Su, B.; Daggupati, P.; Donnelly, C.; Fekete, B.; Flörke, F.; Gosling, S.N.; Hoffmann, P.; et al. Sources of uncertainty in hydrological climate impact assessment: A cross-scale study. *Environ. Res. Lett.* **2018**, *13*, 15006. [CrossRef]
18. Quante, L.; Willner, S.N.; Middelanis, R.; Levermann, A. Regions of intensification of extreme snowfall under future warming. *Sci. Rep.* **2021**, *11*, 16621. [CrossRef]

19. Danco, J.F.; DeAngelis, A.M.; Raney, B.K.; Broccoli, A.J. Effects of a Warming Climate on Daily Snowfall Events in the Northern Hemisphere. *J. Clim.* **2016**, *29*, 6295–6318. [CrossRef]
20. Fontrodona Bach, A.; van der Schrier, G.; Melsen, L.A.; Klein Tank, A.M.G.; Teuling, A.J. Widespread and Accelerated Decrease of Observed Mean and Extreme Snow Depth Over Europe. *Geophys. Res. Lett.* **2018**, *45*, 12312–12319. [CrossRef]
21. Walsh, J.E.; Ballinger, T.J.; Euskirchen, E.S.; Hanna, E.; Mård, J.; Overland, J.E.; Tangen, H.; Vihma, T. Extreme weather and climate events in northern areas: A review. *Earth-Sci. Rev.* **2020**, *209*, 103324. [CrossRef]
22. Lin, W.; Chen, H. Changes in the spatial–temporal characteristics of daily snowfall events over the Eurasian continent from 1980 to 2019. *Int. J. Climatol.* **2022**, *42*, 1841–1853. [CrossRef]
23. Clifton, C.F.; Day, K.T.; Luce, C.H.; Grant, G.E.; Safeeq, M.; Halofsky, J.E.; Staab, B.P. Effects of climate change on hydrology and water resources in the Blue Mountains, Oregon, USA. *Clim. Serv.* **2018**, *10*, 9–19. [CrossRef]
24. Santolaria-Otín, M.; Zolina, O. Evaluation of snow cover and snow water equivalent in the continental Arctic in CMIP5 models. *Clim. Dyn.* **2020**, *55*, 2993–3016. [CrossRef]
25. Kraaijenbrink, P.D.A.; Stigter, E.E.; Yao, T.; Immerzeel, W.W. Climate change decisive for Asia's snow meltwater supply. *Nat. Clim. Chang.* **2021**, *11*, 591–597. [CrossRef]
26. Notarnicola, C. Overall negative trends for snow cover extent and duration in global mountain regions over 1982–2020. *Sci. Rep.* **2022**, *12*, 13731. [CrossRef]
27. Ford, C.M.; Kendall, A.D.; Hyndman, D.W. Effects of shifting snowmelt regimes on the hydrology of non-alpine temperate landscapes. *J. Hydrol.* **2020**, *590*, 125517. [CrossRef]
28. Ford, C.M.; Kendall, A.D.; Hyndman, D.W. Snowpacks decrease and streamflows shift across the eastern US as winters warm. *Sci. Total Environ.* **2021**, *793*, 148483. [CrossRef]
29. Tomczyk, A.M.; Bednorz, E.; Szyga-Pluta, K. Changes in air temperature and snow cover in winter in poland. *Atmosphere* **2021**, *12*, 68. [CrossRef]
30. Szwed, M.; Pińskwar, I.; Kundzewicz, Z.W.; Graczyk, D.; Mezghani, A. Changes of snow cover in Poland. *Acta Geophys.* **2017**, *65*, 65–76. [CrossRef]
31. Solon, J.; Borzyszkowski, J.; Bidłasik, M.; Richling, A.; Badora, K.; Balon, J.; Brzezińska-Wójcik, T.; Chabudziński, Ł.; Dobrowolski, R.; Grzegorczyk, I.; et al. Physico-geographical mesoregions of Poland: Verification and adjustment of boundaries on the basis of contemporary spatial data. *Geogr. Pol.* **2018**, *91*, 143–170. [CrossRef]
32. Beck, H.E.; Zimmermann, N.E.; McVicar, T.R.; Vergopolan, N.; Berg, A.; Wood, E.F. Present and future köppen-geiger climate classification maps at 1-km resolution. *Sci. Data* **2018**, *5*, 180214. [CrossRef] [PubMed]
33. Cornes, R.C.; van der Schrier, G.; van den Besselaar, E.J.M.; Jones, P.D. An Ensemble Version of the E-OBS Temperature and Precipitation Data Sets. *J. Geophys. Res. Atmos.* **2018**, *123*, 9391–9409. [CrossRef]
34. IMGW-PIB Water Yearbook 2021 (Rocznik hydrologiczny 2021). Available online: https://danepubliczne.imgw.pl/data/dane_pomiarowo_obserwacyjne/Roczniki/Rocznikhydrologiczny/RocznikHydrologiczny2021.pdf (accessed on 28 October 2022).
35. IMGW-PIB Bulletin of the National Hydrological and Meteorological Service (Biuletyn Państwowej Służby Hydrologiczno-Meteorologicznej). Available online: https://danepubliczne.imgw.pl/data/dane_pomiarowo_obserwacyjne/Biuletyn_PSHM/Biuletyn_PSHM_2021_ROCZNY.pdf (accessed on 28 October 2022).
36. Jennings, K.S.; Winchell, T.S.; Livneh, B.; Molotch, N.P. Spatial variation of the rain-snow temperature threshold across the Northern Hemisphere. *Nat. Commun.* **2018**, *9*, 1148. [CrossRef]
37. Arciniega-Esparza, S.; Breña-Naranjo, J.A.; Pedrozo-Acuña, A.; Appendini, C.M. HYDRORECESSION: A Matlab toolbox for streamflow recession analysis. *Comput. Geosci.* **2017**, *98*, 87–92. [CrossRef]
38. Aksoy, H.; Wittenberg, H. Nonlinear baseflow recession analysis in watersheds with intermittent streamflow. *Hydrol. Sci. J.* **2011**, *56*, 226–237. [CrossRef]
39. Mann, H.B. Nonparametric Tests Against Trend. *Econometrica* **1945**, *13*, 245–259. [CrossRef]
40. Kendall, M.G. A New Measure of Rank Correlation. *Biometrika* **1938**, *30*, 81–93. [CrossRef]
41. Sen, P.K. Estimates of the Regression Coefficient Based on Kendall's Tau. *J. Am. Stat. Assoc.* **1968**, *63*, 1379–1389. [CrossRef]
42. Radziejewski, M.; Kundzewicz, Z.W. Detectability of changes in hydrological records. *Hydrol. Sci. J.* **2004**, *49*, 39–51. [CrossRef]
43. Greene, C.A.; Thirumalai, K.; Kearney, K.A.; Delgado, J.M.; Schwanghart, W.; Wolfenbarger, N.S.; Thyng, K.M.; Gwyther, D.E.; Gardner, A.S.; Blankenship, D.D. The Climate Data Toolbox for MATLAB. *Geochem. Geophys. Geosystems* **2019**, *20*, 3774–3781. [CrossRef]
44. Kruskal, W.H.; Wallis, W.A. Use of Ranks in One-Criterion Variance Analysis. *J. Am. Stat. Assoc.* **1952**, *47*, 583–621. [CrossRef]
45. Kundzewicz, Z.W.; Robson, A.J. Change detection in hydrological records—A review of the methodology. *Hydrol. Sci. J.* **2004**, *49*, 7–19. [CrossRef]
46. Kunkel, K.E.; Robinson, D.A.; Champion, S.; Yin, X.; Estilow, T.; Frankson, R.M. Trends and Extremes in Northern Hemisphere Snow Characteristics. *Curr. Clim. Chang. Rep.* **2016**, *2*, 65–73. [CrossRef]
47. The Copernicus Climate Change Service (C3S). Available online: https://climate.copernicus.eu/boreal-winter-season-1920-was-far-warmest-winter-season-ever-recorded-europe-0 (accessed on 20 October 2022).

48. Czarnecka, M.; Nidzgorska-Lencewicz, J. The occurrence of atmospheric thaw in Poland over the last 50 years. *Geogr. Pol.* **2013**, *86*, 327–340. [CrossRef]
49. Bogdanowicz, E.; Karamuz, E.; Romanowicz, R.J. Temporal changes in flow regime along the river vistula. *Water* **2021**, *13*, 2840. [CrossRef]

resources

MDPI

Article

Forecasting Monthly River Flows in Ukraine under Different Climatic Conditions

Renata Graf [1,*] and Viktor Vyshnevskyi [2]

1 Department of Hydrology and Water Management, Institute of Physical Geography and Environmental Planning, Adam Mickiewicz University, Bogumiła Krygowskiego Street 10, 61-680 Poznań, Poland
2 Faculty of International Relations, National Aviation University, Liubomyra Huzara Avenue 1, 03058 Kyiv, Ukraine
* Correspondence: renata.graf@amu.edu.pl; Tel.: +48-61-829-6259

Abstract: River-flow forecasts are important for the management and planning of water resources and their rational use. The present study, based on direct multistep-ahead forecasting with multiple time series specific to the XGBoost algorithm, estimates the long-term changes and forecast monthly flows of selected rivers in Ukraine. In a new, applied approach, a single multioutput model was proposed that forecasts over both short- and long-term horizons using grouped or hierarchical data series. Three forecast stages were considered: using train and test subsets, using a model with train-test data, and training with all data. The historical period included the measurements of the monthly flows, precipitation, and air temperature in the period 1961–2020. The forecast horizons of 12, 60, and 120 months into the future were selected for this dataset, i.e., December 2021, December 2025, and December 2030. The research was conducted for diverse hydrological systems: the Prut, a mountain river; the Styr, an upland river; and the Sula, a lowland river in relation to the variability and forecasts of precipitation and air temperature. The results of the analyses showed a varying degree of sensitivity among rivers to changes in precipitation and air temperature and different projections for future time horizons of 12, 60, and 120 months. For all studied rivers, variable dynamics of flow was observed in the years 1961–2020, yet with a clearly marked decrease in monthly flows during in the final, 2010–2020 decade. The last decade of low flows on the Prut and Styr rivers was preceded by their noticeable increase in the earlier decade (2000–2010). In the case of the Sula River, a continuous decrease in monthly flows has been observed since the end of the 1990s, with a global minimum in the decade 2010–2020. Two patterns were obtained in the forecasts: a decrease in flow for the rivers Prut (6%) and the Styr (12–14%), accompanied by a decrease in precipitation and an increase in air temperature until 2030, and for the Sula River, an increase in flow (16–23%), with a slight increase in precipitation and an increase in air temperature. The predicted changes in the flows of the Prut, the Styr, and the Sula rivers correspond to forecasts in other regions of Ukraine and Europe. The performance of the models over a variety of available datasets over time was assessed and hyperparameters, which minimize the forecast error over the relevant forecast horizons, were selected. The obtained RMSE parameter values indicate high variability in hydrological and meteorological data in the catchment areas and not very good fit of retrospective data regardless of the selected horizon length. The advantages of this model, which was used in the work for forecasting monthly river flows in Ukraine, include modelling multiple time series simultaneously with a single model, the simplicity of the modelling, potentially more-robust results because of pooling data across time series, and solving the "cold start" problem when few data points were available for a given time series. The model, because of its universality, can be used in forecasting hydrological and meteorological parameters in other catchments, irrespective of their geographic location.

Keywords: river flow; XGBoost algorithm; trends; multistep-ahead forecasting; climate variability; Ukraine

Citation: Graf, R.; Vyshnevskyi, V. Forecasting Monthly River Flows in Ukraine under Different Climatic Conditions. *Resources* **2022**, *11*, 111. https://doi.org/10.3390/resources11120111

Academic Editors: Diego Copetti and Demetrio Antonio Zema

Received: 11 October 2022
Accepted: 26 November 2022
Published: 30 November 2022

1. Introduction

Some of the most popular scientific issues of modern hydrology are climate variability and change and its impact on the water regime, which is of primary importance for states with generally scarce water resources. Modelling and forecasting the time series of river flows are essential elements in the assessment of the hydrological regime and the management of water resources [1–3]. The results of analyses of multiyear flow measurement series enable an assessment of the reaction rivers to supply factors or their limitation. The decomposition of the time series of flows allows us to capture the change trends, seasonality, and a random factor, which is most often a factor disturbing the forecasting of time series [4]. The results of the forecasts, among others, are important in preventing floods and droughts and reducing the effects of their occurrence [5].

Typically, a statistical and deterministic approach is used in hydrological forecasting. Forecasting models fall into two main groups: the physics-based numerical models and the data-driven prediction models [6]. Physics-based models perform mathematical modelling to simulate dynamic processes, such as floods. Data-driven models are also widely used to model and forecast the flow or rainfall–runoff relationship [7]. It has been shown that data-driven approaches can also achieve a comparable performance to that of physics-based models in predicting flows and extreme hydrological phenomena. According to Tu et al. [8] data-driven models, especially machine-learning models, are becoming alternative approaches to hydrological and hydraulic models. Predictive machine-learning algorithms are used in many studies in the field of applied and data-driven hydrology [9–11]. Shen et al. [12] emphasize that machine-learning models provide improvements to nonlinear modelling over other data-driven techniques. The effectiveness of artificial neural network (ANN) models in hydrological forecasting often exceeds the effectiveness of traditional conceptual or mathematical models based on the modelling of complex hydrological processes [13].

The modelling of temporal flow series is usually performed for both short-term forecasting [14] and long-term forecasting and often for high or low flows [15]. The concept of a short-term real-time daily unsteady flow forecast refers to the relationship between the instantaneous states (water gauge compounds), also without taking into account rainfall data [14,16,17]. Among the long-term forecasting methods, both qualitative and quantitative methods are used [18]. Quantitative methods (referred to as traditional methods) often include statistical methods, correlation, and regression analysis [19–23]. There are many methods available for predicting river flow, including process-based models, which, however, require a lot of data [24]. Models based on computational intelligence and machine-learning algorithms are gaining popularity [6]. Most often, good flow forecast results are obtained by comparing several models [16,25]. In the development of flow forecasts and the assessment of their quality, the conventional linear autoregressive relationship (AR), ANN models (e.g., three-layer feedforward neural network), recursive neural networks (RNN)), and a number of hybrid models are used. Accurate river-flow forecasts are obtained with the RNN model, which often also has the greatest ability to generalize results, showing similar forecast quality in independent tests.

Among the neural networks, the model for the short term works well for river-flow prediction [26–29]. The ANN models have also been used to model rainfall-runoff, river-flow, and flood forecasting, among others, by Imrie et al. [26] for selected catchments in the UK; by Kim and Barros [30] for selected rivers in Pennsylvania (the US); and by Toth et al. [25] for rivers in the Apennines (Italy). Comparisons of prediction and prognostic models of river flows are often made on the basis of autoregressive techniques, neural networks and adaptive neuro-fuzzy inference systems (ANFIS) [31,32]. In the short-term to long-term forecasting of river flows, heuristic optimization algorithms hybridized with ANFIS work well. All developed hybrid algorithms significantly outperformed the traditional ANFIS model performance for all prediction horizons, which is a major advantage compared with the classical black box machine-learning models [33,34]. The performance of neural network models in river-flow forecasting has been compared with linear regression models [35] and stochastic models [36–38]. Forecasts have been made for one river by using different models, or assessments have been made by using different methods (alternative methods) for contrasting catchments. The forecasts are based on flows observed in the cross sections of the river system with a daily delay and without taking into account any rainfall data. Rainfall-runoff simulation modelling is also used, taking into account RNNs [39] as well as deep learning with a long short-term memory (LSTM) network approach [13,40]. The LSTMs are a type of recursive neural networks capable of learning sequence dependencies in sequence prediction problems. Toth et al. [25] have conducted a comparison of short-term rainfall prediction models for real-time Floyd forecasting.

Multimodel data fusion is a tool commonly used in river-flow forecasting. Abrahart and See [16] have conducted data fusion for two different catchments using arithmetic averaging, the probabilistic method (in which the best method is used to generate the current forecast), the last time step model, two different neural network operations, and two different soft computing methodologies. Each site demonstrated several options and potential benefits of using data fusion tools to produce better estimates of hydrological forecasts.

With the improvement of data-driven modelling in hydrological applications, team-learning methods (e.g., gradient boosted decision tree, or GBDT) are widely used [41]. Ensemble machine-learning methods are employed to solve problems related to simulation and prediction in hydrology, including resampling methods (bagging, boosting, and dagging), model averaging, and stacking [42]. Boosting methods (e.g., boosting, adaboost, and extreme gradient boosting) are becoming more and more effective at modelling and forecasting runoff, flooding, and drought and at predicting ice phenomena in rivers [43]. They are used in hydrological modelling to achieve better performance by combining many weaker models into a stronger model [44]. After developing a new streamflow forecasting model based on the modular model, Ni et al. [45] confirmed that models based on extreme gradient gain (XGBoost) give satisfactory forecasts. The results of these studies for the Yangtze River monthly flows indicate that XGBoost is applicable to river-flow forecasting and generally performs better than the support vector machine (SVM). Extreme gradient boosting (XGBoost) has been used to predict flood stages in a data-driven flood alert system (FAS) for a flood-prone watershed in Houston (Texas) [41]. It has been shown that an XGBoost-based FAS can operate continuously to automatically detect flooding, without needing activate an external start-up trigger, as is usually required in conventional event-based warning systems. Flood warning systems built on predictive models constitute a proactive method of flood risk assessment and management [46].

Forecasting changes in the water-balance and hydrological regime is particularly important and valuable for countries struggling with a shortage of water resources and the effects of climate variability and change on such resources. The impact of the human factor is largely uncontrolled, which is noted in Ukraine, where hostilities have been taking place since February 2022. Water resources in Ukraine are becoming increasingly scarce, with the water infrastructure being increasingly exposed to war. In Ukraine, the subject of river-flow forecasting in relation to multiyear measurement series or high (flood) or low flows is undertaken on a different scale. In many studies conducted on Ukrainian rivers, forecasting refers to maximum flows and spring floods, in which the authors rely on the traditional approach, i.e., the use of quantitative methods [47]. In contrast, Khrystiuk and Gorbachova [15] have conducted long-term forecasting on extraordinary spring floods by means of the commensurability method on the Dnipro River near Kyiv, Ukraine. The method used was the Weng Wen-Bo information method, which is a qualitative forecasting method that makes it possible to identify the periods and specific years in which subsequent extraordinary spring river floods may occur. The impact of climate on water resources has also been studied for selected river catchments in Ukraine, e.g., for the Western Bug River in Western Ukraine [48], this by using the water-balance Turk model and the regional model (REMO) [49], climate change scenarios [50] and ecohydrological modelling [51]. Didovets et al. [52] and Vyshnevskyi and Donich [53] have studied one climate change impact on regional water runoff in the Carpathian region, and Loboda and Kozlov [54] rated the water resources of Ukrainian rivers by using average statistical models for the period 2021–2050.

The aim of the research was to determine long-term changes and forecast the flows of selected rivers located in different regions of Ukraine during the period 1961–2020, taking into account changes in precipitation and air temperature. The research covered the following rivers: the Prut (a mountain river), the Styr (an upland river), and the Sula (a lowland river), taking into account the forecast horizons of 12, 60, and 120 months. Monthly river-flow forecasting was carried out in three stages, using input data sets specific to the XGBoost algorithm: (1) forecasting by train and test subsets, (2) forecasting using a model with train-test data, and (3) forecasting by training with all data. The algorithm used allowed direct multistep-ahead forecasting with multiple (grouped or hierarchical) time series, which is important in the case of taking limited measurements of daily river flows. The research results can be used for the management and rational planning of water resources in the studied regions of Ukraine.

2. Materials and Methods

2.1. Study Area

Ukraine is located in the south-eastern part of Europe. The larger part of the country (about 95%) is covered by plains, while the rest is mountainous. For researching long-term changes and predicting river flow in Ukraine, three medium-size river catchments with an area of 500–15,000 km^2 were selected: the Prut River in the mountains (in the Carpathians) and the Styr and Sula rivers on the plains (Figure 1). Three main requirements governed the selection of these rivers: the location within river basins of at least two meteorological stations, a long-term observation period, and a slight impact of human activity on river runoff [55]. We may add that the last demand is especially important in the case of Ukraine, because an important influencing factor, namely irrigation, is present in the south of the country.

In the upper course the Prut River up to the Yaremche hydrological station (with a catchment area of 597 km^2) is a typical mountain river. Moreover, the source of the river is located near the Hoverla mountain, which is the highest in Ukraine (with an elevation of 2061 m) (Figure 1). A major part of the catchment area is covered by forests (primarily spruce). The climate of the catchment area of this river essentially depends on elevation. At the Pozhezhevska meteorological station, at a height of 1451 m a.s.l., the mean annual air temperature is equal to 3.5 °C, the mean air temperature in January is minus 5.8 °C, and the mean air temperature in August (the warmest month) is 13.3 °C. On the other hand, the air temperature at the Yaremche meteorological station (at an elevation of 531 m, near the Yaremche hydrological station) is significantly higher: in the period 1991–2020, the mean annual air temperature equalled 7.9 °C. The coldest month is January (minus 2.2 °C), and the warmest one is July (17.8 °C). Under cold conditions, the snow cover in the upper part of the mountains may persist until early June. The mean annual precipitation in the catchment area is above 1000 mm; at the Pozhezhevska meteorological station, it is 1548 mm; and at the Yaremche station, it is 1013 mm. There are waterfalls on the river, including the Probiy waterfall, several kilometres upstream from Yaremche hydrological station. At the same time, there is no pond or any other water reservoir upstream the station. The water regime of the Prut River depends mainly on the snow cover depth and heavy rains. As a rule, the largest river flow is observed in April, during the snow-melting season. Meanwhile, the largest water discharges are caused by heavy rains in the summer period. The main flow characteristics of the Prut River (Yaremche water gauge) are shown in Table 1.

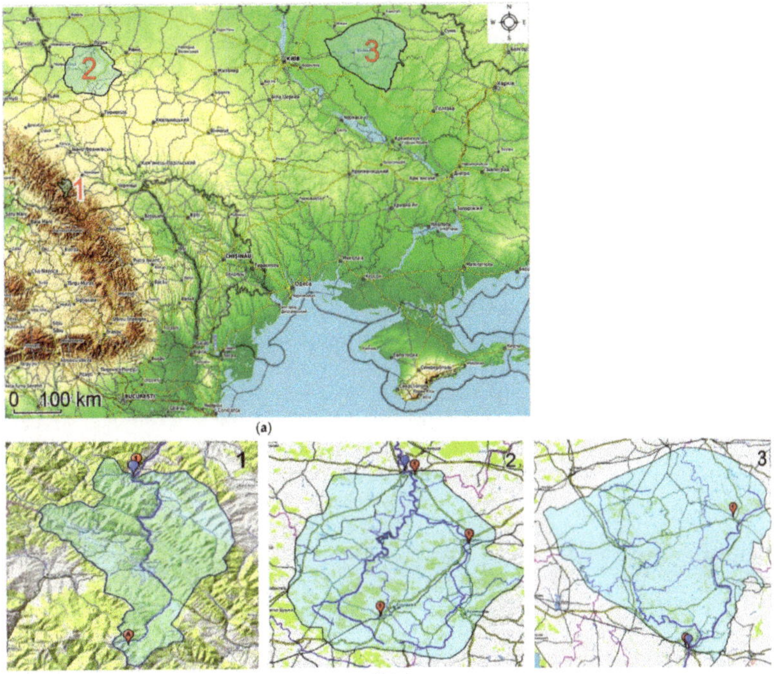

Figure 1. The location of the catchment areas of the studied rivers (**a**), the hydrological (marked in blue) and meteorological (marked in red) stations (**b**): 1, the Prut–Yaremche, 2, the Styr–Lutsk, and 3, the Sula–Lubny. Note that the name of the catchment area includes the name of the river and the water gauge stations (catchment closing cross section).

Table 1. The main characteristics of the discharge conditions of the studied rives, data modified, according to [55].

River and Water Gauge Stations	Catchment Area [km^2]	Discharge [m^3/sec]			
		Long-Term Period Observation			1961–2020
		Mean Annual (Period)	Observed Maximum (Date)	Observed Minimum (Date)	Mean Annual
Prut–Yaremche	597	12.4 (1951–2020)	1530 (08.06.1969)	0.68 (24.01.1976)	12.7
Styr–Lutsk	7200	30.5 (1923–2020)	876 (08.04.1932)	2.26 (18–19.09.2019)	31.2
Sula–Lubny	14,200	27.4 (1936–2020)	1140 (21.04.1942)	0.38 (21.09.2019)	26.4

The Styr River is located in the north-western part of Ukraine (Figure 1). It is a tributary of the Pripyat River, and the Pripyat River is the largest tributary of the Dnipro River. The upper, southern part of the catchment area upstream features the town of Lutsk, on Volyn-Podilsk Upland, while the lower part is on the Polesian Lowland. The climate of the catchment area is moderate with cool winters and warm summers. During the period 1991–2020, the mean annual air temperature at the Lutsk meteorological station was equal to 8.5 °C; the mean air temperature in January was minus 2.9 °C, whereas in July, it was 19.7 °C. The air temperature at other stations, located upstream, was a little higher (by 0.1–0.2 °C). The precipitation in the catchment area upstream from the town of

Lutsk amounts to an average of about 600–700 mm. The water regime of the Styr River at the Lutsk hydrological station (its catchment area is 7200 km^2) is rather stable. The largest discharges observed in April and May are about twice as large as those observed at low water in August and September. In the upper course of the river, there are some reservoirs and many ponds. The reservoirs are rather small, and they do not significantly impact the water regime. The same can be said about ponds, many of which are older than 100 years. The main flow characteristics of the Styr River (Lutsk water gauge) are shown in Table 1.

The Sula River catchment is on the Dnipro Lowland with plain relief (north-east Ukraine), in the forest steppe zone, which actually has little forest (Figure 1). The climate of the catchment area is moderate, with cool winters and warm, sometimes hot, summers. During 1991–2020, the mean annual air temperature at the Lubny meteorological station was equal to 8.6 °C, the mean air temperature in January was minus 4.1 °C, and the mean air temperature in July was 21.3 °C. The air temperature at another station, Romny, located upstream, was lower by approximately 0.5 °C. The precipitation in the catchment area upstream from the town of Lubny amounts to an average of about 600 mm. The water regime of the river, comparable to the Styr River, is much more unstable. The mean discharges at the Lubny hydrological station (with a catchment area of 14,200 km^2), observed in April and May, are approximately 10 times larger than those observed at low water in September. The flow regulation of the Sula River is small. The main flow characteristics of the Sula River (Lubny) are shown in Table 1.

2.2. Dataset Characteristics

The dataset consisted of the average monthly river flows, precipitation, and air temperatures from the period 1961–2020, acquired for three studied catchment areas. The monthly river-flow data were analysed for the Prut–Yaremche, the Styr–Lutsk, and the Sula–Lubny hydrological stations, and for the meteorological stations (Figure 1b), data on the average monthly air temperature and precipitation were analysed. For the Prut River basin, data were used from the Pozhezhevska meteorological station, in the upper part of the river basin, and from the Yaremche meteorological station, near the water gauge. For the Styr River catchment, data from three meteorological stations were used: Brodu (in the south), Dubno (in the east), and Lutsk(near the hydrological station). For the Sula River basin, data from the Romny (in the north-east part of the river catchment) and Lubny (in the south) meteorological stations were used. The time series consisted of the average monthly data for all points of observation.

2.3. Descriptive Statistics

Descriptive statistics of distribution of river-flow variables showed that the threshold values of skewness and kurtosis were exceeded (Table 2); therefore, for the purposes of further analysis, nonparametric measures of central tendency, i.e., Median (Quartiles 1–3), were used. As for the distribution of precipitation and air temperature, the skewness values did not exceed 2.0 and the kurtosis −7.0 (Table 3); hence, the nonparametric measures were used for further analysis (mean and standard deviation).

Table 2. Descriptive statistics of river flow (m^3/s) grouped for the period 1961–2020 (N = 720).

River	M	SD	Mdn	IQR	Min	Max	Skew.	Kurt.
Prut–Yaremche	12.66	10.54	9.37	12.12	1.2	82.5	1.75	4.51
Styr–Lutsk	31.15	17.82	26.95	15.73	4.03	140	2.43	8.78
Sula–Lubny	26.40	32.49	16,00	22.36	0.71	286	3.52	17.66

Explanations: N—sample size; M—mean; SD—standard deviation; Mdn—median; IQR—interquartile range (IQR = Q3 − Q1); Min—minimum value; Max—maximum value; Skew.—skewness; Kurt.—kurtosis.

Table 3. Global descriptive statistics of precipitation and air temperature grouped by river catchment area for the period 1961–2020 (N = 720).

River Catchment	M	SD	Mdn	Min	Max	Skew.	Kurt.
			Precipitation [mm]				
Prut	102.4	60.3	92.5	3.5	458.5	1.38	3.82
Styr	52.6	33.1	45.5	1.67	197.67	1.25	1.70
Sula	51.1	30.7	45.15	0.75	174.65	0.96	0.89
			Air temperature				
Prut	7.4	7.5	7.8	−11.9	19.9	−0.21	−1.16
Styr	8.5	8.5	8.4	−14.0	22.0	−0.19	−1.16
Sula	8.6	8.4	8.4	−16.0	24.7	−0.15	−1.25

Explanations as in Table 1. The time series consisted of the average monthly data for all points of observation.

Compared with the Prut and Styr rivers, the Sula River was characterized by a high standard deviation of monthly flows from the average value, which means that the monthly flows showed high variability over the analysed period (Table 2). The large variance in the flow series was also confirmed by the fact that the highest IQR (22.36) to median ratio (second quartile Q2) was also determined for the Sula–Lubny and the lowest for the Styr–Lutsk.

Table 3 shows that the catchment area of the Prut River had the highest mean precipitation (about 100 mm per month); for the other two rivers' catchments, the precipitation was approximately 50 mm. The degree of variability was similar for all rivers and represented approximately 60% of the average value. The mean air temperature at the Yaremche station during 1961–2020 was about 7.4 °C, and for the other two rivers, the temperature was in the range 8.4–8.5 °C (Table 3). The degree of variability was similar for all rivers and was about 90–100% of the average value.

2.4. Forecast Algorithm

River-flow and meteorological variables forecasting was carried out in several stages, as shown in flowchart (Figure 2). The algorithm used allowed direct multistep-ahead forecasting with multiple (grouped or hierarchical) time series by using the built-in methods of the "R-forecast ML" package [56] inspired by Bergmeir et al. [57]. The applied direct forecasting approach included the following steps: (1) building a single multioutput model that simultaneously forecasts over short- and long-term forecast horizons, (2) assessing the generalization performance of the model over a variety of available data sets over time, and (3) selecting the hyperparameters that minimize forecast error over the relevant forecast horizons and re-train.

The description of the algorithm is broken down into individual steps in Figure 2.

The first step was to create model training and forecasting datasets with lagged, grouped, dynamic, and static features. For this purpose, the create_lagged_df() function of the "R-forecast ML" package was used. The basic idea behind creating custom-feature lags was to improve model accuracy by removing noisy or redundant features in high-dimensional training data. For this purpose, three horizons were defined with forecasts for 1:12, 1:60, and 120 months into the future. A numerical vector indicating the lags in the dataset rows for the creation of the lagged features (lookback) was defined at levels of decreasing range (features from 1 to 36 months, from 54 to 78 months, and from 120 to 126 months in the past) with monthly frequency. The variables month and year were defined as dynamic variables with no static variables in the dataset. Based on multiple time series, the variable "river_id" identified the groups (hierarchies) that were used as model features but were not lagged. The same procedure was performed for the test data set in the model and forecast stage.

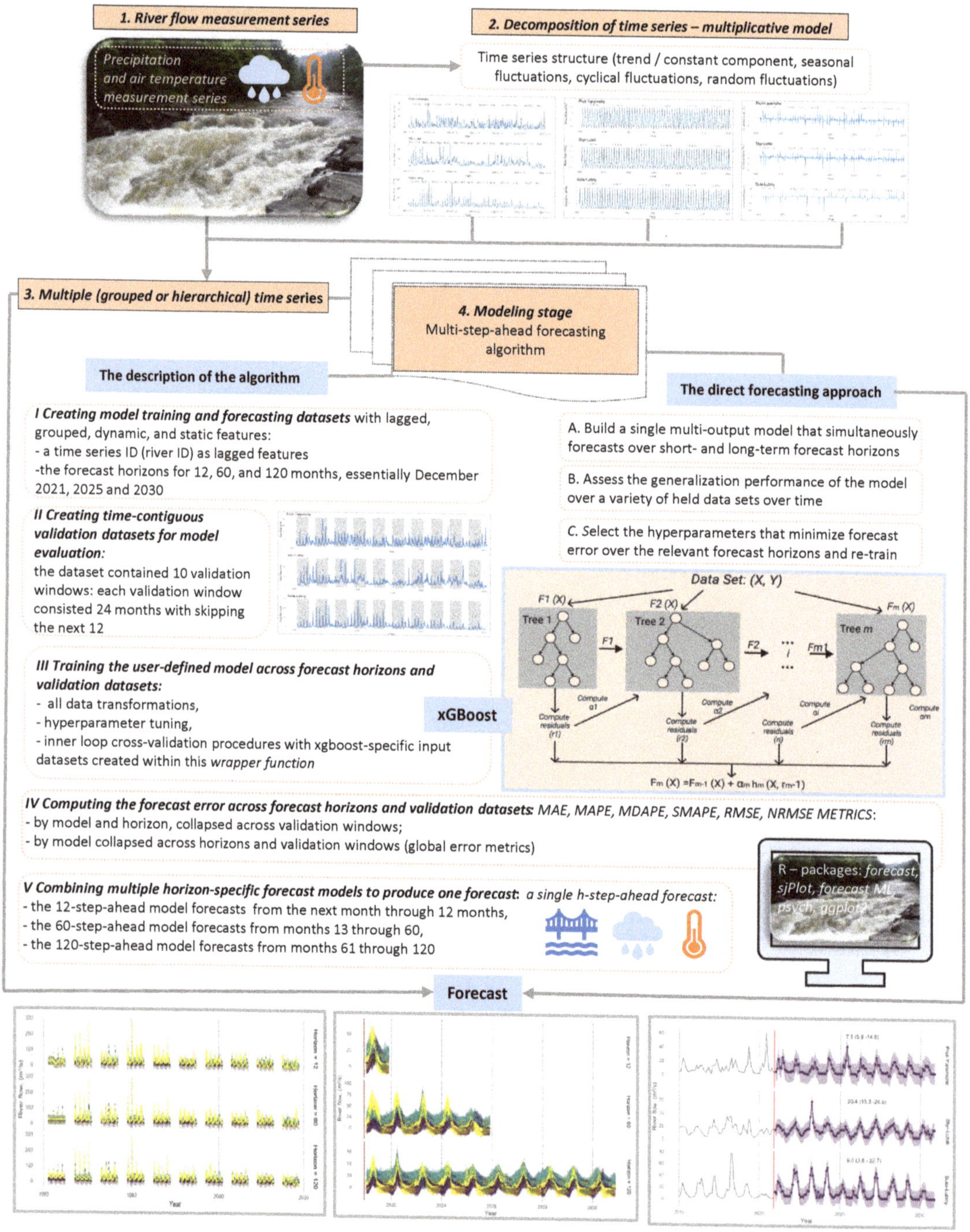

Figure 2. General flowchart of a multistep-ahead forecasting algorithm.

The next step involved creating time-contiguous validation datasets for model evaluation. Given that measurements are taken monthly, each validation window consisted of 24 months, skipping the next 12 months. As a result, the dataset contained 10 validation windows. Custom validation subsets with skipped river-flow data are shown in Figure S1 (Supplementary Materials). A similar approach was used for precipitation and air temperature.

Training the user-defined model across forecast horizons and validation datasets was carried out in the next step. The user-defined model consisted of all data transformations, hyperparameter tuning, and inner loop cross-validation procedures with XGBoost-specific [58] input datasets created within this wrapper function.

The next step involved computing the forecast error across forecast horizons and validation datasets (Figure S1 in Supplementary Materials). Estimating forecast error metrics on the test datasets was based on the calculation of the following: mean absolute error (MAE), mean absolute percentage error (MAPE), median absolute percentage error (MDAPE), symmetrical mean absolute percentage error (SMAPE), root mean squared error (RMSE), and normalized root mean squared error by standard deviation (NRMSE), which have been discussed in detail in the works of, i.a., Bahrami-Pichaghchi and Aghelpour [59], and Aghelpour and Norooz-Valashedi [60].

The final step involved combining the multiple direct-horizon forecast models to produce a single h-step-ahead forecast. Prediction intervals were estimated by calculating the $\alpha/2$ and 1-$\alpha/2$ percentiles of the simulated data for each prediction horizon by using bootstrap residuals and recursive multivariate prediction. The one-step-ahead prediction error was defined as:

$$e_t = y_t - \hat{y}_{t|t-1} + \epsilon \quad (1)$$

where y_t represents the actual prediction, $\hat{y}_{t|t-1}$ represents the prediction conditioned by previous prediction, ϵ represents the estimation error predictions, which were simulated by including the residuals (sampling from the collection of errors from past prediction iterations). The expected variance was represented by the difference in predictions.

The approach was to combine forecasts across models in such a way that short-term models would produce short-term forecasts and long-term models would produce long-term forecasts. This implies that for 120 months, the ahead forecast included (1) the 12-step-ahead model forecasts from the next month through 12 months, (2) the 60-step-ahead model forecasts from months 13 through 60, and (3) the 120-step-ahead model forecasts from months 61 through 120.

Analyses were performed with the statistical language R (version 4.1.1) [61].

XGBoost Algorithm

For modelling and forecasting river flows and meteorological variables, an XGBoost was chosen as it outperforms other machine-learning (ML) algorithms, such as artificial neural networks (ANN) or support vector regression (SVR), and statistical models in the case of a multilevel predictive average [62]. The process is called gradient boosting because it makes use of an algorithm to minimize losses when adding new models (Figure 3). Training is an iterative process: new trees are added that predict the residuals or errors of the previous trees, and they are subsequently combined with the previous trees to make the final prediction. XGBoost minimizes a regularized (L1 and L2) objective function that combines a convex loss function (based on the difference between predicted and target outputs) and a penalty term for model complexity.

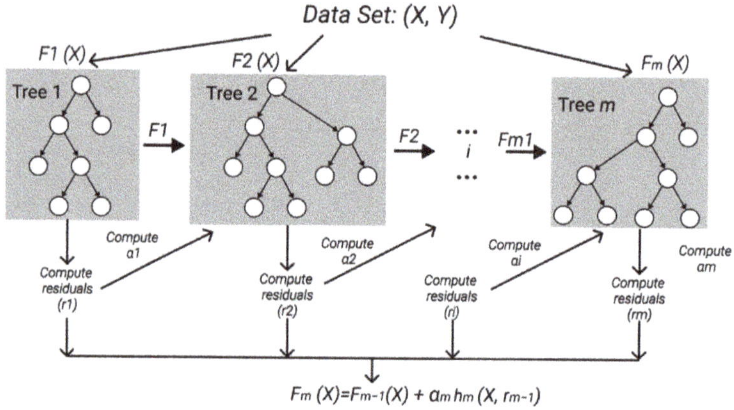

Figure 3. Schematic principle of the XGBoost algorithm for the training process (the figure was prepared on the basis of XGBoost documentation [63]).

In Figure 3, αi and ri are regularization parameters and residuals computed in the ith tree separately. Here, hi is a function that is trained to predict results, and ri uses X for the ith tree.

To compute αi, the computed residuals used ri and estimated the following:

$$\mathrm{arg}min_{\propto} = \sum_{i=1}^{m} L\,(Y_i,\ F_{i-1}(X_i) + \alpha h_i\,(X_i,\ r_i)) \tag{2}$$

where $L\,(Y, F\,(X))$ is a differentiable loss function. $L(\theta) = \sum_i (y_i - \hat{y}_i)^2$, where L represents the training loss function, θ represents model parameters, y represents the actual value, and $\hat{y}$ represents the predicted value.

Using a sampling procedure, 80% of the sample was qualified for the training set and 20% for the test set. The training of the model was performed by using eXtreme Gradient Boosting Training with the following parameters of tree booster: maximum depth of a tree = 3; number of threads = 2; maximum number of boosting iterations = 20; control the learning rate ϵ = 0.3; minimum loss reduction required to make a further partition on a leaf node of the tree γ = 0.5; minimum sum of instance weight (hessian) needed in a child = 5; and evaluation metric *rmse*. For task parameters, the regression with squared loss was used as an objective function with early stopping rounds = 5. The regression with squared loss is referred to as:

$$L(\theta) = \sum_i (y_i - \hat{y}_i)^2 \tag{3}$$

where L represents the training loss function, θ represents model parameters, y represents the actual value, $\hat{y}$ represents the predicted value.

3. Results

3.1. River-Flow Dynamics in Relation to Changes in Precipitation and Air Temperature over the Decades 1961–2020

For a graphical representation of the flow distribution during the study period, see Figure 4. From the data provided, one can follow the dynamics of flow over the decades. In the case of the Prut River at the Yaremche station, variable dynamics was evident over the study period. The smallest values of flow were recorded in the period 1960–1970, the largest in the period 2000–2010. In the last decade, a significant decrease in flow was observed at the level of the local minimum for 50 years (median = 8.0 m^3/s). A similar situation was observed in the case of the Styr River (Lutsk water gauge), where the maximum flow values recorded over 2000–2010 changed into the minimum flows during the entire observation period recorded over 2010–2020. In the last decade, the median was 23.7 m^3/s (with the median of 26.9 m^3/s for the period 1961–2020), which is 7.1 m^3/s lower than that in the 2000–2010 decade. Similarly, in the case of the Sula River (Lubny water gauge), the period of increase in the flow between 1960 and 1990 was followed by a continuous decline, with a global minimum recorded in the last decade.

The trend dynamics of river-flow amounts over the decades is shown in Figure 5. In the case of the Prut–Yaremche, an alternating trend was observed, with periods of growth (1960–1980, 1990–2010) alternating with periods of decline (1980–1990, 2010–2020). The last decade was characterized by the sharpest decline during the study period, when the flow reached the global minimum. A similar situation was observed in the case of the Styr–Lutsk, where the flow decreased by more than 25% in 2010–2020. In the case of the Sula–Lubny, the period of increase in flow (1960–1990) was followed by a period of decrease at the end of the study period, when the global flow minimum was reached in 2010–2020.

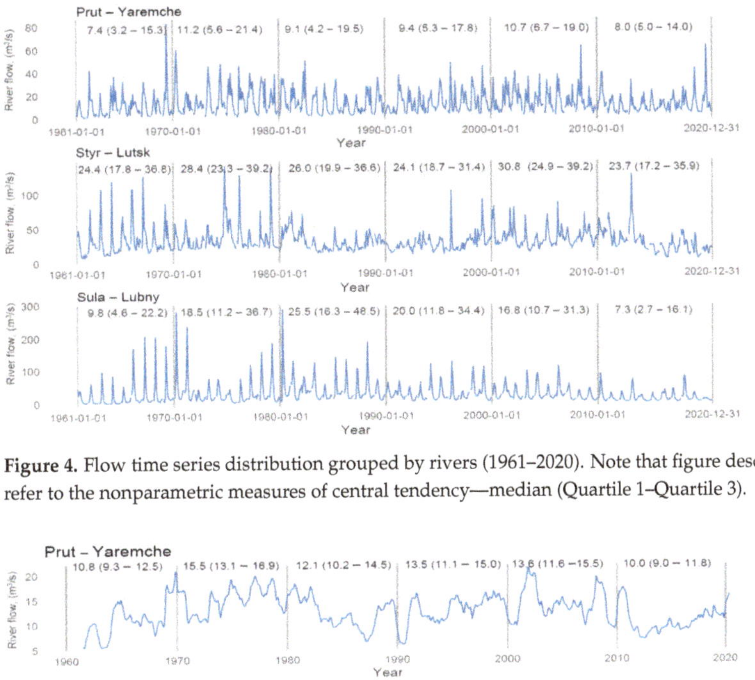

Figure 4. Flow time series distribution grouped by rivers (1961–2020). Note that figure descriptions refer to the nonparametric measures of central tendency—median (Quartile 1–Quartile 3).

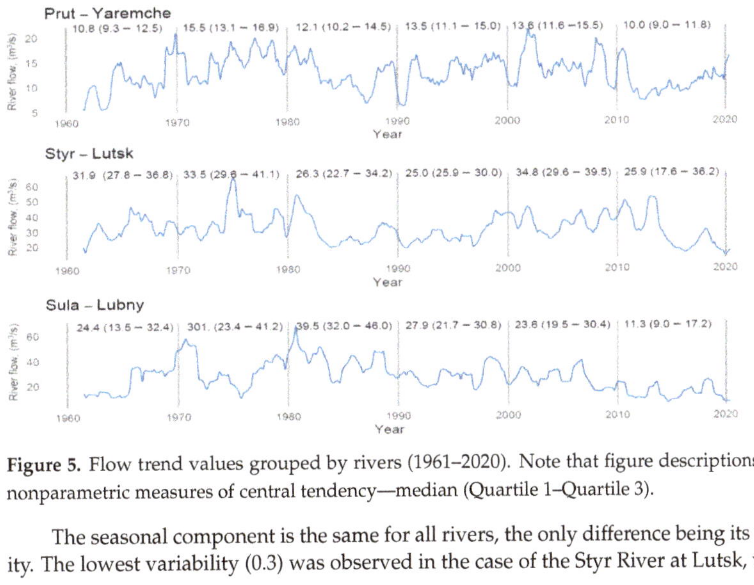

Figure 5. Flow trend values grouped by rivers (1961–2020). Note that figure descriptions refer to nonparametric measures of central tendency—median (Quartile 1–Quartile 3).

The seasonal component is the same for all rivers, the only difference being its variability. The lowest variability (0.3) was observed in the case of the Styr River at Lutsk, whereas the highest in the case of the Sula–Lubny (0.9). The dynamics of the seasonal component of river flows in the period 1961–2020 is shown in Figure S2 (in Supplementary Materials). The level of the random component was similar for all rivers. Variability in the case of rivers Prut–Yaremche and Sula–Lubny was also similar, but it was lower for the Styr River at Lutsk. In all cases, a slow decline in the average component and variability was observed throughout the study period, with local or global minima in the last decade, except the Sula–Lubny, where a wave of variation was observed. The dynamics of the random component of river flow is shown in Figure S3. In the case of the Prut River catchment, the lowest precipitation was recorded between 1980 and 1990, and it was followed by an increase in precipitation with a global maximum observed in the last decade 2010–2020 (Figure S4). In the case of the Styr River catchment, a slight variation in the dynamics of precipitation was observed: the period of increasing precipitation alternated with the period of decreasing precipitation. In the Sula River catchment, following a clear increase in precipitation in the 1970s and a corresponding stable level during the next 40 years, a decrease in precipitation below 50 mm was recorded in 2010–2020, which corresponds to the local minimum for 50 years.

The trend dynamics of precipitation amounts over the decades is shown in Figure S5 (in Supplementary Materials). The precipitation data indicate that in the case of the Prut River catchment, the declining trend period in the 1960s–1990s was replaced by an increasing trend with the global maximum in the last decade of the study period. In the case of the Styr River at Lutsk, there was a slight variation in the trend over the period with increases/decreases during one or two decades with the global average at the end of the study period. In the Sula River catchment, maximum values of the precipitation trend were observed in the period 1970–2000, which was followed by a period of a downward trend with a local minimum in the last decade of the study period. The seasonal component is the same for the Prut River and the Styr River catchments. With the same average seasonality, the Sula River catchment was characterized by the lowest variability of the three rivers. The dynamics of the seasonal component of precipitation is demonstrated in Figure S6. The mean of the random component was the same for all the rivers' catchments, the degree of variability increasing slightly in the second half of the study period, reaching local or global maxima. The lowest variability was observed for the Prut River and the highest for the Sula River catchment. The dynamics of the random component of precipitation is demonstrated in Figure S7.

For all three river catchment areas, the lowest average air temperatures were recorded in the first decade of the study period. The study period was characterized by an almost-continuous increase in average air temperatures. The value of the increase ranged from 1.6 °C to 2.1 °C with global maxima in the last decade (Figure S8 in Supplementary Materials). The trend dynamics of air temperature over the decades is presented in Figure S9. The data show a continuous upward trend for each river's catchment from the beginning of the observation to the end of the study period. Global trend maxima were reached in the last decade. The degree of variability for all river catchments was characterized by a period of increase, with the maximum in the period 1980–1990, followed by a decrease in variability to local or global minima. The seasonal component of air temperature was the same for the Prut River and Styr River catchments. With the same average seasonality, the Prut River catchment was characterized by the lowest variability (Figure S10). The level of the random component was similar for all the rivers' catchments, the variability was also similar (Figure S11). In all cases, a slow decline in the average component and variability was observed throughout the study period, with global minima in the last decade.

3.2. Forecasting Monthly River Flows and Meteorological Variables

River-flow, precipitation, and air temperature forecasting was performed in three stages: using train (80%) and test (20%) subsets, using a model with train-test data, and training with all data.

3.2.1. River-Flow Forecasting

The fitting of the data within the selected validation subgroups is presented in Figure S12 (in Supplementary Materials), where the dashed lines represent the actual data not fitted by the model. Measures of data fit in terms of prediction errors can be found in Tables S1 and S2 (in Supplementary Materials). The data show that the fitting errors are quite large. This is best seen in the normalized RMSE (NRMSE), which was calculated as the ratio of the RMSE to the global SD of flow for each river, separately. Global NRMSE values of 0.78, 0.82, and 0.61 tell us that the model does not fit the data well enough to trust such predictive results (>0.5), but it is better than the usual data randomization based on global M and SD (<1). The absence of the overfitting and underfitting (the RMSE value is similar at each fitting step for the training and test sets) of the data is evidence for judiciously selecting model parameters. A significant RMSE value at each fitting step indicates high variability in the data that cannot be reasonably well fitted by retrospective data, regardless of the selected horizon length.

Given the above limitations, the prediction of river flow for the next approach was performed by using the training set and the test set for individual rivers in periods of 12, 60, and 120 months. The aggregated prediction plots are shown in Figure 6.

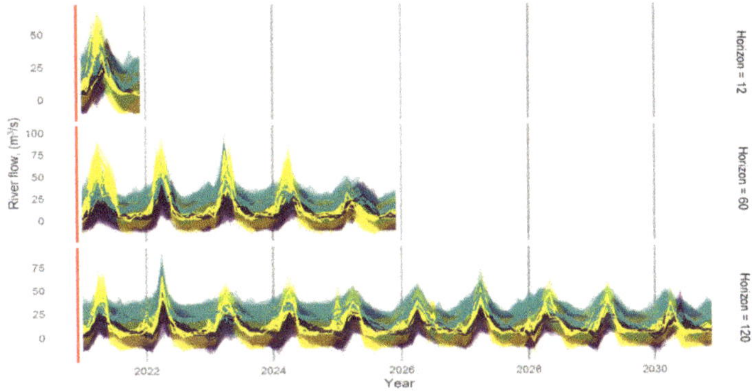

Figure 6. 12-, 60-, and 120-step-ahead forecasts of river flows (purple line: the Prut–Yaremche, turquoise line: the Styr–Lutsk, yellow line: the Sula–Lubny).

The predictions of river flow within each horizon are shown in Table 4.

Table 4. Forecast by river-flow model with train-test data grouped by horizon.

Horizo (Months)	River	N	Mdn	IQR
12	Prut–Yaremche	120	8.3	5.6–18.6
60		600	6.8	4.6–14.6
120		1200	8.0	5.5–15.7
12	Styr–Lutsk	120	17.9	15.0–22.1
60		600	19.8	14.3–25.5
120		1200	20.8	16.7–26.6
12	Sula–Lubny	120	9.6	5.9–21.8
60		600	8.5	4.7–20.4
120		1200	8.5	5.1–17.9

Explanations: N—sample size; Mdn—median; IQR—interquartile range.

An analogous prediction was made, taking into account all the data in the form of a test set. Figure 7 shows combined multiple direct-horizon forecast models for producing a 120-month step-ahead forecast grouped by "river id".

The predictions for the entire set are essentially consistent with the predictions in Table 4. Based on the results given in Figure 7, it can be concluded that the flow level of the Prut–Yaremche will remain the same or will decrease by about 6% in the period 2021–2030 (compared with the flow level in the period 2010–2020). In the case of the Styr River at Lutsk, the flow level will decrease by about 12–14%. In contrast, the flow of the Sula–Lubny in 2021–2030 is expected to be 1.2–1.7 m^3/s higher than in 2010–2020, which corresponds to a projected increase of 16–23%.

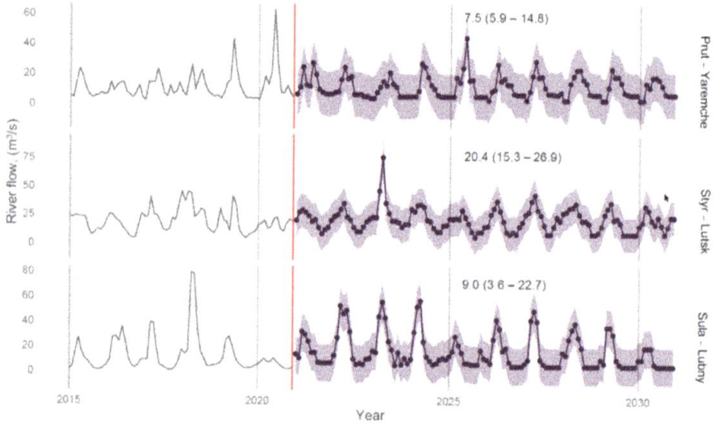

Figure 7. Combined plot with partial actual data (2015–2020) and predicted data (2021–2030) for the river-flow model trained with all data. Explanations: The left side of the graph shows a portion of the actual data (2015–2020) and the red vertical bar shows the boundaries of the actual data and the beginning of the forecast data. The data in purple represents the forecast data. The purple circles reflect the values with a monthly step, and the filled background represents the prediction intervals.

3.2.2. Precipitation Forecasting

In the case of precipitation forecasting, the fitting of the data within the selected validation subgroups is shown in Figure S13 (in Supplementary Materials), where the dashed lines represent the actual data not fitted by the model.

The data in Tables S3 and S4 (in Supplementary Materials) show that the fitting errors of prediction are quite large. This is best seen in the normalized RMSE (NRMSE), which was calculated as the ratio of the RMSE to the global SD of precipitation for each river catchment, separately. Global NRMSE values of 0.79, 0.86, and 0.96 indicate that the model does not fit the data well enough to trust such predictive results (>0.5) but are better than the usual data randomization based on global M and SD (<1). The absence of the overfitting and underfitting of the data (the RMSE value is similar at each fitting step for the training and test sets) is evidence for judiciously selecting model parameters.

Given the above limitations, as in the case of river-flow forecasting, precipitation forecasting using the training set and the test set approach were performed for individual rivers in periods of 12, 60, and 120 months. The aggregated forecast plots are shown in Figure S14. The predictions of precipitation values within each horizon are shown in Table 5.

Table 5. Forecast by precipitation model with train-test data grouped by horizon.

Horizon (Months)	River Catchment	N	M	SD
12	Prut	120	100.3	32.0
60		600	103.3	33.6
120		1200	103.0	37.0
12	Styr	120	51.2	13.3
60		600	52.3	14/7
120		1200	53.3	15.2
12	Sula	120	50.3	10.1
60		600	53.6	12.8
120		1200	52.3	11.8

Explanations: N—sample size, M—mean, SD—standard deviation.

An analogous prediction (as for river-flow forecast) was made, taking into account all the data in the form of a test set. Precipitation predictions for the 120-month period for each river catchment are shown in Figure 8.

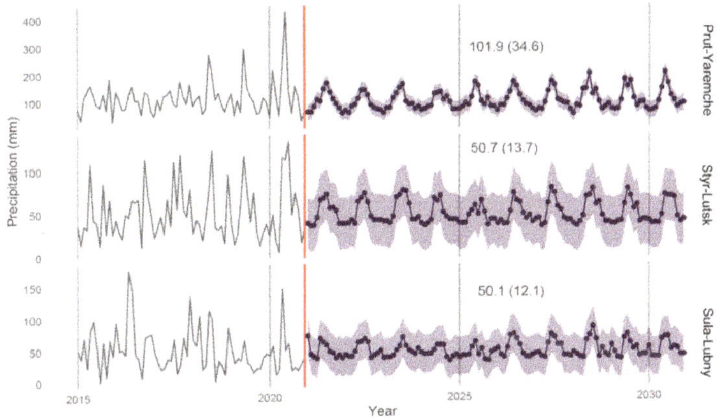

Figure 8. Combined plot with partial actual data (2015–2020) and predicted data (2021–2030) for the precipitation model trained with all data. Explanations are the same as those for Figure 7.

Based on the results provided in Figure 8, it can be concluded that in the period 2021–2030, the amount of precipitation in the Prut River catchment will decrease by about 7–10% (compared with the precipitation level in 2010–2020). In the case of the Styr River catchment, the amount of precipitation will remain at the level of the last decade or slightly decrease (by 1–2 mm). For the Sula River basin, the forecast shows an increase in average precipitation (by 1–3 mm) compared with the last decade.

3.2.3. Air Temperature Forecasting

In air temperature forecasting, the fitting of the data within the selected validation subgroups is shown in Figure S15 (in Supplementary Materials), where the dashed lines represent the actual data, almost all of which were fitted by the model.

The data in Tables S5 and S6 (in Supplementary Materials) show that the fitting errors of prediction were rather moderate. This is best seen in the normalized RMSE (NRMSE), which was calculated (similar to river flows and precipitation) as the ratio of the RMSE to the global SD of temperature for each river, separately. Global NRMSE values of 0.28–0.30 tell us that the model provides a 3.3–3.5 times better prediction of the data than in the case of the usual data randomization based on global M, SD. A low RMSE value at each fitting step indicates that the data can be reasonably well fitted by retrospective data, regardless of the horizon length chosen.

Given the above limitations, air temperature forecasting for the approach with the training set and the test set was performed for individual rivers in periods of 12, 60, and 120 months. The aggregated forecast plots are shown in Figure S16.

The predictions of air temperature values within each horizon are shown in Table 6.

An analogous prediction was made taking into account all the data in the form of a test set. Air temperature predictions for the 120-month period for each river catchment are shown in Figure 9.

Based on the results provided in Figure 9, it can be concluded that the air temperature in each individual river basin will increase by 0.1 °C to 0.4 °C. In the case of the Styr River and the Sula River catchments, a trend of further variability decrease is observed. Air temperature variability for the Prut River basin will remain unchanged in the period 2021–2030.

Table 6. Forecast by air temperature model with train-test data grouped by horizon.

Horizon (Months)	River Catchment	N	M	SD
12		120	9.12	6.98
60	Prut	600	9.16	7.33
120		1200	9.02	7.50
12		120	9.73	7.61
60	Styr	600	9.28	8.15
120		1200	9.21	8.19
12		120	9.70	8.22
60	Sula	600	9.28	8.77
120		1200	9.12	8.81

Explanations: N—sample size, M—mean, SD—standard deviation.

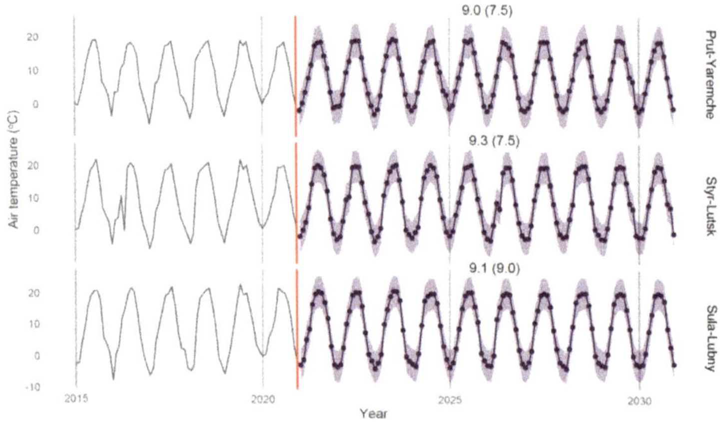

Figure 9. Combined plot with partial actual data (2015–2020) and predicted data (2021–2030) for the air temperature model trained with all data. Explanations are the same as those for Figure 7.

4. Discussion

The results of the monthly flow forecasts made for three rivers in Ukraine show a different fit of the model to the data, which is underlined by the measures of the model's performance. To some extent, similar patterns of the long-term variability of the monthly flow were obtained. However, forecasts of flows were different, which to a greater or lesser extent is related to the multiyear variability and different forecasts of precipitation and air temperature.

4.1. Long-Term Changes of Monthly River Flow under Different Precipitation and Air Temperature Conditions

In all the studied rivers, a variable dynamic of flows was observed in the years 1961–2020, with a clearly marked decrease in monthly flows in the last decade of 2010–2020. The last decade of low flows on the Prut–Yaremche and the Styr–Lutsk rivers was preceded by their noticeable increase in the earlier 2000–2010 decade. On the other hand, in the case of the Sula–Lubny river, a continuous decrease in monthly flows had been observed since the end of the 1990s (after a 30-year period of flow increase since the beginning of the 1960s) with a global minimum in the 2010–2020 decade. These changes are confirmed by the dynamics of flow trends, showing alternating periods of increase and decrease in flows over the decades of the analysed multiannual period (Figure 5).

The variability of river flows is partly related to the trends of precipitation and air temperature in the studied catchments. The decrease in flows observed on the Prut River in 1980–1990 corresponds to the lowest precipitation recorded in the 1980s–1990s. On the other hand, the observed small increase in precipitation during 1961–2020 was accompanied by a small decrease in river flow. One of the reasons for this discrepancy was a significant increase of air temperature and a corresponding increase of evaporation from the catchment area, which is also documented by the research of Vyshnevskyi and Donich [53]. Similar tendencies of climate parameters and river-flow changes were obtained for the Styr and the Sula rivers. Despite a small increase of precipitation, water runoff tends to decrease. This decrease is bigger than that of the Prut River. In our opinion, such decrease of flow of plain rivers may be explained by two factors: an increase of water intake during warm weather (in particular due to the irrigation of farming land) and an increase of evaporation from ponds and reservoirs. As shown in the paper by Vyshnevskyi [64], the dependence between evaporation and air temperature is nonlinear: a small increase in air temperature results in a significant increase of evaporation.

For all the studied rivers, a similarity was identified in the seasonal fluctuation component and the level of the random component of monthly flows. A detailed analysis of random flow fluctuations for selected rivers has not been carried out; therefore, research on these issues should be continued in the future.

4.2. Forecasting of Monthly River Flow with Precipitation and Air Temperature Forecasts

Results of the forecasts for 2021–2030 for the studied catchments showed that the dominant pattern is as follows: a decrease in flows is accompanied by a decrease in precipitation and an increase in air temperature (the Prut and the Styr). The differences relate to the magnitude of the flow change of approximately 6% compared with the 2010–2020 decade for the mountainous Prut River (or even no change in flow) and about 14% for the Styr River, predominantly in the uplands. On the other hand, the forecasts for the Sula River provide for an increase in flows (16–23%), with an increase in precipitation and air temperature, which is what makes it different from other rivers. It may be assumed that with an increase in precipitation in the Sula River catchment area, soil retention (underground retention) will also become more important in shaping the river flows, which will contribute to levelling the differences between high spring flows and low summer–autumn flows. The predicted changes in the flows of the Prut, the Styr, and the Sula rivers correspond to the forecasts in other regions of Ukraine and Europe. Over the past decades, climate change has been observed in Europe, and it has been emphasized that these changes will be stronger in the future [65–70]. Many studies on the impact of climate variability and change on water resources in Europe, using hydrological models on a continental and global scale, predict lower flows, with increasing air temperature and lower or no significant changes in precipitation [71–74]. However, the various types of changes are not evenly distributed either in time or in space. Didovets et al. [69] analysed changes in precipitation and temperature until the end of the 20th century in the catchment areas of Ukraine and determined their impact on the water availability of main river basins by using global hydrological models, which were compared with global climate models

(GCM) in two representative concentrations pathways scenarios: RCP 2.6 and RCP 8.5. A historical period (1971–2000) and two future periods (2041–2070 and 2071–2100) were considered, showing that changes in mean annual precipitation range between −14% and +10% and changes in the mean annual river discharge range from −30% to +6%, depending on RCP. The largest decrease in the mean annual runoff was forecast for the Pripyat, the Southern Bug, and the Dniester basins, reaching up to −30% by the end of the century in RCP 8.5, which was also confirmed by the present current study for the Styr River, which is a tributary of the Pripyat River.

4.3. Assessment of the Forecasts of Monthly River Flow

In the present research (regarding the Prut, Styr, and Sula rivers), attention was drawn to a significant RMSE value at each fitting step, which indicates high variability of the data that cannot be reasonably well fitted by retrospective data, regardless of the selected horizon length. This leads to the conclusion that the monthly data frequency for flow is too low to produce good fitting results and that it is worthwhile to use the daily frequency. Comparing the obtained results with other forecasts yields an emphasis on the fact that uncertainty regarding the forecasting of hydrological characteristics is still high, regardless of the location of the rivers and the adopted research methods. The variability of the river-flow time series can be influenced by various parameters, such as air temperature, precipitation, and evaporation, which makes accurate estimations and predictions of the flow almost impossible, the more so as it is further complicated by complexity, nonstationarity, and the nonlinearity of the river-flow phenomenon [75–77].

An analysis of the homogeneity and stationarity of an average monthly flows of Ukrainian rivers (305 water gauge stations, from the beginning of the observation until 2010), carried out with the use of hydrogenetic methods [78], has shown that most of the observation series are homogeneous and stationary. The exceptions include rivers with significant anthropogenic impact (river regulations, water intakes, deforestation, etc.). Observation series with a full cycle of long-term cyclic fluctuations (dry and wet phase) are stationary, whereas other observation series are quasistationary. Many studies indicate a large diversity of predictions and forecasts, especially those obtained for large rivers (often transboundary) and a high uncertainty as to the predicted hydrological effects in the future due to the visible climate changes and their impact on the hydrological characteristics of rivers, also in Ukraine [15,78]. The importance of studies on the strength of simulated changes and the effects on river flows has also been confirmed [79–82].

A staged forecasting process used in this study takes into account the gradient enhancement approach of the XGBoost model. In the approach applied to Ukrainian rivers, a single multioutput model was proposed that simultaneously forecasts over short- and long-term horizons. The generalization performance of the model over a variety of available data sets over time was assessed, and the hyperparameters which minimize the forecast error over the relevant forecast horizons were selected. The algorithm used allowed direct multistep-ahead forecasting with multiple time series (grouped or hierarchical) inspired by Bergmeir et al. [57]. As suggested by Sanders et al. [41] and Osman et al. [83], XGBoost models use all input data directly, without any preprocessing or selection, and a proper selection of the input fields is essential to improve the performance of XGBoost models. Usually, unavoidable errors may result from introduced model limitations or the nonstationarity of hydrological processes [84]. An additional difficulty may be the quality of hydrological and meteorological data, another being the calibration of parameters [85,86]. Yang et al. [87] suggest including additional hydrological information in river-flow forecasts to improve model performance. On the other hand, research conducted by Sanders et al. [41] shows that using more information on the catchment area and data from point measurements can lead to a reduction in the performance of forecasting models. According to Aghelpour et al. [88], consideration of the climatic indices (e.g., Southern Oscillation Index, Global Mean Temperature Index, North Pacific pattern, Pacific Decadal Oscillation, North Atlantic

Oscillation) as inputs in prediction models of river flows in Iran may improve monthly flow prediction by 24% on average.

Complex river-flow prediction models typically require multiple hydrological and climatological parameters as input, many of which may not be available for some locations, and the complexity of the flow processes makes it difficult to use physical models. Many estimates and forecasts of river flows have well documented that combined models (hybrid models) may show better performance compared with stand-alone models [89–91].

5. Conclusions

Long-term changes were estimated, and a forecast of monthly river flows was made for three rivers (the Prut, the Styr, and the Sula) in Ukraine with reference to the variability and forecasts of precipitation and air temperature. The years 1961–2020 were selected as the historical period, while the forecast horizons of 12, 60, and 120 months into the future were specified for this dataset, i.e., December 2021, December 2025, and December 2030.

The forecasting of monthly river flows was carried out in three stages with the use of input data sets characteristic for the XGBoost algorithm: train (80%) and test subsets (20%), using a model with train-test data, and forecasting by training with all data for forecast horizons (December 2021, December 2025, and December 2030). The adoption of this algorithm allowed direct multistep-ahead forecasting with multiple (grouped or hierarchical) time series, which may be of importance taking into account sometimes-scarce measurements of daily river flows. This is one of the main advantages and a new, unique feature of the adopted forecasting method, which is characterized by high versatility of use, regardless of the spatial location of the river catchment areas. The direct forecasting approach used included the following steps: build a single multioutput model that simultaneously forecasts over short- and long-term forecast horizons, assess the generalization performance of the model over a variety of held data sets over time, and select the hyperparameters that minimize forecast error over the relevant forecast horizons and re-train.

The selected rivers flow through different physical geographical and climatic regions and represent different hydrological systems: the Prut, a mountain river; the Styr, an upland river; and the Sula, a lowland river. These features undoubtedly influenced their different degree of sensitivity and resistance to variability of climatic conditions and possible anthropogenic influences. The conducted analysis showed some differences in the trends of changes in monthly flows in individual decades of the period 1961–2020 and similarities in terms of the seasonal component and random flow fluctuations. An emphasis was put on the last decade, significant for future forecasts, in which a decrease in flow was determined for the Prut and the Styr rivers, whereas an increase was determined for the Sula River. Two patterns were obtained in the forecasts, which partially correspond to the results of flow forecasts obtained for Ukrainian rivers with the use of other methods and models. A decrease in flow, accompanied by a decrease in precipitation and an increase in air temperature until 2030, are expected according to the forecasts for the Prut and Styr rivers, whereas for the Sula River, the forecasts assume an increase in flow, accompanied by an increase in precipitation and air temperature.

The advantages of the model used for forecasting of monthly river flows in Ukraine, i.e., modelling many time series simultaneously using one model, include the simplicity of modelling, potentially more-robust results because of pooling data across time series, and solving the "cold start" problem when few data points are available for a given time series. Unlike the recursive or iterative method of creating forecasts in multiple steps used in traditional forecasting (such as ARIMA), direct forecasting involves creating a series of distinct, horizon-specific models. Although there are several hybrid methods for creating multistep forecasts, the simple, direct forecasting method used avoids the exponentially more-difficult problem of "predicting the predictors" for forecast horizons beyond one step in the future.

The obtained results of the forecasts, although approximate to some extent, provide information on future (in the next decade) changes in monthly flows of Ukrainian rivers (the Prut, the Styr, and the Sula) occurring under various environmental and climatic conditions, and they highlight the need to extend research into assessments of the impact of anthropogenic factors on the long-term dynamics of changes in river flows. In future studies of the daily flows, precipitation and air temperature should test the effectiveness of multistage forecasting with multiple, grouped, or hierarchical time series in the forecasting of hydrological and meteorological parameters. The research results can be used for the management and rational planning of water resources in the studied regions of Ukraine. They can provide the basis for the rational use of river water resources, such as for irrigation purposes, which is certainly a great challenge in the situation of increasingly frequent and longer periods of drought.

Supplementary Materials: The following supporting information can be downloaded at: https://www.mdpi.com/article/10.3390/resources11120111/s1, Figure S1. Custom validation subsets (grey areas) with skipped data (white areas) for river flow, Figure S2. Seasonal flow values grouped by rivers (1961–2020), Figure S3. Random flow values grouped by rivers (1961–2020), Figure S4. Precipitation time series distribution grouped by rivers' catchment areas (1961–2020). Note: Figure descriptions refer to non-parametric measures-mean (standard deviation), Figure S5. Precipitation trend values grouped by rivers (1961–2020), Figure S6. Precipitation seasonal values grouped by rivers (1961–2020), Figure S7. Precipitation random values grouped by rivers (1961–2020), Figure S8. Air temperature time series distribution grouped by rivers' catchment areas (1961–2020). Note: Figure descriptions refer to non-parametric measures-mean (standard deviation), Figure S9. Air temperatur trend values grouped by rivers (1961–2020), Figure S10. Air temperatur seasonal values grouped by rivers (1961–2020), Figure S11. Air temperatur and random values grouped by rivers (1961–2020), Figure S12. River flows predicted (solid line) vs. actual (dashed line) values grouped by horizons (purple line: the Prut River-Yaremche, turquoise line: the Styr River–Lutsk, yellow line: the Sula River-Lubny), Figure S13. Precipitation predicted (solid) vs. actual (dashed) values grouped by horizons (purple line: Prut River catchment. turquoise line: Styr River catchment. yellow line: Sula River catchment), Figure S14. 12, 60 and 120 step-ahead forecasts of precipitation (purple line: the Prut River catchment, turquoise. line: the Styr River catchment, yellow line: the Sula River catchment), Figure S15. Air temperature predicted (solid) vs. actual (dashed) values grouped by horizons (purple line: PrutRiver catchment. turquoise line: Styr River catchment. yellow line: Sula River catchment), Figure S16. 12, 60 and 120 step-ahead forecasts of air temperature (purple line: the Prut River catchment, turquoise line: the Styr River catchment, yellow line: the Sula river catchment). Table S1. Forecast error by river flow forecast horizon, Table S2. Global river flow forecast error, Table S3. Forecast error by precipitation forecast horizon and river catchment, Table S4. Global precipitation forecast error grouped by river catchment, Table S5. Forecast error by air temperature forecast horizon and river catchment, Table S6. Global air temperature forecast error grouped by river catchment.

Author Contributions: Conceptualization, R.G. and V.V.; methodology, R.G.; software, R.G.; data curation, V.V.; writing—original draft preparation, R.G. and V.V.; investigation, R.G.; validation, R.G.; visualization, R.G.; writing—reviewing and editing, R.G. and V.V.; discussion, R.G. and V.V.; supervision, R.G. All authors have read and agreed to the published version of the manuscript.

Funding: This research received no external funding.

Data Availability Statement: The source of data derives from the measurement and observation data of the Hydrometeorological Service in Ukraine.

Acknowledgments: The authors thank their universities for their support. The authors would also like to acknowledge the Hydrometeorological Service in Ukraine for the release of the input database.

Conflicts of Interest: The authors declare no conflict of interest.

References

1. Shmueli, G. To explain or to predict? *Stat. Sci.* **2010**, *25*, 289–310. [CrossRef]

2. Blöschl, G.; Bierkens, M.F.P.; Chambel, A.; Cudennec, C.; Destouni, G.; Fiori, A.; Kirchner, J.W.; McDonnell, J.J.; Savenije, H.H.G.; Sivapalan, M.; et al. Twenty-three Unsolved Problems in Hydrology (UPH)–a community perspective. *Hydrol. Sci. J.* **2019**, *64*, 1141–1158.
3. Papacharalampous, G.A.; Tyralis, H.; Koutsoyiannis, D.; Montanari, A. Quantification of predictive uncertainty in hydrological modelling by harnessing the wisdom of the crowd: A large-sample experiment at monthly timescale. *Adv. Water Resour.* **2020**, *136*, 103470. [CrossRef]
4. Montanari, A. Hydrology of the Po River: Looking for changing patterns in river discharge. *Hydrol. Earth Syst. Sci.* **2012**, *16*, 3739–3747. [CrossRef]
5. Steirou, E.; Gerlitz, L.; Apel, H.; Sun, X.; Merz, B. Climate influences on flood probabilities across Europe. *Hydrol. Earth Syst. Sci.* **2019**, *23*, 1305–1322. [CrossRef]
6. Hussain, F.; Wu, R.-S.; Wang, J.-X. Comparative Study of Very Short-Term Flood Forecasting Using Physics-Based Numerical Model and Data-Driven Prediction Model. *Nat. Hazards* **2021**, *107*, 249–284. [CrossRef]
7. Yaseen, Z.M.; Ebtehaj, I.; Bonakdari, H.; Deo, R.C.; Mehr, A.D.; Melini, W.H.; Mohtar, W.; Diop, L.; El-shafie, A.; Singh, V.P. Novel approach for streamflow forecasting using a hybrid ANFIS-FFA model. *J. Hydrol.* **2017**, *554*, 263–276. [CrossRef]
8. Tu, H.; Wang, X.; Zhang, W.; Peng, H.; Ke, Q.; Chen, X. Flash Flood Early Warning Coupled with Hydrological Simulation and the Rising Rate of the Flood Stage in a Mountainous Small Watershed in Sichuan Province, China. *Water* **2020**, *12*, 255. [CrossRef]
9. Sivakumar, B.; Jayawardena, A.W.; Fernando, T.M.K.G. River flow forecasting: Use of phase-space reconstruction and artificial neural networks approaches. *J. Hydrol.* **2002**, *265*, 225–245. [CrossRef]
10. Chau, K.W. A split-step particle swarm optimization algorithm in river stage forecasting. *J. Hydrol.* **2007**, *346*, 131–135. [CrossRef]
11. Abbas, S.A.; Xuan, Y. Development of a new quantile-based method for the assessment of regional water resources in a highly-regulated river basin. *Water Resour. Manag.* **2019**, *33*, 3187–3210. [CrossRef]
12. Shen, C.; Laloy, E.; Elshorbagy, A.; Albert, A.; Bales, J.; Chang, F.-J.; Ganguly, S.; Hsu, K.-L.; Kifer, D.; Fang, Z.; et al. HESS Opinions: Incubating Deep-Learning-Powered Hydrologic Science Advances as a Community. *Hydrol. Earth Syst. Sci.* **2018**, *22*, 5639–5656.
13. Hu, C.; Wu, Q.; Li, H.; Jian, S.; Li, N.; Lou, Z. Deep Learning with a Long Short-Term Memory Networks Approach for Rainfall-Runoff Simulation. *Water* **2018**, *10*, 1543. [CrossRef]
14. Siuta, T.J. Modelowanie serii czasowych przepływów w krótkoterminowej prognozie hydrologicznej. *Acta Sci. Polonorum. Form. Circumiectus* **2020**, *19*, 3–14. [CrossRef]
15. Khrystiuk, B.; Gorbachova, L. Long-term Forecasting of Extraordinary Spring Floods by Commensurability Method on the Dnipro River Near Kyiv City, Ukraine. *J. Environ. Res. Eng. Manag. EREM* **2019**, *75*, 74–81.
16. Abrahart, R.S.; See, L. Multi-model data fusion for River flow forecasting; an evaluation of six alternative methods based on two contrasting catchment. *Hydrol. Earth Syst. Sci.* **2002**, *6*, 655–670. [CrossRef]
17. Zeynoddin, M.; Bonakdari, H.; Azari, A.; Ebtehaj, I.; Gharabaghi, B.; Riahi Madavar, H. Novel hybrid linear stochastic with non-linear extreme learning machine methods for forecasting monthly rainfall a tropical climate. *J. Environ. Manag.* **2018**, *222*, 190–206. [CrossRef]
18. Peng, Z.; Zhang, L.; Yin, J.; Wang, H. Commensurability-Based Flood Forecasting in Northeastern China. *Pol. J. Environ. Stud.* **2017**, *26*, 2689–2702. [CrossRef]
19. Apel, H.; Thieken, A.H.; Merz, B.; Blöschl, G. Flood risk assessment and associated uncertainty. *Nat. Hazards Earth Syst. Sci.* **2004**, *4*, 295–308. [CrossRef]
20. Khrystiuk, B.F. The forecasting of the average, maximum and minimum for a ten-day period of water discharges on Upper Danube. *Proc. Ukr. Hydrometeorol. Inst.* **2012**, *262*, 206–220. (In Ukrainian)
21. Khrystiuk, B.; Gorbachova, L.; Koshkina, O. The impact of climatic conditions of the spring flood formation on hydrograph shape of the Desna River. *Meteorol. Hydrol. Water Manag.* **2017**, *5*, 63–70. [CrossRef]
22. Shevnina, E. Methods of long-range forecasting of dates of the spring flood beginning and peak flow in the estuary sections of the Ob and Yenisei rivers. *Russ. Meteorol. Hydrol.* **2009**, *34*, 51–57. [CrossRef]
23. Scitovski, R.; Maričić, S.; Scitovski, S. Short-term and long-term water level prediction at one river measurement location. *Croat. Oper. Res. Rev. (CRORR)* **2012**, *3*, 80–90.
24. Sharma, P.; Machiwal, D. Streamflow forecasting. In *Advances in Streamflow Forecasting*; Sharma, P., Machiwal, D., Eds.; Elsevier: Amsterdam, The Netherlands, 2021; pp. 1–50.
25. Toth, E.; Brath, A.; Montanari, A. Comparison of short-term rainfall prediction models for real-time flood forecasting. *J. Hydrol.* **2000**, *239*, 132–147. [CrossRef]
26. Imrie, C.E.; Durucan, S.; Korre, A. River flow prediction Rusing artificial neural networks: Generalisation beyond the calibration range. *J. Hydrol.* **2000**, *233*, 138–153. [CrossRef]
27. Ozgur, K. River Flow Modeling Using Artificial Neural Networks. *J. Hydrol. Eng.* **2004**, *9*, 60–63.
28. Teschl, R.; Randeu, W.L. An Artificial Neural Networkbased Rainfall-Runoff Model Using Gridded Radar Data. In Proceedings of the Third European Conference on Radar in Meteorology and Hydrology (ERAD), Visby, Sweden, 6–10 September 2004.
29. Krzanowski, S.; Wałęga, A. Zastosowanie sztucznych sieci neuronowych do predykcji szeregów czasowych stanów wody i przepływów w rzece. *Acta Sci. Pol. Form. Circumiectus* **2007**, *6*, 59–73.

30. Kim, G.; Barros, A.P. Quantitative flood forecasting using multisensor data and neural networks. *J. Hydrol.* **2001**, *246*, 45–62. [CrossRef]
31. Firat, M.; Güngör, M. River flow estimation using adaptive neuro fuzzy inference system. *Math. Comput. Simul.* **2007**, *75*, 87–96. [CrossRef]
32. Lohani, A.K.; Kumar, R.; Singh, R.D. Hydrological time series modeling: A comparison between adaptive neuro-fuzzy, neural network and autoregressive techniques. *J. Hydrol.* **2012**, *442*, 23–35. [CrossRef]
33. Dehghani, M.; Seifi, A.; Riahi-Madvar, H. Novel forecasting models for immediate-short-term to long-term influent flow prediction by combining ANFIS and grey wolf optimizations. *J. Hydrol.* **2019**, *576*, 698–725. [CrossRef]
34. Riahi-Madvar, H.; Dehghani, M.; Memarzadeh, R.; Gharabaghi, B. Short to Long-Term Forecasting of River Flows by Heuristic Optimization Algorithms Hybridized with ANFIS. *Water Resour. Manag.* **2021**, *35*, 1149–1166. [CrossRef]
35. Achouri, I.; Hani, I.; Bougherira, N.; Djabri, L.; Chaffai, H.; Lallahem, S. River flow model using artificial neural networks. *Energy Proc.* **2015**, *74*, 1007–1014. [CrossRef]
36. Abudu, S.; Cui, C.; King, J.P.; Abudukadeer, K. Comparison of performance of statistical models in forecasting monthly streamflow of Kizil River. China. *Water Sci. Eng.* **2010**, *3*, 269–281.
37. Valipour, M.; Banihabib, M.E.; Behbahani, S.M.R. Comparison of the ARMA, ARIMA, and the autoregressive artificial neural network models in forecasting the monthly inflow of Dez dam reservoir. *J. Hydrol.* **2013**, *476*, 433–441. [CrossRef]
38. Aghelpour, P.; Varshavian, V. Evaluation of stochastic and artificial intelligence models in modeling and predicting of river daily flow time series. *Stoch. Hydrol. Hydraul.* **2020**, *34*, 33–50. [CrossRef]
39. Hsu, K.; Gupta, H.V.; Soroochian, S. Aplication of a recurrent neural network to rainfall-runoff modeling. *Proc. Aesthet. Constr. Environ.* **1997**, *31*, 2517–2530.
40. Kratzert, R.; Klotz, D.; Brenner, C.; Schulz, K.; Herrnegger, M. Rainfall–runoff model ling Rusing Long Short-Term Memory (LSTM) networks. *Hydrol. Earth Syst. Sci.* **2018**, *22*, 6005–6022. [CrossRef]
41. Sanders, W.; Li, D.; Li, W.; Fang, Z.N. Data-Driven Flood Alert System (FAS) Using Extreme Gradient Boosting (XGBoost) to Forecast Flood Stages. *Water* **2022**, *14*, 747. [CrossRef]
42. Zounemat-Kermani, M.; Batelaan, O.; Fadaee, M.; Hinkelmann, R. Ensemble machine learning paradigms in hydrology: A review. *J. Hydrol.* **2021**, *598*, 126266.
43. Graf, R.; Kolerski, T.; Zhu, S. Predicting Ice Phenomena in a River Using the Artificial Neural Network and Extreme Gradient Boosting. *Resources* **2022**, *11*, 12. [CrossRef]
44. Zhang, Z.; Zhang, Q.; Singh, V.P.; Shi, P. River flow modelling: Comparison of performance and evaluation of uncertainty using data-driven models and conceptual hydrological model. *Stoch Environ. Res. Risk Assess.* **2018**, *32*, 2667–2682.
45. Ni, L.; Wang, D.; Wu, J.; Wang, Y.; Tao, Y.; Zhang, J.; Liu, J. Streamflow forecasting using extreme gradient boosting model coupled with Gaussian mixture model. *J. Hydrol.* **2020**, *586*, 124901.
46. Fares, A. *Climate Change and Extreme Events*; Elsevier: San Diego, CA, USA, 2021.
47. Shakirzanova, J. Forecasting of the maximum water flow of the spring flood in basin Dnieper with use of the automated program complexes. *Hydrol. Hydrochem. Hydroecol.* **2011**, *4*, 48–55. (In Ukrainian)
48. Fischer, S.; Pluntke, T.; Pavlik, D.; Bernhofer, C. Hydrologic effects of climate change in a sub-basin of the Western Bug River, Western Ukraine. *Environ. Earth Sci.* **2014**, *72*, 4727–4744. [CrossRef]
49. Snizhko, S.; Kuprikov, I.; Shevchenko, O.; Evgen, P.; Didovets, I. Assessment of local water resources runoff in Ukraine by using the water-balance Turk model and the regional model REMO in the XXI century. *Bryansk State Univ. Her.* **2014**, *4*, 191–201.
50. Loboda, N.; Bozhok, Y. Water resources of Ukraine in the XXI century based on climate change scenarios. *Ukr. Hydrometeorol. J.* **2016**, *17*, 114–121.
51. Didovets, I.; Lobanova, A.; Bronstert, A.; Snizhko, S.; Maule, C.F.C.F.; Krysanova, V. Assessment of climate change impacts on water resources in three representative Ukrainian catchments using eco-hydrological modelling. *Water* **2017**, *9*, 204. [CrossRef]
52. Didovets, I.; Krysanova, V.; Bürger, G.; Snizhko, S.; Balabukh, V.; Bronstert, A. Climate change impact on regional floods in the Carpathian region. *J. Hydrol. Reg. Stud.* **2019**, *22*, 100590.
53. Vyshnevskyi, V.I.; Donich, O.A. change in the Ukrainian Carpathians and its possible impact on river runoff. *Acta Hydrol. Slovaca.* **2021**, *22*, 3–14. [CrossRef]
54. Loboda, N.; Kozlov, M. Assessment of water resources of the Ukrainian rivers according to the average statistical models of climate change trajectories RCP4.5 and RCP8.5 over the period of 2021 to 2050. *Ukr. Hydrometeorol. J.* **2020**, *25*, 93–104. [CrossRef]
55. Vyshnevskyi, V.I.; Kutsiy, A.V. *Long-Term Changes in the Water Regime of Rivers in Ukraine*; Naukova Dumka: Kyiv, Ukraine, 2022; 252p, Available online: https://er.nau.edu.ua/handle/NAU/56293 (accessed on 22 October 2022).
56. Redell, N. forecastML: Time Series Forecasting with Machine Learning Methods; R Package Version 0.9.0,<URL; 2020. Available online: https://CRAN.R-project.org/package=forecastML (accessed on 3 September 2022).
57. Bergmeir, C.; Hyndman, R.J.; Koo, B. A note on the validity of cross-validation for evaluating autoregressive time series prediction. *Comput. Stat. Data Anal.* **2018**, *120*, 70–83. [CrossRef]
58. Chen, T.; Guestrin, C. XGBoost: A Scalable Tree Boosting System. In Proceedings of the 22nd SIGKDD Conference on Knowledge Discovery and Data Mining, San Francisco, CA, USA, 13–17 August 2016.
59. Bahrami-Pichaghchi, H.; Aghelpour, P. An estimation and multi-step ahead prediction study of monthly snow cover area, based on efficient atmospheric-oceanic dynamics. *Clim. Dyn.* **2022**. [CrossRef]

60. Aghelpour, P.; Norooz-Valashedi, R. Predicting daily reference evapotranspiration rates in a humid region, comparison of seven various data-based predictor models. *Stoch. Environ. Res. Risk Assess.* **2022**, *36*, 4133–4155. [CrossRef]
61. R Core Team. *R: A Language and Environment for Statistical Computing*; R Foundation for Statistical Computing: Vienna, Austria, 2021; Available online: https://www.R-project.org/ (accessed on 5 August 2022).
62. Abolghasemi, M.; Hyndman, R.; Garth, T.; Bergmeir, C.H. Machine learning applications in time series hierarchical forecasting. *arXiv* **2019**, arXiv:1912.00370v1.
63. XGBoost Documentation. Available online: https://xgboost.readthedocs.io/en/latest/index.html (accessed on 1 August 2022).
64. Vyshnevskyi, V.I. The impact of climate change on evaporation from the water surface in Ukraine. *J. Geol. Geogr. Geoecol.* **2022**, *31*, 163–170. [CrossRef]
65. Kovats, R.S.; Valentini, R.; Bouwer, L.M.; Georgopoulou, E.; Jacob, D.; Martin, E.; Rounsevell, M.J.-F.S. Europe. In *Climate Change 2014: Impacts, Adaptation, and Vulnerability*; Part B: Regional Aspects; Contribution of Working Group II to the Fifth Assessment Report of the Intergovernmental Panel on Climate Change; Cambridge University Press: Cambridge, UK; New York, NY, USA, 2014; pp. 1267–1326.
66. IPCC. Climate Change and Land, IPCC Special Report on Climate Change, Desertification, Land Degradation, Sustainable Land Management, Food Security, and Greenhouse Gas Fluxes in Terrestrial Ecosystems. Summary for Policymakers. 2018. Available online: https://www.ipcc.ch/site/assets/uploads/2019/08/4.-SPM_Approved_Microsite_FINAL.pdf (accessed on 10 August 2022).
67. Jacob, D.; Kotova, L.; Teichmann, C.; Sobolowski, S.P.; Vautard, R.; Donnelly, C.; Koutroulis, A.G.; Grillakis, M.G.; Tsanis, I.K.; Damm, A.; et al. Climate impacts in Europe under +1.5 °C global warming. *Earth's Futur.* **2018**, *6*, 264–285. [CrossRef]
68. Graf, R.; Wrzesiński, D. Relationship between Water Temperature of Polish Rivers and Large-Scale Atmospheric Circulation. *Water* **2019**, *11*, 1690. [CrossRef]
69. Didovets, I.; Krysanova, V.; Hattermann, F.F.; López, M.R.R.; Snizhko, S.; Schmied, H.M. Climate change impact on water availability of main river basins in Ukraine. *J. Hydrol. Reg. Stud.* **2020**, *32*, 100761.
70. Wrzesiński, D.; Graf, R. Temporal and spatial patterns of the river flow and water temperature relations in Poland. *J. Hydrol. Hydromech.* **2022**, *70*, 12–29. [CrossRef]
71. Gudmundsson, L.; Wagener, T.; Tallaksen, L.M.; Engeland, K. Evaluation of nine large-scale hydrological models with respect to the seasonal runoff climatology in Europe. *Water Resour. Res.* **2012**, *48*, W11504.
72. Van Vliet, M.T.H.; Donnelly, C.; Strömbäck, L.; Capell, R.; Ludwig, F. European scale climate information services for water use sectors. *J. Hydrol.* **2015**, *528*, 503–513. [CrossRef]
73. Donnelly, C.; Greuell, W.; Andersson, J.; Gerten, D.; Pisacane, G.; Roudier, P.; Ludwig, F. Impacts of climate change on European hydrology at 1.5, 2 and 3 degrees mean global warming above preindustrial level. *Clim. Change* **2017**, *143*, 13–26. [CrossRef]
74. Mentaschi, L.; Vousdoukas, M.; Besio, G.; Feyen, L. alphaBetaLab: Automatic estimation of subscale transparencies for the Unresolved Obstacles Source Term in ocean wave modelling. *SoftwareX* **2019**, *9*, 1–6. [CrossRef]
75. Bayazit, M. Nonstationarity of Hydrological Records and Recent Trends in Trend Analysis: A State-of-the-art Review. *Environ. Process.* **2015**, *2*, 527–542. [CrossRef]
76. Adnan, R.M.; Mostafa, R.R.; Elbeltagi, A.; Yaseen, Z.M.; Shamsuddin, S.; Kisi, O. Development of new machine learning model for streamflow prediction: Case studies in Pakistan. *Stoch. Environ. Res. Risk Assess.* **2022**, *36*, 999–1033. [CrossRef]
77. Meira-Neto, A.A.; Niu, G.; Roy, T.; Tyler, S.; Troch, P.A. Interactions between snow cover and evaporation lead to higher sensitivity of streamflow to temperature. *Commun. Earth Environ.* **2020**, *1*, 1–7.
78. Gorbachova, L. Place and role of hydro-genetic analysis among modern research methods runoff. *Proc. Ukr. Hydrometeorol. Institute* **2015**, *268*, 73–81. (In Ukrainian)
79. Frieler, K.; Lange, S.; Piontek, F.; Reyer, C.P.O.; Schewe, J.; Warszawski, L.; Zhao, F.; Chini, L.; Denvil, S.; Emanuel, K.; et al. Assessing the impacts of 1.5 °C global warming-Simulation protocol of the Inter-Sectoral Impact Model Intercomparison Project (ISIMIP2b). *Geosci. Model Dev.* **2017**, *10*, 4321–4345. [CrossRef]
80. Haddeland, I.; Heinke, J.; Biemans, H.; Eisner, S.; Flörke, M.; Hanasaki, N.; Konzmann, M.; Ludwig, F.; Masaki, Y.; Schewe, J.; et al. Global water resources affected by human interventions and climate change. *Proc. Natl. Acad. Sci. USA* **2014**, *111*, 3251–3256. [CrossRef]
81. Hattermann, F.F.; Krysanova, V.; Gosling, S.N.; Dankers, R.; Daggupati, P.; Donnelly, C.; Flörke, M.; Huang, S.; Motovilov, Y.; Buda, S.; et al. Cross-scale intercomparison of climate change impacts simulated by regional and global hydrological models in eleven large river basins. *Clim. Chang.* **2017**, *141*, 561–576. [CrossRef]
82. Hattermann, F.F.; Vetter, T.; Breuer, L.; Su, B.; Daggupati, P.; Donnelly, C.; Fekete, B.; Florke, F.; Gosling, S.N.; Hoffmann, P.; et al. Sources of uncertainty in hydrological climate impact assessment: A cross-scale study. *Environ. Res. Lett.* **2018**, *13*, 015006. [CrossRef]
83. Osman, A.I.A.; Najah Ahmed, A.; Chow, M.F.; Feng Huang, Y.; El-Shafie, A. Extreme Gradient Boosting (Xgboost) Model to Predict the Groundwater Levels in Selangor Malaysia. *Ain Shams Eng. J.* **2021**, *12*, 1545–1556. [CrossRef]
84. Vieux, B.E.; Cui, Z.; Gaur, A. Evaluation of a Physics-Based Distributed Hydrologic Model for Flood Forecasting. *J. Hydrol.* **2004**, *298*, 155–177. [CrossRef]
85. Nayak, P.C.; Sudheer, K.P.; Rangan, D.M.; Ramasastri, K.S. Short-Term Flood Forecasting with a Neurofuzzy Model: Flood Forecasting with a Neurofuzzy Model. *Water Resour. Res.* **2005**, *41*, W04004. [CrossRef]

86. Hosseiny, H.; Nazari, F.; Smith, V.; Nataraj, C. A Framework for Modeling Flood Depth Using a Hybrid of Hydraulics and Machine Learning. *Sci. Rep.* **2020**, *10*, 8222. [CrossRef] [PubMed]
87. Yang, T.; Asanjan, A.A.; Welles, E.; Gao, X.; Sorooshian, S.; Liu, X. Developing reservoir monthly inflow forecasts using artificial intelligence and climate phenomenon information. *Water Resour. Res.* **2017**, *53*, 2786–2812. [CrossRef]
88. Aghelpour, P.; Bahrami-Pichaghchi, H.; Varshavian, V.; Graf, R.; Gholami Sefidkouhi, M.A.; Khoshravesh, M. Evaluating the impact of large-scale climatic indices as inputs for forecasting monthly river flow in Mazandaran Province, Iran. *Pure Appl. Geophys.* **2022**, *179*, 1309–1331. [CrossRef]
89. Feng, Z.-K.; Niu, W.-J.; Tang, Z.-Y.; Xu, Y.; Zhang, H.-R. Evolutionary Artificial Intelligence Model via Cooperation Search Algorithm and Extreme Learning Machine for Multiple Scales Nonstationary Hydrological Time Series Prediction. *J. Hydrol.* **2021**, *595*, 126062.
90. Kilinc, H.C.; Haznedar, B. A Hybrid Model for Streamflow Forecasting in the Basin of Euphrates. *Water* **2022**, *14*, 80. [CrossRef]
91. Wang, X.; Wang, Y.; Yuan, P.; Wang, L.; Cheng, D. An Adaptive Daily Runoff Forecast Model Using VMD-LSTM-PSO Hybrid Approach. *Hydrol. Sci. J.* **2021**, *66*, 1488–1502. [CrossRef]

resources

MDPI

Article

Predicting Ice Phenomena in a River Using the Artificial Neural Network and Extreme Gradient Boosting

Renata Graf [1,*], Tomasz Kolerski [2] and Senlin Zhu [3]

1 Department of Hydrology and Water Management, Institute of Physical Geography and Environmental Planning, Adam Mickiewicz University, 61-680 Poznań, Poland
2 Faculty of Civil and Environmental Engineering, Gdańsk University of Technology, 80-233 Gdańsk, Poland; tomasz.kolerski@pg.edu.pl
3 College of Hydraulic Science and Engineering, Yangzhou University, Yangzhou 225009, China; slzhu@yzu.edu.cn
* Correspondence: renata.graf@amu.edu.pl; Tel.: +48-61-829-6259

Abstract: Forecasting ice phenomena in river systems is of great importance because these phenomena are a fundamental part of the hydrological regime. Due to the stochasticity of ice phenomena, their prediction is a difficult process, especially when data sets are sparse or incomplete. In this study, two machine learning models—Multilayer Perceptron Neural Network (MLPNN) and Extreme Gradient Boosting (XGBoost)—were developed to predict ice phenomena in the Warta River in Poland in a temperate climate zone. Observational data from eight river gauges during the period 1983–2013 were used. The performance of the model was evaluated using four model fit measures. The results showed that the choice of input variables influenced the accuracy of the developed models. The most important predictors were the nature of phenomena on the day before an observation, as well as water and air temperatures; river flow and water level were less important for predicting the formation of ice phenomena. The modeling results showed that both MLPNN and XGBoost provided promising results for the prediction of ice phenomena. The research results of the present study could also be useful for predicting ice phenomena in other regions.

Keywords: river freezing; Multilayer Perceptron Neural Network (MLPNN); Extreme Gradient Boosting (XGBoost); predictor variables; balanced accuracy; Poland

Citation: Graf, R.; Kolerski, T.; Zhu, S. Predicting Ice Phenomena in a River Using the Artificial Neural Network and Extreme Gradient Boosting. *Resources* **2022**, *11*, 12. https://doi.org/10.3390/resources11020012

Academic Editor: Diego Copetti

Received: 22 November 2021
Accepted: 24 January 2022
Published: 26 January 2022

1. Introduction

Prediction of ice phenomena in rivers is an important element of hydrological regime analysis [1] and the assessment of the risk of ice jam type floods [2]. The changing thermal conditions of river waters during the winter season and the nature of river ice may significantly change the hydro-ecological and socio-economic aspects of the functioning of the river ecosystem.

Due to the stochastic nature of ice phenomena, their prediction is difficult, especially when data sets for rivers are sparse or incomplete. An additional complication is the scale of the event (local and regional scales) and the influence of numerous factors on the process of river freezing, e.g., meteorological (e.g., air temperature, solar radiation, wind velocity) [3,4], hydrological (e.g., flow rate, inflow and outflow conditions) [5,6], the complexity of interactions between hydroclimatic factors [7,8], hydraulic (e.g., trough cross-section geometry, river bathymetry, water table drop) [9] and thermodynamic factors (e.g., water temperature and thermal conductivity) [10,11]. Relations between river freezing and features of the hydrological regime, including flow, water state, and water temperature, are usually complex and non-linear, and are also spatially heterogeneous due to the variability of environmental conditions. In addition to the process that determines the number of occurrences of a given phenomenon, there is also a dichotomous process determining whether it has a chance of occurring in a given period [12]. This task is further complicated by the fact that ice phenomena occur in three phases: freezing of the river (first symptom of ice), permanent ice cover, and the disappearance phase when an ice floe is formed and related phenomena appear, such as ice jams, which often lead to winter floods. However, the full freezing cycle is not always recorded for rivers.

The analysis of time series relating to ice phenomena allows for the determination of the frequency and duration of their occurrence and the tendency of changes over time, and also for an assessment of the ice phases, which provides a good background for the characteristics of the freezing process in many regions. However, it is not sufficient for their prediction and forecasting [13]. Although Shulyakovskii [14] has developed a manual for forecasting the freezing of inland rivers and lakes, there are few studies related to this topic, especially works dealing with the prediction of ice phenomena at various stages of their occurrence. The problem most frequently discussed is the prediction of ice jams on rivers and their consequences in the form of ice jam floods. The theoretical model of river ice jams was developed by Uzuner and Kennedy [15]. Existing forecasts of ice extent are most often based on the location of the 0 °C isotherm [16]. Good results in this regard have also been obtained from observations of river ice ranges carried out with the use of satellites. Remote sensing is useful for the monitoring of ice characteristics, such as different types and thicknesses of ice or ice cover, and for tracking the progress of the breakup of ice jams, which can help predict the location and timing of ice blockages [17,18]. However, the results of field studies and analyses of satellite images do not always provide accurate data for forecasting ice and scenarios of changes in ice dynamics [19].

Prediction models for ice phenomena are usually limited to the empirical or the stochastic due to the difficulties in applying deterministic models. The methods used to predict ice phenomena (e.g., ice jams) include empirical single-variable threshold analyses, logistic regression [2,20], and discriminant function analysis [21]. Many numerical models have been developed to simulate ice formation on rivers [22,23]. According to Wang et al. [24] and Beltaos [25], a better understanding of physical processes has increased the possibility of developing more accurate numerical models of ice jams and ice jam floods in rivers, e.g., the public-domain river-ice RIVICE model [10], the DynaRice model, a two-dimensional coupled hydrodynamic and ice dynamic model [23], and hydraulic models [19]. An interesting ice jam flood forecasting system that considers requirements for the real-time predictions of water, ice, and sediment transport, was developed for the lower Odra River [26]. The prediction of ice phenomena was also carried out using teleconnection indices, as presented by Sutyrina [27] in relation to spring ice phenomena in lakes and reservoirs (including for Lake Baikal).

In the prediction of ice phenomena, machine learning methods are used less frequently, although they have already been utilized widely in forecasting time series of hydrometeorological data [28–30]. Artificial neural networks (ANNs) have been used to forecast freezing conditions in rivers [31,32] and predict ice jams [4,33]. For example, Chokmani et al. [34] estimated the thickness of ice using artificial neural networks (ANNs), while Hu et al. [35] predicted the disappearance of ice phenomena using hybrid artificial neural networks. Furthermore, fuzzy logic systems have provided favorable results as regards ice phenomena forecasting and its effects [20,36]. For example, Zhao [3] predicted the breakup date of flood ice using a wavelet neural network (WNN) model. Whereas Yan and Ding [37] proposed a predictive model of ice formation based on a dynamic fuzzy neural network (D-FNN) in combination with a particle swarm optimization (PSO) algorithm. The significant advantages of artificial neural networks over standard statistical classification methods consist in their ability to adapt to data of different formats and configurations [32,34]. Ensemble machine learning methodologies, including resampling methods (bagging, boosting, and dagging), model averaging, and stacking, are used in the solving of problems related to simulation and prediction in hydrology [38,39].

The increase in the amount of hydrological and meteorological data makes it more difficult not only to select the methods for their analysis but also to choose predictive and prognostic models so as to maintain both their legibility and accuracy [40]. For the integrated management of an aquatic ecosystem, it is necessary to determine how the thermal–ice regime of the river will develop and change in the future, considering global climate change and local conditions in particular [41]. The identification of the most important hydrological and thermal variables influencing the course of ice phenomena in a river may result in more accurate forecasts of the freezing process.

The main goal of the present study is to predict ice phenomena in a river with the use of the Multilayer Perceptron Neural Network (MLPNN) and Extreme Gradient Boosting (XGBoost) algorithms, which belong to the group of machine learning methods. MLPNN is one of the most widely used ANN models in the field of hydrology [4,7,8]. According to Zounemat-Kermani et al. [42], the boosting methods (e.g., boosting, AdaBoost, and Extreme Gradient Boosting) are becoming more and more effective for modeling and forecasting water quality, runoff, sediment transport, groundwater, flooding, and drought. One of the advantages of XGBoost compared to neural networks is the ability to assess the importance of predictors in the model, and in this study, by employing the XGBoost model, we can assess the dominant factors controlling the dynamics of ice phenomena in the studied river. The objective of the present research was to show the predictability of the selected models and explain spatial differences in terms of the predictors: air temperature (Ta), water temperature (Tw), water level (H), and river flow (Q), as well as the ice phenomenon of the previous day and of the month of occurrence of the phenomenon. The predictions will be carried out using the example of the Warta River in Poland (Central Europe), which is a river of great economic significance and considerable natural value. The results of the study are important for determining the range of intensification of thermal and hydrological ice phenomena variables and the conditions under which their reduction will occur.

2. Study Area

The Warta River is a tributary of the Odra River and the third longest (808 km) river in Poland (Figure 1). Its catchment area (area 54,500 km^2) is characterized by a significant diversity of topography and terrain and climatic and hydrological conditions [43]. Within the Warta Water Region there are three main types of relief: old-glacial in the southern part, young-glacial in the northern and central parts, and upland, south of Wielun.

The catchment belongs to nine out of 28 climatic regions designated in Poland by Woś [44]. The average annual air temperature ranges from 7.5 °C in the north to 8.5 °C in the west. In the coldest month—January—the average temperature ranges from −1.2 °C (in the west) to −2.5 °C (in the southeast). Annual rainfall totals in the study area are diverse and range from 520 mm in the Kujawy region (in the northeast) to 675 mm in the south. A regional differentiation of features of the hydrological regime has been observed along the analyzed section of the Warta [45]—from a medium-developed (the upper and lower course of the river) to a highly-developed (along the section from Nowa Wies (Nowa Wieś Podgórna to Poznan) nival regime (Figure 1). The average dates of appearance of ice phenomena on the Warta River, as well as the dates of their disappearance, vary. Research by Graf et al. [46] for the 1991–2010 observation series showed that the earliest ice phenomena occurred in the third decade of December (about 45% of the total number of observations) and the latest in the first decade of January. The disappearance of ice phenomena is usually observed from the end of January to the end of March [47], while about 30% of observations are made in the third decade of February. Most days with ice phenomena on the Warta River are in January (41% of observations), and the most common form of ice is frazil ice (46% of phenomena) and ice cover (30%).

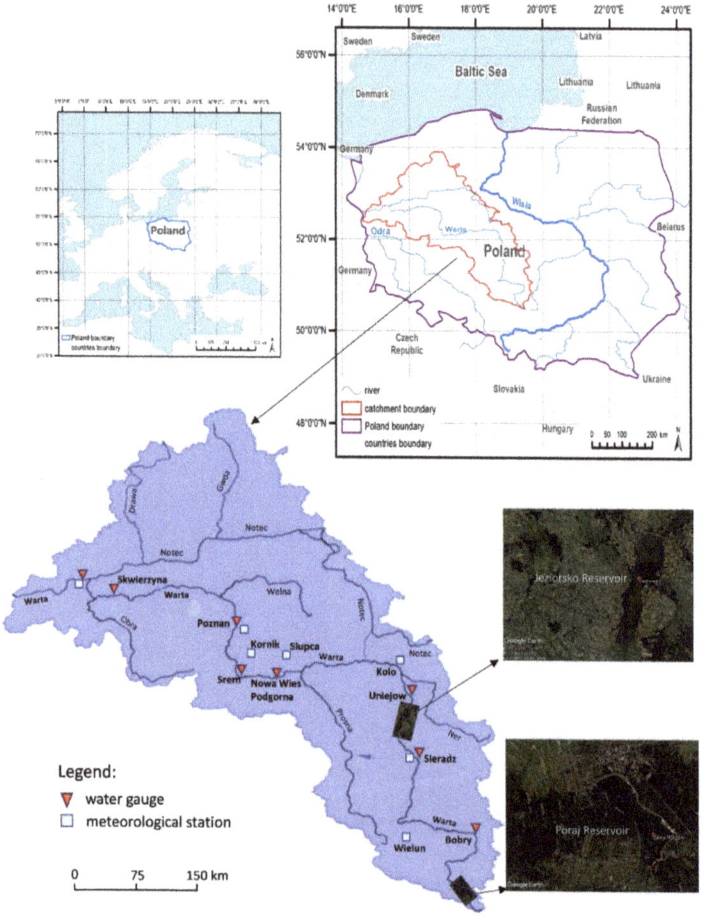

Figure 1. Study area and the locations of water gauges and meteorological stations of the Institute of Meteorology and Water Management—National Research Institute (IMGW-PIB, Warsaw, Poland).

3. Materials and Methods

Predictions of ice phenomena and their numerical descriptions were performed based on daily data on the number of occurrences (the number of days on which the phenomena were observed) and the nature of ice phenomena, and on air temperature (Ta), water temperature (Tw), water levels (H), and river flow (Q) for the years 1983–2013 from the Central Database of Historical Data of the Institute of Meteorology and Water Management—National Research Institute (IMGW-PIB) in Warsaw, Poland (Figure 1). The observation series includes data for the period after 1980, when changes in water temperature in rivers and further consequences, including the lower incidence of ice phenomena, were revealed in response to the sudden climate change associated with changes in CRS (climate regime shift). The regime shift of the late 1980s is a well-documented example of CRS in Poland [48].

Use was made of data from eight water gauges on the Warta River (Bobry, Sieradz, Uniejow, Nowa Wies (Nowa Wieś Podgórna), Srem, Poznan, Skwierzyna, and Gorzow Wielkopolski) and seven meteorological stations (Wielun, Sieradz, Koło, Słupca, Kornik, Poznan, and Gorzow Wielkopolski). Data have been presented in relation to the hydrological year, which in Poland lasts from 1 November until 31 October.

3.1. Classification of Ice Phenomena

The full ice cycle of the river includes forms of ice phenomena observed within the IMGW-PIB water gauge network: frazil ice, border ice, border ice and frazil ice, frazil ice jam, ice cover, ice floes, ice floes and border ice, ice floes and frazil ice, and ice jams. For modeling and predicting ice phenomena, these were grouped into three basic categories: (1) river freeze-up, (2) stable ice cover, and (3) breakup of ice cover—the disappearance of ice (Table 1). The joining into classes is not random, and indeed follows from the order in which ice phenomena appear on the river depending on the thermal and hydrological conditions of the winter season. Each observed ice phenomenon was assigned to the date of occurrence (month and year).

Table 1. Classification and grouping of ice phenomena.

Class	Ice Phenomena	Ice Phase of the River
1 class	Frazil ice Border ice Border ice and frazil ice Frazil ice jam	I phase— Freeze-up
2 class	Ice cover	II phase—Ice cover
3 class	Ice floe Ice floe and border ice Ice floe and frazil ice Ice jam	III phase— Breakup and ice deterioration

In each of the mentioned phases, characteristic fluvial processes occur as a result of the appearance of various forms of ice. Strictly defined forms, such as frazil ice jams or ice jams, are ephemeral forms. According to data from the IMGW-PIB, in the analyzed period, the Warta River had only several days with frazil ice jams and ice jams, which accounted for 0.1% of all observed ice phenomena. Frazil ice jam embolism occurred on the Warta River in Skwierzyna (three days with jams) and Poznan (one day with jams). Ice jams on the Warta occurred in Srem (three days with jams) and Uniejow (two days with jams).

It should be emphasized that ice jams are not a common event in the Warta River. They are among the characteristic features of the river's morphology, which makes the reaches of the river susceptible to the formation of frazil ice [46]. In addition, climatic change serves to significantly decrease the intensity of ice jams by increasing the temperature in the vicinity of supercooled water and thereby prohibiting the formation of ice jams. Furthermore, the Warta River is strongly impacted by anthropogenic activity. Regarding the features of its morphology, the Warta River consists of various bed slopes, from mild to steep, which follow each other in a way that affects the pattern of ice formation. Surface ice is observed mainly over the milder sloped beds. Less surface ice is observed over the steeper slopes, while more suspended frazil particles are present. These ice particles may accumulate under the cover of the following flat sections, forming hanging dams at the inlet of the low-sloped sector. There is no specific data for the Warta River regarding hanging dams that would allow us to distinguish between the ice cover itself and frazil depositions with a greater degree of certainty.

3.2. Data Preparation

Predictions of ice phenomena were performed based on daily data on the number of occurrences (number of days with the phenomena) and the nature of ice phenomena and on air temperature (Ta), water temperature (Tw), water levels (H), and river flow (Q), as well as ice phenomena of the previous day (the day before occurrences of ice phenomena from classes 1, 2, or 3, or from class "none") and the month of individual phenomena (six months of the hydrological winter half-year XI-IV). The choice of input variables was not accidental. The hydroclimatic factors and thermal conditions are important predictors for the process of ice phenomena formation.

To improve the predictability of the tested models and accelerate the process of simulation convergence (in particular as regards artificial neural networks (ANNs)), before inputting the data into Multilayer Perceptron Neural Network (MLPNN) the variables were normalized by converting their values into standardized values (so-called Z-scores) [29]:

$$z_i = \frac{x_i - \underline{x}}{sd_x} \tag{1}$$

where x_i is the i-th value of x, $\underline{x}$ is the mean of x, and sd_x is the standard deviation of x. The resulting variable z has a mean of zero and a standard deviation of one while retaining all the properties of the original variable. The following variables were transformed: Ta, Tw, H, Q, the day of the month (mon.), and the year (Y).

Additionally, ice phenomena of the previous day, encoded in four columns with the one-hot method (zero for the absence of a given phenomenon, and one when it occurred), and the month of individual phenomena, also encoded with the one-hot method, were introduced into the models. Encoding ice phenomena using the one-hot method with the addition of labels allows the assignment of your own characteristics to specific phenomena, showing the similarities between phenomena or the features that make them different. The ice phenomena data used are treated as categorical variables (also called nominal variables), that is, they represent the types of data that can be broken down into groups. In the tested example, three classes were distinguished (Table 1). However, the categories cannot be ordered from highest to lowest. In the classification methods these variables—as target variables (the ones we want to predict)—are usually converted to numerical form using one-hot coding.

3.3. Descriptive Statistics of the Frequency of Ice Phenomena

The research methodology included several stages. The development of predictive models for ice phenomena on the river was preceded by a statistical description of ice phenomena and their changes in the studied period.

The statistical description of ice phenomena included an analysis of the frequency of ice phenomena in the set of analyzed data, which was determined separately for each measuring station and assuming the classification of ice phenomena into three classes, as presented in Table 1. The next stage concerned the analysis of ice phenomena as sequential phenomena. For this purpose, cross-tables were made comparing ice phenomena from the current day with phenomena from the previous day. In the last step, the relationships between the classes of ice phenomena and air temperature, as well as water temperature, water level, and river flow, were analyzed. For this purpose, box and violin plots were made for the distribution of these parameters for each class of ice phenomenon. The violin plot is a combination of a box plot and a density plot, thus showing more details of data distribution, especially the kernel density distribution [49]. As a result, the problem of overlapping the traditional density plot, which is difficult to identify, is eliminated. Wider sections of the graph signify the higher probability of occurrence of certain values, while narrower sections denote lesser probability. According to Hintze and Nelson [50], the violin plot is used to visualize quantitative and qualitative data, including those that do not conform to the normal distribution, and to define the data structure. Like box plots, violin plots are used to present a comparison of variable distribution (or sample distribution) across different categories.

3.4. Prediction Models

Figure 2 depicts the stages of research activities in brief. The current research was carried out in three stages in total. To begin with, the data that had been cleaned, standardized, and adapted to the needs was referred to as prepared data. The second step was to use the R tools to test model predictions using both XGBoost and MLPNN methods. In the prediction of ice phenomena, the following formula was used (according to the choice of the MLPNN and XGboost algorithms):

$$\begin{gathered} ice\_0 + ice\_1 + ice\_2 + ice\_3 \sim \\ Ta + Tw + Q + H + D + Y + \\ day\_before0 + day\_before1 + day\_before2 + day\_before3 + \\ mo1 + mo2 + mo3 + mo4 + mo5 + mo6 \end{gathered} \quad (2)$$

where ice_0 means no ice phenomena, ice_1–3 is the classes of ice phenomena (classes 1–3 adopted on the basis of the classification and grouping of ice phenomena presented in Table 1), Ta is air temperature, Tw is water temperature, Q is river flow, H is water level, D is day of the month, Y is day of the year, day_before0 is day before the day with no ice phenomena (class "none"), day_before1–3 means the day before occurrences of ice phenomena from classes 1, 2, 3 and mo1–6 means months in the winter half-year (November– April, according to the hydrological year).

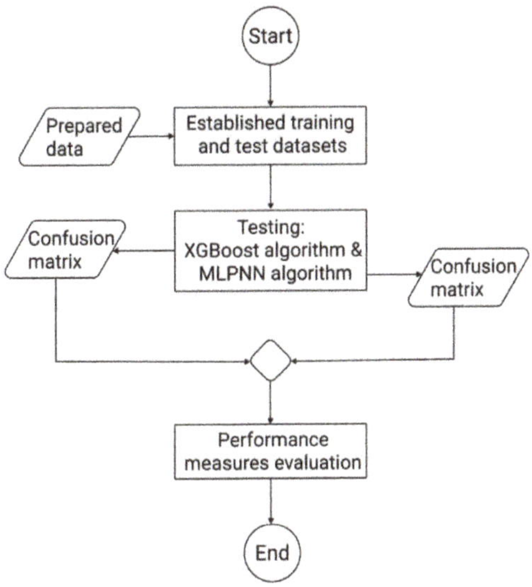

Figure 2. Stages of research activities.

The training and test sets were created using the stratified sampling algorithm, with the year and month variables functioning as layers. The process of determining datasets is detailed in the description Evaluating the Predictions. The confusion matrices were formed on the basis of the second stage of the activity and were then used as inputs for the third stage, which involved evaluating the performance of the XGBoost and MLPNN methods.

3.4.1. The Multilayer Perceptron Neural Network (MLPNN)

The most commonly used type of neural network method is the multi-layered perception method. In this method, the signal is passed to a one-way loop-free input-to-output network. Neither neuron acts on itself. This architecture is referred to as feed-forward, and consists of multiple inputs, hidden layers, and an output, as shown in Figure 3.

The first model used to predict ice phenomena was the Multilayer Perceptron Neural Network (MLPNN), which included an input layer, one hidden layer, and an output layer, and is one of the most widely used ANN models in the field of hydrology [4,7,8]. The input layer, which comprises the predictors, does not perform any calculations. The hidden layer is made of artificial neurons. A single hidden neuron 'collects' activations from each neuron of the input layer and calculates the weighted sum of the input variables. Each hidden layer neuron is connected to each input layer neuron. The hidden layer neurons then perform a non-linear transformation of the weighted sums using an activating function and pass the results to the output layer, which in this application is represented by ice phenomena. A neural network of this type with an output variable Y and containing n neurons in the hidden layer can be expressed as follows [51]:

$$Y = f_2\left[\sum_{j=1}^{n} w_{jk}\left[f_1\left(\sum_{j=1}^{n} x_i w_{ij} + \delta_j\right)\right] + \delta_0\right] \tag{3}$$

where x_i is the value of the input variable i, w_{ij} is the weight (synapse) between the input variable i and the hidden neuron j, δ_j is the bias of the hidden neuron j, f_1 is the sigmoidal function constituting the activation function for hidden neurons, w_{jk} is the synapse between the hidden neuron j and the output neuron k (here $k = 4$), f_2 is also the activation sigmoid function, and δ_0 is the bias of the output layer neuron. The use of the sigmoidal function as an activation function for neurons of the output layer ensured that the predictions would be obtained from the model.

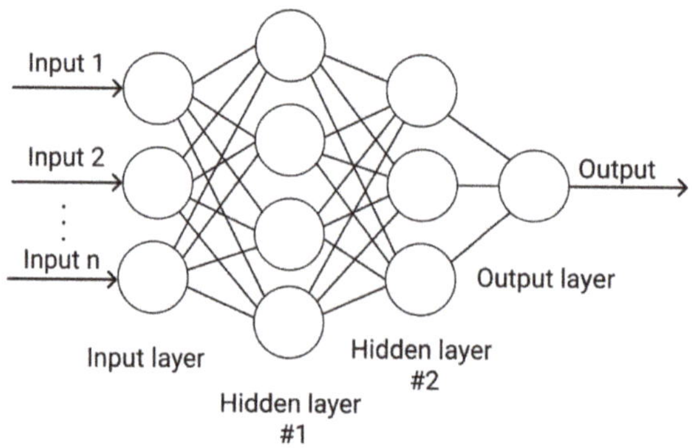

Figure 3. Feed-forward multilayer perceptron architecture.

To estimate the weights and biases, the neural net package [52] and implemented elastic back propagation [53] were used. Cross entropy was used as a function of cost. Models with three, four, five, and six neurons in the hidden layer were calculated for each station.

3.4.2. The Extreme Gradient Boosting (XGBoost) Model

The second model tested was the Extreme Gradient Boosting (XGBoost) implemented by Chen et al. [54]—also in the form of the XGBoost library for the R platform. The gradient boosting machine is a team learning technique based on decision trees. A decision tree generates an output variable estimate based on optimized predictor thresholds that divide the data into multiple groups. The gradient boosting algorithm in each subsequent step aims to reduce the prediction error of the previous step. Technically, in each subsequent step the algorithm estimates the parameters of the model whose purpose is to predict the residuals (prediction errors) of the model estimated in the previous step. The objective function (J) in round t (step t) is given by Equation (4) [54]:

$$J^{(t)} = \sum_{i=1}^{n} l(y_i,\ \hat{y}_i) + \sum_{k=1}^{K} \Omega(f_k) \tag{4}$$

where: l is the training loss, Ω is regulations, f_k is the function of the K–tree. In this study, y_i is the observed ice phenomena and $\hat{y}_i$ is the obtained final prediction value.

In the present study, decision trees with a maximum depth of five nodes were used. Formally, a tree is any consistent acyclic graph, i.e., a graph that does not contain cycles. The multi-class log loss function was used as the cost function. Predictions of $Y^{(t)}$ from the model for iteration t are obtained from Equation (5) [39,54]:

$$Y^{(t)} = \sum_{k=1}^{t} f_k(X) = Y^{(t-1)} + f_t(X) \quad (5)$$

where X is the predictor or set of predictors and f_k is the function that returns the predicted values of the predictors. The second part of the equation shows explicitly that the algorithm prediction in the iteration t is the sum of predictions from the $t-1$ iteration and the new predictions from the t iteration. In XGBoost, the function f_k consists of classification and regression trees that enable the modeling of arbitrary nonlinear relations and the prediction of variables of any nature (Figure 4).

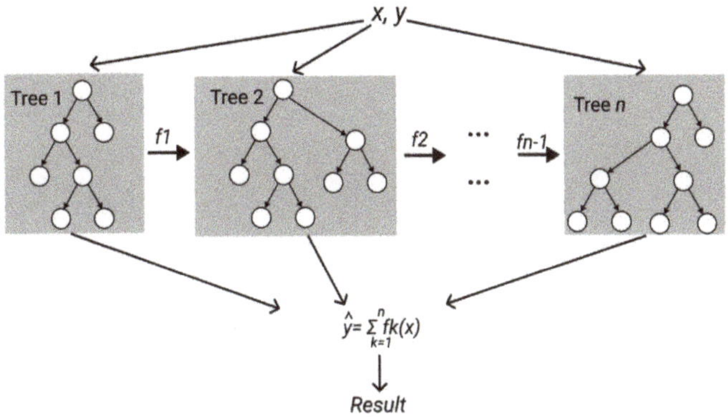

Figure 4. A general architecture for XGBoost.

One of the advantages of XGBoost compared to neural networks is the ability to assess the importance of predictors in the model. The importance of a predictor for regression and classification trees in the gradient boosting algorithm is defined as the profit that the predictor contributes to the entire model by using it to create successive branches of the tree. In this study, by employing the XGBoost model, we can assess the dominant factors controlling the dynamics of ice phenomena in the studied river.

3.5. Evaluating the Predictions

To assess the predictive power of the tested models, cross-validation and four goodness-of-fit metrics were used. Cross-validation was performed by training the models on the available data (training data) and then calculating predictions and goodness-of-fit metrics for the data on which the algorithms were not trained (test data). The XGBoost model was taught on 70% of the training set, and the prediction model was tested on 30% of the test set. The ANN model was taught on the first 50% of the sample, and the prediction model was tested on the remaining 50%. The test and training sets were created using the stratified sampling algorithm. The year and month variables were used in the form of layers. This was done specifically so that the training and test sets had a comparable number of observations within each year and month included in the analysis. Such divisions are in line with the general practice of evaluating machine learning algorithms [51].

The test and training sets were created using the stratified sampling algorithm, with the year and month variables functioning as layers. As a result, the test and random sets had a comparable number of observations within each year and month.

Four metrics were used as goodness-of-fit metrics, calculated separately for each class of ice phenomena: sensitivity, specificity, precision, and weighted validity [55]. For ease of interpretation of these statistics, consider the following cross tables (Table 2), where the letters A–D represent the counts:

Table 2. Goodness-of-fit metrics.

Prediction	Observation	
	Phenomenon	**No Phenomenon**
Phenomenon	A (TP)	B (FP)
No phenomenon	C (FN)	D (TN)

Explanation: A—TP (True-Positive), the number of true positive predictions, i.e., correctly classified examples from the selected class, B—FP (False-Positive), the number of false-positive predictions, i.e., examples incorrectly assigned to the selected class when in fact they do not belong to it, the so-called false alarm, C—FN (False-Negative), the number of false-negative predictions, i.e., misclassified examples from this class, i.e., a negative decision while the example is positive (the so-called error of miss), D—TN (True-Negative), the number of truly negative predictions, i.e., examples correctly not assigned to the selected class (correct rejection).

The statistics used are defined by the formulas:

$$\text{Sensitivity} = A/(A + C) \tag{6}$$

$$\text{Specificity} = D/(D + B) \tag{7}$$

$$\text{Precision} = A/(A + B) \tag{8}$$

$$\text{Balanced Accuracy} = (\text{Sensitivity} + \text{Specificity})/2 \tag{9}$$

Sensitivity TPR (True-Positive Rate) is a measure of "reach" (coverage, "reaching") that indicates the percentage of the positive class that has been covered by a positive prediction [56]. Specificity TNR (True-Negative Rate) is a measure of "coverage" that indicates the percentage of the negative class being covered by the negative prediction. Theoretically, Sensitivity (TPR) and Specificity (TNR) are independent measures, however. in practice increasing sensitivity often leads to a decrease in specificity [55]. Precision, referred to as the Positive Predictive Value (PPV), is a measure of precision that indicates how confidently we can trust positive predictions, i.e., the percentage of positive predictions that are positive. The confidence interval for the three distinguished measures is built based on the Clopper–Pearson method for a single proportion. Accuracy is the proportion of correct predictions with a set of test data. It is the ratio of the number of correct predictions to the total number of input samples. In turn, Balanced Accuracy is the arithmetic mean of the recall for each class. The closer the value is to 1, the better the prediction. However, exactly 1 indicates a problem that may be typically labeled as over-fitting. For highly unbalanced classification problems, as in the case of the analyzed data, balanced accuracy is particularly useful, because this statistic depends on both the level of correct prediction of a phenomenon and the level of prediction of the absence of a phenomenon.

Data analyses and operations were performed using the R 4.02 statistical environment [57]. The analyses and the necessary data restructuring, as well as the visualization of the data and the results of the analyses, were performed using the basic functions of the R environment and dedicated libraries for a given type of algorithm. The libraries used are cited in the corresponding analysis.

4. Results

4.1. *Probability of Occurrence of Ice Phenomena*

The frequency of ice phenomena on the Warta River in the analyzed period has been presented in Table 3. At the majority of measuring stations, ice phenomena from class 1 were observed on slightly more than 10% of days, while the frequency of occurrence of phenomena from class 2 varies from about 1.5% to over 8%. Ice phenomena from class 3 (breakup of ice cover—disappearance of freezing) were the least frequently observed. At each measuring station, this class was observed on less than 1% of days.

Table 3. The frequency of ice phenomena.

Class of Ice Phenomena		Bobry	Sieradz	Uniejow	Nowa Wies	Srem	Poznan	Skwierzyna	Gorzow Wlkp.
1	Nr. of days	518	275	287	417	404	735	449	626
	(%)	10.99	5.8	12.18	10.00	11.73	15.60	10.33	13.29
2	Nr. of days	82	393	130	309	259	69	354	278
	(%)	1.74	8.34	5.52	7.41	7.52	1.46	8.14	5.90
3	Nr. of days	2	45	4	12	4	45	42	15
	(%)	0.04	0.96	0.17	0.29	0.1	0.95	0.97	0.32
No *	Nr. of days	4110	3997	1935	3431	2777	3864	3503	3793
	(%)	87.22	84.86	82.13	82.30	80.63	81.99	80.57	80.50

* No—no ice phenomena.

The probability of occurrence of ice phenomena in specific months of the year (in the cold semester of the hydrological year) has been presented in Figure 5. The probability of occurrence of ice phenomena from class 1 is highest in the months of December and January. Class 2 events are most likely to occur in January and February, whereas the greatest probability of the breakup of ice cover and disappearance of freezing (class 3) is associated with the month of January; with February in Sieradz, and with March in Poznan.

The results of the analysis of ice phenomena as sequential phenomena have been presented in Figure 6. The cross tables compare ice phenomena from the current day with the phenomena of the previous day. It was noted that each class was most often preceded by a phenomenon from its class. Additionally, there was often no ice at all at the river stations the day before the occurrence of ice phenomena from class 1. In a small percentage of days, ice phenomena from class 1 preceded class 2 events. Class 3 occurrences were regularly preceded by phenomena from classes 1 and 2 (Figure 6).

4.2. The Relationship between Ice Phenomena and Hydrological Conditions and Thermal Variables

The assessment of the relationship between various classes of ice phenomena and thermal conditions and hydrological factors has been presented in the form of violin plots of the distribution of these parameters for each class of the phenomenon. Figure 7 shows the differentiation of the variables with respect to the water gauges. For each water gauge station on the Warta River, the differentiation of the occurrence of ice phenomena in relation to air (Ta) and water (Tw) temperatures as well as water level (H) and flow (Q) was presented. As drawn, the graphs indicate certain regularities of occurrence of ice phenomena on the river.

The phenomena from the first stage of ice (border ice, frazil ice) are characteristic of the conditions of poor cooling of the water and mild flow, i.e., for the months of November and December. Although frazil ice requires a significant subcooling of the water and an effective dissipation of the heat of solidification, it can form particularly abundantly during strong, cold winds, even if the air temperature drop is insignificant (even at a few degrees below 0 °C). The analysis of the data showed that the ice phenomena from the first phase occur even at a water temperature of the Warta River of 0.2–0.8 °C, and ice cover is maintained at a water temperature of 0.2 °C and at negative air temperatures, which is understandable. In the case of lowland rivers, which also include the Warta River, the ice cover expansion phase, due to low flows and falls, lasts the longest, and its formation is favored by the persistence of negative air temperatures for a long time [46]. The period of ice cover disappearance as a result of an increase in air temperature occurs on the river in stages, as a result of which an ice floe is created that moves downstream (ice procession). The flow of ice floes in the river usually accelerates the cracking of the ice cover caused by the rise in the water level in the spring.

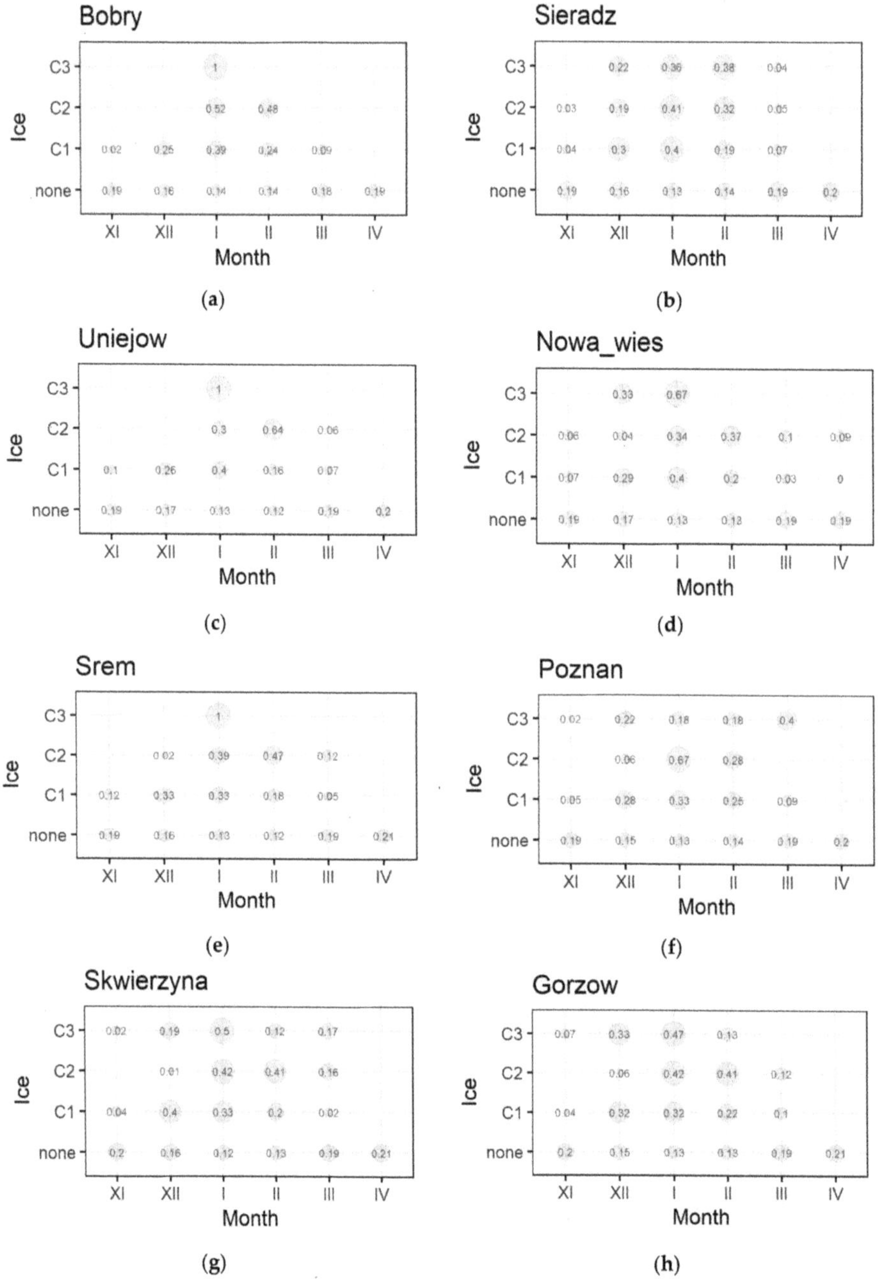

Figure 5. Probability of occurrence of ice phenomena (classes 1–3 and none) as a function of the month for the water gauge stations on the Warta River. Note: Water gauge stations are labeled in the order (**a**–**h**), according to their location on the river (from upper to lower course).

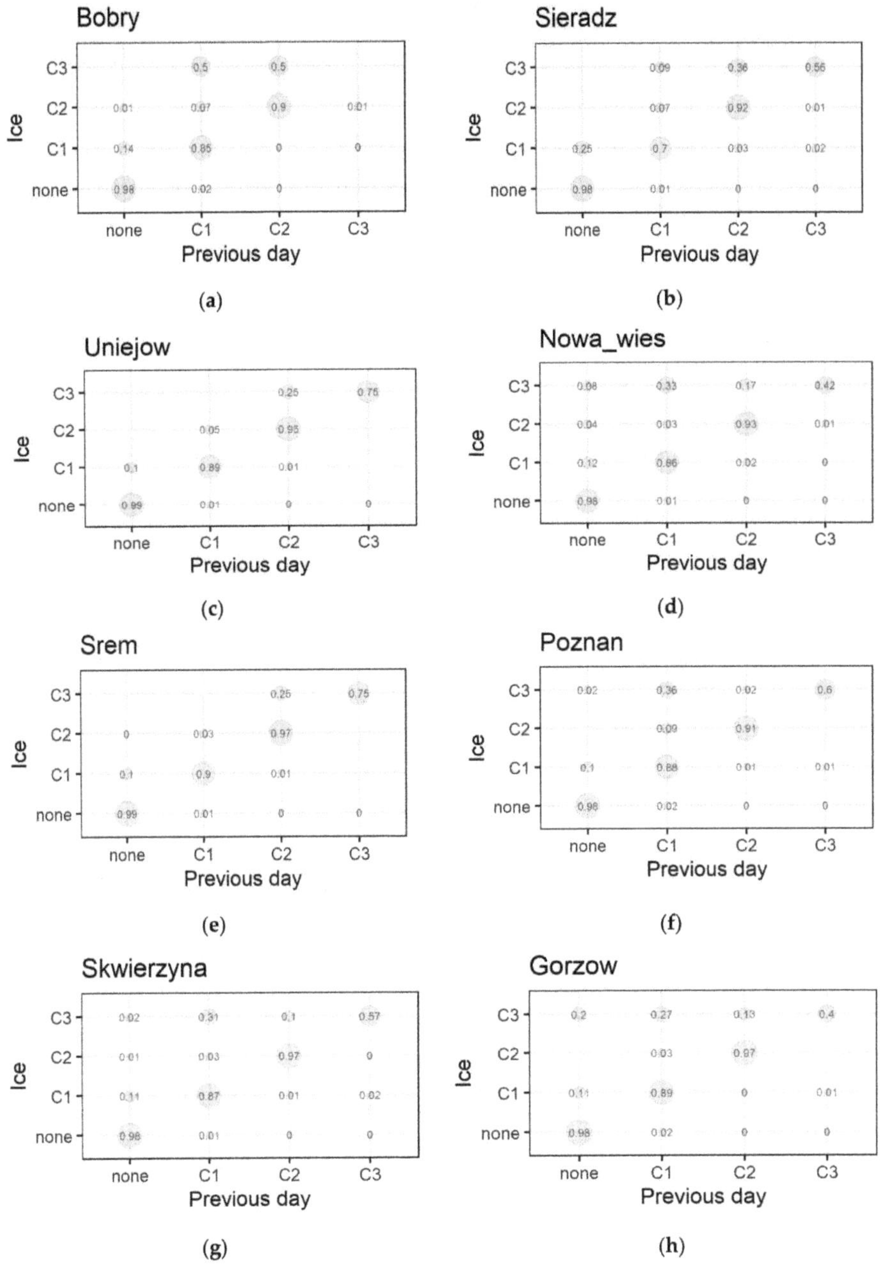

Figure 6. Probability of the order of occurrence of ice phenomena classes (none, C1, C2, C3) as a function of the ice phenomenon of the previous day for the water gauge stations on the Warta River. Note: Water gauge stations are labeled in the order (**a**–**h**), according to their location on the river (from upper to lower course).

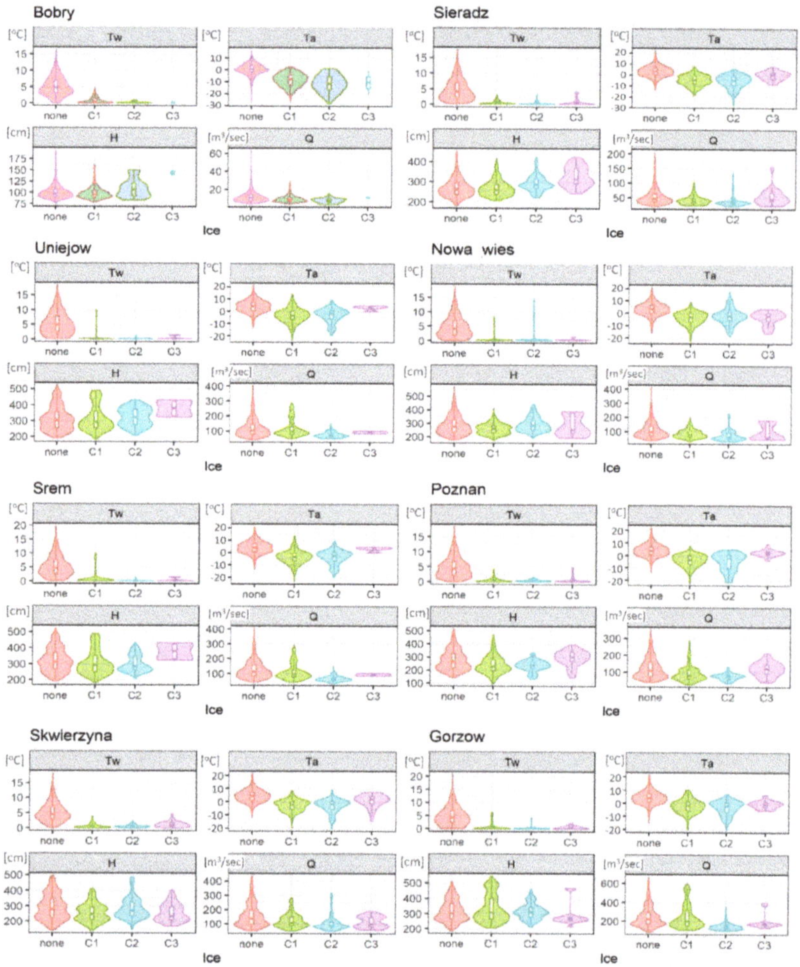

Figure 7. Distributions of the relationship between classes of ice phenomena (none, C1, C2, C3) and air and water temperatures (Ta, Tw), the water level (H), and river flow (Q).

The distribution of ice phenomena concerning water temperature has a distinct character. In this case, the distribution for phenomena classes 1–3 is unimodal and has one high "peak" at very low water temperatures, which indicates the typical regularity of the occurrence of the first ice on the river. Considering the ice cover, the distribution partly takes the form of a slanting distribution with a long tail, which can be seen in the graph for the water gauges of Uniejow, Nowa Wies, and Srem (Figure 7).

As regards the relationship between ice phenomena and air temperature, distribution becomes more diverse depending on the class of the phenomenon and the location of the observation post. For class 1, the distribution is predominantly unimodal. For the majority of measuring stations, distribution is asymmetric and has features of skewed distribution. At the Sieradz and Skwierzyna stations, this distribution shows a tendency to bimodality, which would suggest the presence of two characteristic periods of air temperature and thus favor an increase in the probability of occurrence of ice phenomena from the first phase in these locations. In the case of class 2 (permanent ice cover), distributions at almost all stations are unimodal with a clear skew towards very low air temperatures, which strongly suggests that the probability of ice cover is related to the accumulation of days with negative air temperature. The exception is the Poznan water gauge, for which distribution has features similar to the bimodal distribution (Figure 7). For class 3 (breakdown of the ice cover), the distribution has features typical of unimodal distribution and is focused on an air temperature ranging from 0 °C to a few degrees above zero. This form of relationship is typical of most water gauges on the Warta River. Finally, as regards the water gauges in Nowa Wies and Skwierzyna, the skewness of the distribution increases, and this points to an increase in outliers.

The distribution of ice phenomena from class 1 (formation of ice phenomena) concerning the water level displayed predominantly bimodal features (Figure 7). In the case of the Bobry and Sieradz stations, the distribution shows features of asymmetry and develops as a skewed distribution. In the case of distributions with two or more mods, the widest sections of the violin diagram indicate the greatest probability of observing ice phenomena on the river at a low and medium water level. However, additional periods with a specific water level on the Warta River (states above the average) at which ice phenomena will occur under favorable river thermal conditions are not excluded. The bimodal distribution indicates that the distribution of ice phenomena in this relationship is unstable or very variable. Distribution displays similar features in the case of the relation of the ice cover (class 2) to the state of the water, which is also bimodal at most stations (Figure 7). The distribution shows similar features as regards the relation between the ice cover (class 2) and the state of the water, which, too, is bimodal at the majority of water gauges. The exceptions here are Sieradz and Poznan, for which a unimodal distribution with a specific skewness has been identified. For ice phenomena from class 3, the relationship with the water level shows different types of distributions: unimodal (Poznan and Gorzow Wlkp.) and biomodal (Nowa Wies and Skwierzyna). In the case of Sieradz, the distribution is flat, while in Gorzow Wlkp. it is strongly skewed.

The distribution of the relationship between ice phenomena from class 1 and river flow is bimodal at Nowa Wies, Poznan, and Skwierzyna and unimodal at other stations (Figure 7). In the case of Uniejow, Srem, and Gorzow Wlkp., distribution is also strongly skewed. The greatest probability of occurrence of ice phenomena in the initial period of the Warta River's freezing is associated with the low flow of the river. In the case of permanent ice cover (class 2), the distribution is unimodal at all water gauges, except for Nowa Wies, where it exhibits features of bimodality. This means that the distribution of ice in this relationship is relatively stable along the entire river. However, as regards ice phenomena from class 3, the distribution at certain stations has unimodal (Uniejow, Srem, and Gorzow Wlkp.) or bimodal (Sieradz, Nowa Wies, and Skwierzyna) features.

4.3. Predicting Ice Phenomena

The results of predictive modeling have been presented for three sections of the Warta River: the upper course (Bobry, Sieradz, and Uniejow water gauges)—Table 4; the middle course (Nowa Wies, Srem, and Poznan water gauges)—Table 5; and the lower course (Skwierzyna and Gorzow Wlkp. water gauges)—Table 6. In most of the analyzed instances, the predictive power of the tested models was comparable, and the differences in metrics between models were inconsiderable.

In the upper section of the Warta River (Bobry station), the MLPNN with four hidden units (NN4) was the best among the models, as indicated by the highest values of "balanced accuracy" (BA) statistics for ice phenomena from class 2 (BA = 0.971), and for the "no ice phenomena" class (BA = 0.933), and the second-highest value of statistics for class 1 (BA = 0.913) in the test set (Table 4). The XGBoost model predicted ice phenomena to a comparable extent. It exhibited a similar "balanced accuracy" profile, but one slightly infe-rior to NN models. Class 3 was too small in terms of abundance for the model to success-fully learn the relationship between the class and the predictors in this dataset. For the Sieradz station, it is difficult to identify the model with the highest predictive power (Table 4). The XGBoost model and the NN3–NN5 models successfully predicted each class of ice phenomena. From the NN models, the model with four hidden units was the most sensitive to the rarely occurring class 3 (balanced accuracy BA = 0.76), while predictions for class 1 were less accurate (BA = 0.834). Among all the models used in the work, the XGBoost model best predicted ice phenomena from class 1 (BA = 0.923) but demonstrated the weakest prediction of phenomena from classes 2 (BA = 0.954) and 3 (BA = 0.64) (Table 4). The predictive power of the tested models for the Uniejów station was different depending on the class of phenomena (Table 3). At this location, the XGBoost model achieved the highest values of balanced accuracy in the test set for ice phenomena from classes 1 (BA = 0.951) and 2 (BA = 0.984). At the same time, the NN5 model was the only one to correctly predict ice phenomena from class 3 (BA = 0.998).

Table 4. Results of predictive modeling of ice phenomena for the Warta River (upper course).

Water Gauge	Model	Class	Training Set				Test Set			
			Sensitivity	Specificity	Precision	Balanced Accuracy	Sensitivity	Specificity	Precision	Balanced Accuracy
Bobry	XGBoost	* No	0.986	0.892	0.985	0.939	0.984	0.852	0.978	0.918
		1	0.87	0.987	0.888	0.928	0.83	0.982	0.852	0.906
		2	0.93	0.997	0.87	0.964	0.872	1	0.971	0.936
		3	-	1	-	-	0	1	-	0.5
	NN3	No	0.988	0.907	0.987	0.947	0.982	0.861	0.979	0.922
		1	0.874	0.987	0.892	0.931	0.811	0.983	0.863	0.897
		2	0.978	0.998	0.917	0.988	0.919	0.994	0.723	0.957
		3	1	1	1	1	0	1	0	0.5
	NN4	No	0.988	0.928	0.989	0.958	0.98	0.886	0.984	0.933
		1	0.905	0.988	0.905	0.947	0.844	0.982	0.844	0.913
		2	0.933	0.998	0.913	0.966	0.946	0.995	0.761	0.971
		3	1	1	1	1	0	1	-	0.5
	NN5	No	0.988	0.921	0.988	0.954	0.98	0.876	0.982	0.928
		1	0.909	0.987	0.898	0.948	0.863	0.979	0.833	0.921
		2	1	1	1	1	0.791	0.998	0.872	0.894
		3	0	1	-	0.5	0	1	-	0.5
	NN6	No	0.998	0.934	0.99	0.966	0.982	0.855	0.979	0.918
		1	0.928	0.997	0.972	0.962	0.831	0.981	0.844	0.906
		2	0.927	1	1	0.963	0.829	0.997	0.829	0.913
		3	-	1	-	-	0	1	-	0.5
Sieradz	XGBoost	No	0.992	0.919	0.985	0.955	0.983	0.936	0.989	0.96
		1	0.765	0.987	0.807	0.876	0.869	0.977	0.675	0.923
		2	0.934	0.989	0.895	0.962	0.912	0.995	0.943	0.954
		3	0.2	1	0.8	0.6	0.28	1	0.875	0.64
	NN3	No	0.983	0.965	0.993	0.974	0.973	0.917	0.986	0.945
		1	0.88	0.983	0.767	0.932	0.774	0.974	0.644	0.874
		2	0.961	0.992	0.917	0.977	0.957	0.993	0.922	0.975
		3	0.37	0.997	0.556	0.683	0.444	0.999	0.8	0.722
	NN4	No	0.991	0.928	0.987	0.959	0.98	0.915	0.985	0.948
		1	0.739	0.991	0.836	0.865	0.686	0.982	0.701	0.834
		2	0.99	0.994	0.934	0.992	0.959	0.987	0.869	0.973
		3	0.591	0.997	0.684	0.794	0.522	0.999	0.8	0.76
	NN5	No	0.997	0.929	0.987	0.963	0.984	0.852	0.976	0.918
		1	0.77	0.996	0.919	0.883	0.559	0.986	0.696	0.772
		2	0.976	0.992	0.917	0.984	0.952	0.991	0.904	0.972
		3	0.643	1	1	0.821	0.412	0.997	0.467	0.704
	NN6	No	0.992	0.947	0.991	0.969	0.982	0.888	0.98	0.935
		1	0.871	0.993	0.89	0.932	0.684	0.983	0.715	0.833
		2	0.99	0.992	0.919	0.991	0.954	0.984	0.851	0.969
		3	0.273	0.999	0.75	0.636	0	1	-	0.5
Uniejów	XGBoost	No	0.986	0.922	0.983	0.954	0.993	0.941	0.988	0.967
		1	0.873	0.983	0.879	0.928	0.912	0.99	0.926	0.951
		2	0.938	0.996	0.938	0.967	0.969	0.999	0.984	0.984
		3	0	1		0.5	0	1	-	0.5
	NN3	No	0.994	0.971	0.994	0.983	0.981	0.892	0.976	0.936
		1	0.951	0.992	0.945	0.972	0.818	0.98	0.854	0.899
		2	0.969	0.999	0.984	0.984	0.97	0.995	0.914	0.982
		3	1	1	1	1	0	1	-	0.5
	NN4	No	0.993	0.977	0.995	0.985	0.979	0.927	0.984	0.953
		1	0.974	0.989	0.931	0.982	0.881	0.978	0.838	0.93
		2	0.919	1	1	0.96	0.956	0.996	0.942	0.976
		3	1	1	1	1	0	1	-	0.5
	NN5	No	0.998	0.958	0.991	0.978	0.98	0.835	0.965	0.908
		1	0.954	0.997	0.98	0.976	0.756	0.978	0.816	0.867
		2	0.951	1	1	0.975	0.855	0.999	0.983	0.927
		3	1	1	1	1	1	0.997	0.333	0.998
	NN6	No	0.999	0.986	0.997	0.993	0.99	0.888	0.976	0.939
		1	0.98	0.999	0.993	0.989	0.804	0.989	0.91	0.897
		2	1	1	1	1	0.938	0.993	0.884	0.966
		3	1	1	1	1	0	0.999	0	0.5

* No means no ice phenomena.

Table 5. Results of predictive modeling of ice phenomena for the Warta River (middle course).

Water Gauge	Model	Class	Training Set				Test Set			
			Sensitivity	Specificity	Precision	Balanced Accuracy	Sensitivity	Specificity	Precision	Balanced Accuracy
Nowa Wies	XGBoost	* No	0.988	0.904	0.979	0.946	0.984	0.91	0.981	0.947
		1	0.855	0.988	0.896	0.922	0.867	0.983	0.846	0.925
		2	0.948	0.995	0.936	0.972	0.896	0.994	0.926	0.945
		3	0	1	-	0.5	0	1	-	0.5
	NN3	No	0.989	0.917	0.982	0.953	0.98	0.89	0.976	0.935
		1	0.86	0.99	0.899	0.925	0.819	0.985	0.864	0.902
		2	0.956	0.994	0.933	0.975	0.94	0.991	0.887	0.965
		3	0	1		0.5	0	1	-	0.5
	NN4	No	0.984	0.933	0.986	0.958	0.986	0.842	0.966	0.914
		1	0.901	0.985	0.872	0.943	0.805	0.986	0.868	0.896
		2	0.929	0.997	0.96	0.963	0.838	0.994	0.921	0.916
		3	0.4	0.998	0.333	0.699	0.143	1	0.5	0.571
	NN5	No	0.979	0.95	0.989	0.965	0.981	0.91	0.981	0.946
		1	0.93	0.981	0.844	0.955	0.827	0.985	0.861	0.906
		2	0.956	0.997	0.968	0.977	0.906	0.991	0.888	0.949
		3	0.625	1	1	0.812	0.25	0.996	0.111	0.623
	NN6	No	0.987	0.966	0.992	0.976	0.972	0.919	0.983	0.946
		1	0.967	0.989	0.911	0.978	0.898	0.973	0.791	0.936
		2	0.957	0.998	0.975	0.977	0.865	0.994	0.921	0.93
		3	1	1	1	1	0.167	0.998	0.2	0.582
Srem	XGBoost	No	0.99	0.936	0.984	0.963	0.991	0.935	0.985	0.963
		1	0.895	0.99	0.921	0.942	0.903	0.989	0.917	0.946
		2	0.969	0.997	0.962	0.983	0.945	0.999	0.984	0.972
		3	0	1		0.5	0	1	-	0.5
	NN3	No	0.992	0.982	0.996	0.987	0.976	0.927	0.982	0.951
		1	0.967	0.993	0.949	0.98	0.876	0.977	0.829	0.926
		2	0.984	0.998	0.969	0.991	0.955	0.998	0.977	0.977
		3	0.667	1	1	0.833	0	0.999	0	0.499
	NN4	No	0.997	0.962	0.991	0.98	0.99	0.894	0.976	0.942
		1	0.94	0.997	0.981	0.968	0.82	0.99	0.912	0.905
		2	1	1	1	1	0.931	0.996	0.945	0.963
		3	1	1	1	1	0.333	0.998	0.25	0.666
	NN5	No	0.999	0.957	0.99	0.978	0.988	0.905	0.977	0.947
		1	0.929	0.997	0.979	0.963	0.874	0.986	0.896	0.93
		2	0.977	1	1	0.988	0.877	0.998	0.974	0.938
		3	1	1	1	1	0.5	0.998	0.25	0.749
	NN6	No	0.997	0.976	0.994	0.987	0.98	0.912	0.979	0.946
		1	0.95	0.997	0.975	0.974	0.822	0.983	0.869	0.903
		2	0.993	0.999	0.985	0.996	0.992	0.994	0.925	0.993
		3	1	1	1	1	0.333	0.999	0.333	0.666
Poznan	XGBoost	No	0.987	0.922	0.983	0.955	0.984	0.889	0.975	0.937
		1	0.906	0.983	0.906	0.945	0.874	0.98	0.893	0.927
		2	0.935	0.999	0.906	0.967	0.895	0.999	0.919	0.947
		3	0.567	0.998	0.81	0.782	0.267	0.998	0.5	0.632
	NN3	No	0.985	0.935	0.986	0.96	0.981	0.919	0.981	0.95
		1	0.914	0.985	0.914	0.949	0.89	0.976	0.876	0.933
		2	0.909	0.998	0.882	0.954	0.917	0.999	0.943	0.958
		3	0.81	0.998	0.773	0.904	0.417	0.996	0.5	0.706
	NN4	No	0.984	0.93	0.983	0.957	0.98	0.956	0.991	0.968
		1	0.919	0.981	0.905	0.95	0.918	0.974	0.86	0.946
		2	0.829	0.999	0.935	0.914	0.853	1	0.967	0.926
		3	0.867	1	1	0.933	0.4	0.995	0.333	0.697
	NN5	No	0.989	0.962	0.992	0.976	0.981	0.903	0.978	0.942
		1	0.946	0.989	0.941	0.967	0.864	0.976	0.871	0.92
		2	0.97	0.999	0.941	0.984	0.806	0.999	0.906	0.902
		3	0.833	0.997	0.741	0.915	0.524	0.995	0.478	0.759
	NN6	No	0.99	0.956	0.99	0.973	0.98	0.901	0.979	0.94
		1	0.949	0.988	0.937	0.969	0.886	0.976	0.869	0.931
		2	0.912	0.999	0.939	0.955	0.857	1	1	0.929
		3	0.75	0.999	0.9	0.875	0.429	0.997	0.529	0.713

* No means no ice phenomena.

Table 6. Results of predictive modeling of ice phenomena for the Warta River (lower course).

Water Gauge	Model	Class	Training Set				Test Set			
			Sensitivity	Specificity	Precision	Balanced Accuracy	Sensitivity	Specificity	Precision	Balanced Accuracy
Skwierzyna	XGBoost	* No	0.987	0.941	0.985	0.964	0.987	0.908	0.979	0.948
		1	0.891	0.985	0.872	0.938	0.863	0.982	0.844	0.922
		2	0.969	0.996	0.959	0.982	0.963	0.998	0.975	0.98
		3	0.381	0.999	0.727	0.69	0.143	1	1	0.571
	NN3	No	0.983	0.956	0.989	0.97	0.983	0.964	0.991	0.973
		1	0.913	0.98	0.84	0.946	0.936	0.98	0.84	0.958
		2	0.95	0.999	0.983	0.974	0.931	0.998	0.982	0.965
		3	0.542	0.996	0.619	0.769	0.389	0.996	0.438	0.692
	NN4	No	0.99	0.977	0.994	0.984	0.983	0.917	0.98	0.95
		1	0.952	0.99	0.92	0.971	0.832	0.978	0.813	0.905
		2	0.979	1	0.995	0.989	0.952	0.997	0.963	0.974
		3	0.842	0.998	0.762	0.92	0.391	0.997	0.562	0.694
	NN5	No	0.993	0.968	0.993	0.981	0.99	0.887	0.972	0.938
		1	0.925	0.991	0.921	0.958	0.796	0.987	0.882	0.891
		2	0.984	0.999	0.989	0.991	0.942	0.995	0.947	0.969
		3	0.625	0.997	0.625	0.811	0.385	0.996	0.556	0.69
	NN6	No	0.995	0.986	0.997	0.99	0.98	0.934	0.984	0.957
		1	0.981	0.991	0.925	0.986	0.903	0.979	0.842	0.941
		2	0.995	1	1	0.997	0.97	0.995	0.941	0.982
		3	0.55	1	1	0.775	0.182	0.999	0.571	0.59
Gorzow Wlkp.	XGBoost	No	0.984	0.938	0.985	0.961	0.982	0.9	0.976	0.941
		1	0.911	0.981	0.877	0.946	0.866	0.982	0.883	0.924
		2	0.973	0.999	0.986	0.986	0.977	0.998	0.963	0.987
		3	0	1	-	0.5	0	1	-	0.5
	NN3	No	0.986	0.929	0.982	0.958	0.982	0.907	0.978	0.944
		1	0.907	0.985	0.904	0.946	0.875	0.98	0.869	0.927
		2	0.98	1	0.993	0.99	0.969	0.998	0.969	0.983
		3	0	1	-	0.5	0	1	-	0.5
	NN4	No	0.981	0.946	0.987	0.963	0.98	0.926	0.982	0.953
		1	0.92	0.978	0.865	0.949	0.885	0.978	0.863	0.932
		2	0.986	1	0.993	0.993	0.964	0.996	0.937	0.98
		3	0	1	-	0.5	0	1	-	0.5
	NN5	No	0.987	0.959	0.989	0.973	0.972	0.908	0.979	0.94
		1	0.937	0.986	0.914	0.961	0.858	0.974	0.83	0.916
		2	0.986	1	0.993	0.993	0.912	0.995	0.912	0.953
		3	0.444	0.999	0.571	0.722	0.333	0.997	0.222	0.665
	NN6	No	0.989	0.947	0.987	0.968	0.973	0.913	0.979	0.943
		1	0.929	0.986	0.914	0.958	0.882	0.974	0.838	0.928
		2	0.986	1	1	0.993	0.949	0.998	0.97	0.974
		3	0.333	1	1	0.667	0.167	0.999	0.25	0.583

* No means no ice phenomena.

In the middle course of the Warta River, for the Nowa Wies water gauge, the NN5 model turned out to be the best at predicting ice phenomena (Table 5). This model performed well for ice phenomena from classes 1 (BA = 0.906) and 2 (BA = 0.949), comparable with other models, while at the same time being the most sensitive for class 3 (BA = 0.623). However, the best performance in predicting class 1 events was achieved by the NN6 model (BA = 0.936), and the best performance for class 2 by the NN3 model (BA = 0.965). The XGBoost model achieved similar performance to the NN5 model as regards the prediction of phenomena from class 2 (BA = 0.945). In the case of the Srem water gauge, it is difficult to indicate the best model (Table 5). A neural network model NN5 best predicted the ice phenomena from class 3 (BA = 0.749). The NN6 model showed the best prediction for class 2 (BA = 0.993), and class 1 events were best predicted by XGBoost (BA = 0.946). Nevertheless, it is the neural network model with five hidden units (NN5) that seems to have the most balanced prediction profile for all classes of ice phenomena. For the Poznan water gauge (Table 5), the predictive power of the tested models was comparable. The NN3 model can be viewed as the best for predicting ice phenomena at this location because it predicted classes 1 (BA = 0.933) and 2 (BA = 0.958) best and was the third most effective in predicting the level of class 3 events (BA = 0.706). At the same time, the NN5 model turned out to be the best for ice phenomena from class 3 (BA = 0.759).

For the lower course of the Warta River, in the Skwierzyna profile, the predictive power of the tested models was comparable (Table 6). The NN3 model appears to present the most balanced predictive profile. This model best predicted the occurrence of ice phenomena from class 1 (BA = 0.958) and also the absence of river freezing (BA = 0.973). The best prediction of ice phenomena from class 2 was achieved by the NN6 model (BA = 0.982), with the results of the XGBoost model being comparable (BA = 0.98). The NN3 model also displayed good predictability of phenomena from classes 2 (BA = 0.965) and 3 (BA = 0.692). Its class 3 prediction performance is comparable to that of the NN4 model, for which BA = 0.694 was determined. For the Gorzow Wlkp. water gauge, one of the better predictive models for ice phenomena was the NN4 model (BA = 0.932 for class 1, BA = 0.98 for class 2) (Table 6). The NN5 model predicted classes 1 and 2 comparably to the other models, and at the same time was the most sensitive in terms of predicting ice phenomena from class 3 (BA = 0.665). The XGBoost model predicted the phenomena from group 2 best (BA = 0.987), similarly to the NN3 model (BA = 0.983).

4.3.1. Spatial Differences in Model Performance

Among the NN models used, the best predictions were given by the NN5 (eight-fold confirmation of the best prediction) and NN4 models (seven-fold) (Table 7). The XGBoost model also has high predictive power, and the model turned out to be the best in predicting ice phenomena from classes 1 and 2. In three cases, its performance was comparable with those of the NN models. The phenomena from the initial stage of freezing (class 1) were best predicted by the XGBoost model. On the other hand, the disintegration of the ice cover and accompanying ice phenomena were best predicted by the NN5 model (at five water gauge stations). No dependence of the models' performance on the location of water gauges (Table 7) was observed, although as regards predictions of ice phenomena in the upper section of the Warta River (Bobry, Sieradz, Uniejow stations), the XGBoost model and the NN4 and NN5 models proved to be superior.

Table 7. Models with the best prediction of ice phenomena on the Warta River.

Water Gauge	No Ice	Class 1	Class 2	Class 3
Bobry	NN4	NN5	NN4 XGBoost	* No results
Sieradz	XGBoost	XGBoost	NN3	NN4
Uniejow	XGBoost	XGBoost	XGBoost	NN5
Nowa Wies	XGBoost	NN5	NN5 XGBoost	NN5
Srem	XGBoost	XGBoost	NN6	NN5
Poznan	NN4	NN3	NN3	NN5
Skwierzyna	NN3	NN3	NN6 XGBoos	NN4
Gorzow Wlkp.	NN4	NN4	XGBoost	NN5

* No results—no results from the learned relations between the class of the phenomenon and predictors.

Ice phenomena predictions for the river along its middle section (stations in Nowa Wies, Srem, and Poznan) were made most reliably by the XGBoost and NN5 models (Nowa Wies and Srem) and the NN3–NN5 models (Poznan) (Table 7). For the prediction of ice phenomena along the lower section of the Warta, superior performance was demonstrated by the NN models, taking into account the lower efficiency of the XGBoost model.

The most difficult prediction was that for ice phenomena in the decay phase and the formation of ice floes and, consequently, ice jams. Due to the lowest frequency of observations, there were problems with their prediction in the case of the Bobry station. In this case, no results were obtained from the relations determined between the class of ice phenomena and the predictors.

4.3.2. Evaluation of the Importance of Predictors in the Models

The use of XGBoost, as opposed to ANNs, made it possible to assess the importance of predictors in the model. The selected predictor variables were ranked according to the normalized reduction in model error, also known as "variable importance". Figure 8 shows the most important predictor variables in the final model: water and air temperature, hydrological conditions (water level and river flow), and data for the "day before", month, and year. The results of this analysis indicate that for each measuring station the most important predictor of ice phenomena is the type of ice phenomenon the day before the identification of a given event, with water temperature and air temperature coming next. In the case of the stations in Uniejow and Srem, water temperature is the second most important predictor of the occurrence of ice phenomena.

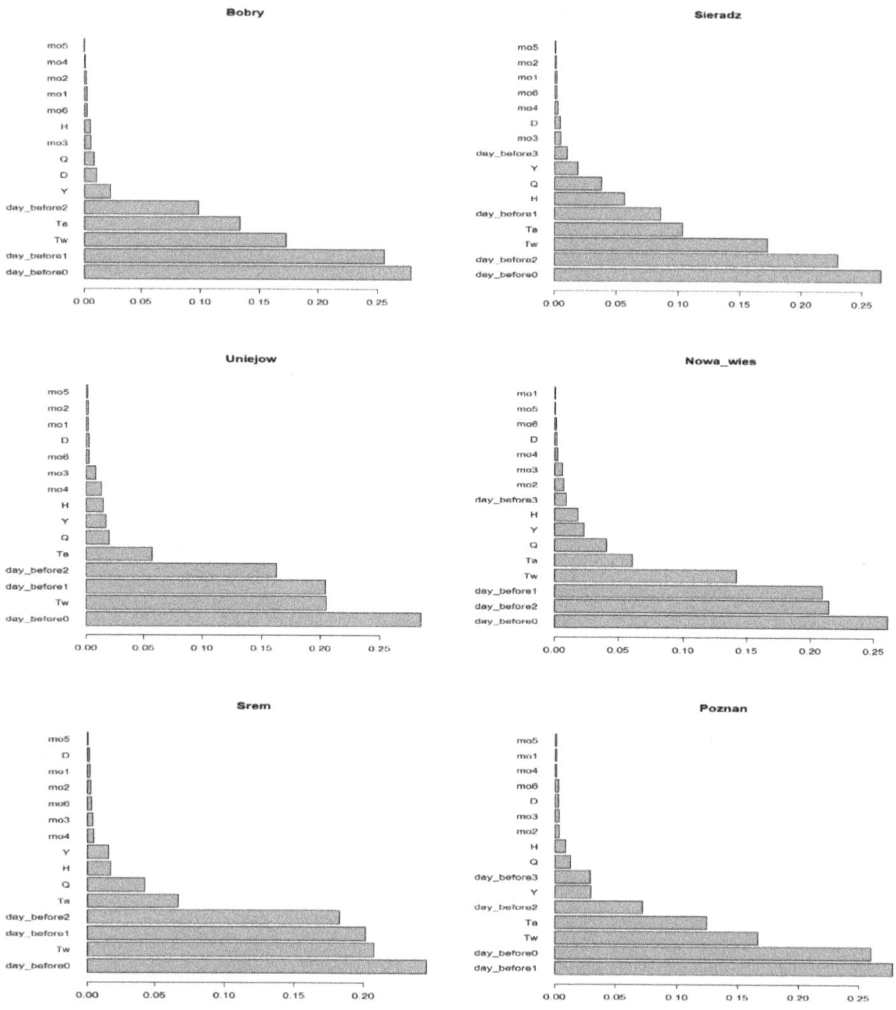

Figure 8. *Cont.*

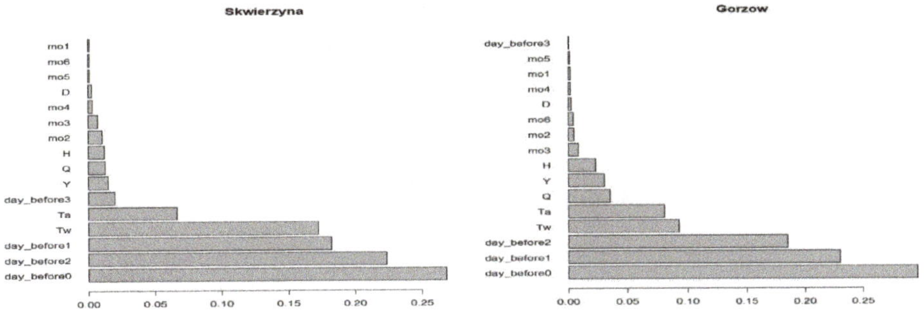

Figure 8. Relative importance of predictors in the XGBoost model (profit values).

These results suggest that when looking for a balance between the complexity of the model and its predictive power, the two most important predictors for the occurrence of ice phenomena on the Warta River should be taken into account, i.e., the nature of the ice phenomenon on the day preceding the observation (especially for class 2 or class 1 events), and water temperature.

5. Discussion

5.1. Selection of Predictors as Input Variables

The predictive modeling of ice phenomena carried out on the example of the Warta River showed that the prediction of their occurrence in different phases and spatial locations gives different results. In this case, the prediction was a difficult process, mainly due to the complexity of interactions between hydroclimatic factors and thermal conditions that contribute to the occurrence of freezing.

In the research conducted on the Warta River, an important assumption was the selection of input variables that affect the accuracy of predictions in the neural network models and XGBoost. A set of daily data were used, these including thermal and hydrological variables, the type of ice phenomenon (group of phenomena) on the day preceding water gauge observations, and the month of their occurrence. The premises confirming the correctness of their choice are the results of studies of the ice regimes of rivers in Poland, including those conducted on the Vistula River [13,58], Oder River [59], on the rivers of the Baltic coastal zone [60,61], Bug River [62], and Warta River [46,63,64]. The selection of input variables significantly affects the performance of ANN models [7], however, it is often arbitrary [8].

5.2. The Most Important Predictor Variables in the Final Model

The results of the predictive models that we developed for the Warta River showed that all the input parameters (predictors) that were taken into account had some significance for the formation of ice phenomena from different classification groups. However, under the thermal conditions established for the reference period (research period 1984–2013), hydrological parameters—river flow and water level—were less important for the process of ice phenomena formation. The research established that ice phenomena occurred irregularly and periodically in the studied period and that the structure of freezing along the river course was diversified. The phenomena from class 1 were predominant, i.e., from the freezing phase of the river, represented by frazil ice and border ice, which is now a typical feature of the ice regime of most rivers in Poland [47].

The results of modeling confirmed that the most important predictors in the analyzed case were the nature of the phenomenon on the day preceding the observation (most often class 2 or class 1), as well as water temperature, and then air temperature (Figure 8). Graf [12] examined the dependencies of the trends of ice phenomena in the Noteć River, in western Poland (a tributary of the Warta River), on air and water temperature using regression models for count data and the Zero-Inflated Negative Binomial Model; results showed that the temperature values are the best predictors. In some locations, however, the model predicting the number of ice phenomena—taking into account the relationship with temperature—turned out to be statistically insignificant. Graf and Tomczyk [11] determined that for the Noteć River, a faster increase in accumulated sequences of negative air temperature contributes to an increase in the probability of a permanent ice cover, and the average degree day increase by one degree increases the chance of ice cover on the river in the range of 1.5–6.0% in different water gauges.

The period of intense changes in thermal conditions in the Warta catchment area, e.g., the cold period or sudden spring river supply, can be represented by changes in the models. However, it is also visible in the types of distributions illustrating the relationship between the classes of ice phenomena on the Warta River and hydrological factors and thermal conditions, which has been presented in the violin plots. The violin plots show diverse and complicated relations resulting from the differing variability of hydroclimatic factors and thermal conditions, which determines the nature of the distribution (Figure 5). Conditions conducive to the emergence of ice phenomena are not always the same in every location on the river, which is the result of local conditions, including, e.g., channel morphology and the influence of anthropic pressure.

5.3. The Performance of Predictive Models

The models performed promisingly in predicting the occurrence of ice phenomena on the Warta River, and this—in addition to their low demand for computational data resources, speed of operation, and ease of use—makes them particularly attractive. Further, it was found that the ANN approach served its purpose. By using more advanced and specialized network architectures (NN5, NN6), the ability to learn and predict the non-linear behavior of ice phenomena was increased for classes 2 and 3, which were characterized by a lower frequency of occurrence in the Warta River.

In ANN predictive modeling, the use of the sigmoidal function as an activation function for output layer neurons ensured that predictions obtained with the model would be the probabilities of occurrence of a given class, since this function maps real values to the range 0–1. As a result, models with 3, 4, 5, and 6 neurons in the hidden layer were developed, and this made it possible to compare their performance in predicting ice phenomena. Guo et al. [4], basing their work on the ANN theory, used the sigmoidal function as the activation function in the hidden layer for the forecasting of ice jams during river ice breakup. Their results were promising, as they predicted the annual occurrence of ice blockages with an accuracy of 85%, while the projected decay date with the projected ten-day period showed a maximum error of two days.

Concerning the prediction of class 1 and class 2 phenomena in the Warta River (permanent ice cover) and their non-occurrence, ANNs require further improvement, although present results indicate that they are comparable to the XGboost algorithm for predicting group 2 phenomena, i.e., permanent ice cover. The performance of the NN models and the XGBoost algorithm is also comparable for the different water gauge locations on the Warta River, although an overall better fit of XGBoost and NN4/NN5 models was observed for the upper course of the river; XGBoost and NN5/NN3 were most successful for the middle course, while NN models predominated for the lower course. The results of a comparison of both types of models in terms of their suitability for predicting ice phenomena on the Warta River showed a high accuracy of prediction for the XGBoost method, which has not been used on a larger scale in this regard so far.

XGBoost models variable interactions and handles the multi-linearity common to ecological datasets seamlessly [65]. Moreover, XGBoost works faster than many other gradient-increasing algorithms due to the regularization factor and the parallel computing functionality. One of the advantages of this method is its resistance to outliers, which eliminates the need to supplement missing data, and thus, in the case of the Warta River, it allows an increase in the efficiency of the prediction of a given phenomenon, even when eliminating outliers. The obtained results were considered satisfactory, which was confirmed by four model fit measures.

The comparison of results obtained for the Warta River with the ANN models used to predict ice phenomena on other rivers shows their considerable similarity. Most of the predictive and prognostic models developed confirm that the results of ice condition forecasts made with the use of ANNs are satisfactory and consistent with the measured data [37,66,67]. Moreover, the high accuracy of forecasts is indicated, which takes into account factors influencing the formation and disappearance of ice phenomena. Too many simplifications, as made in some models, may lower their prognostic accuracy and limit their usefulness for other rivers [36]. According to Massie [33], neural network classifiers, just like in the case of empirical methods, are most likely location-specific, but it is possible to transfer ANN models to other locations with minimal modifications. However, there are still no solutions for the prediction of phenomena in individual phases of their occurrence using the XGBoost algorithm. In Poland to date, models from the ANN group and the XGBoost algorithm have not been used to predict ice phenomena.

A review of the literature shows that numerous parameters are needed to support models developed for forecasting ice phenomena, most commonly ice jams and the resultant floods, but obtaining this data is sometimes difficult or even impossible. Despite the progress made in forecasting ice processes on rivers, this field still has great research potential; however, it also requires comprehensive observations, the collection and testing of data from stationary measurement networks, and direct field studies [8].

6. Conclusions

In the present study, MLPNN and XGBoost models were developed to forecast ice phenomena on the Warta River in Poland. The results obtained lead to the following conclusions:

(1) Both the MLPNN and XGBoost models produced promising results for the forecasting of ice phenomena, which are presented using the four model fit measures.

(2) For highly unbalanced classification problems, as in the case of the analyzed data, the "Balanced Accuracy" is particularly useful, since this statistic depends on both the level of correct prediction of a phenomenon and the level of prediction of the absence of a phenomenon.

(3) The XGBoost turned out to be the best for predicting freeze-up (class 1) and ice cover (class 2 of ice phenomena), and at three water gauges its performance was comparable with that of the NN models, whereas breakup and ice deterioration (class 3) were best predicted by the NN5 model (at five water gauge stations). No dependence of the performance of individual models on the location of water gauges was observed.

(4) The choice of input variables impacts the accuracy of the models developed. The nature of ice phenomenon on the day preceding the observation, as well as water and air temperature values, are important predictors, while river flow and water level were less important for the process of ice phenomena formation. This information was provided by the XGBoost algorithm.

(5) The forecasting of ice phenomena is complicated due to the complex interactions between their determinants. This is confirmed by the types of distribution (unimodal, bimodal), illustrating the relationship between classes of phenomena on the river and hydroclimatic factors and thermal conditions.

The results of the research conducted here have important implications for forecasting ice phenomena, specifically as regards the application of XGBoost. Preliminary results seem to indicate that XGBoost, as an ensemble machine learning model, works well as a forecasting tool in hydrological research. Though the MLPNN and XGBoost models performed competently, there is still scope for further improvement through additional studies and the construction of hybrid models. Other factors influencing the occurrence of ice phenomena on rivers that would additionally help to improve the accuracy of these models should also be looked at (e.g., channel morphology, the accumulated degree days of frost and thaw, and the rates of change in water level and flow during the freeze-up and breakup periods). Since the present results concern only one river, future research will focus on applying models to rivers in different geographic locations and hydrological regimes to more accurately test the suitability and effectiveness of models.

Author Contributions: Conceptualization, R.G.; funding acquisition, R.G.; methodology, R.G.; project, R.G.; administration, R.G.; investigation, R.G.; software, R.G.; formal analysis, R.G.; validation, R.G.; visualization, R.G.; writing—original draft preparation, R.G; writing—review and editing, R.G., T.K., and S.Z.; supervision, R.G., T.K., and S.Z. All authors have read and agreed to the published version of the manuscript.

Funding: This research received funding from an internal grant "Research subsidy for scientific research activities of the Faculty of Geographical and Geological Sciences of the Adam Mickiewicz University in Poznan, Poland.

Institutional Review Board Statement: Not applicable.

Informed Consent Statement: Not applicable.

Data Availability Statement: Hydrological and meteorological measurement data are available directly from https://hydro.imgw.pl (accessed on 10 July 2021) and https://danepubliczne.imgw.pl (accessed on 15 July 2021) as operational data. The data that support the findings of this study are available from the corresponding author upon reasonable request.

Acknowledgments: The authors thank the Faculty of Geographical and Geological Sciences, Adam Mickiewicz University, Poznan, Poland for their support. The authors would also like to acknowledge the Institute of Meteorology and Water Management—National Research Institute (IMWM-NRI, Warsaw, Poland) for the release of the input database.

Conflicts of Interest: The authors declare no conflict of interest.

References

1. Magnuson, J.J.; Robertson, D.; Benson, B.; Wynne, R.; Livingstone, D.; Arai, T.; Assel, R.; Barry, R.; Card, V.; Kuusisto, E.; et al. Historical trends in lake and river ice cover in the northern hemisphere. *Science* **2000**, *289*, 1743–1746. [CrossRef] [PubMed]
2. Hicks, F.; Beltaos, S. River Ice. In *Cold Region Atmospheric and Hydrologic Studies. The Mackenzie GEWEX Experience*; Woo, M., Ed.; Springer: Berlin/Heidelberg, Germany, 2008.
3. Zhao, L. River Ice Breakup Forecasting Using Artificial Neural Networks and Fuzzy Logic Systems. Ph.D. Thesis, Department of Civil and Environmental Engineering, University of Alberta, Edmonton, AB, Canada, 2012.
4. Guo, X.; Wang, T.; Fu, H.; Guo, Y.; Li, J. Ice-jam forecasting during river breakup based on neural network theory. *J. Cold Reg. Eng.* **2018**, *32*, 04018010. [CrossRef]
5. Beltaos, S.; Prowse, T. River-ice hydrology in a shrinking cryosphere. *Hydrol. Proc.* **2009**, *23*, 122–144. [CrossRef]
6. Lindenschmidt, K.-E.; Huokuna, M.; Burrell, B.C.; Beltaos, S. Lessons learned from past ice-jam floods concerning the challenges of flood mapping. *Int. J. River Basin Manag.* **2018**, *16*, 457–468. [CrossRef]
7. Massie, D.D.; White, K.D.; Daly, S.F. Application of neural networks to predict ice jam occurrence. *Cold Reg. Sci. Technol.* **2002**, *35*, 115–122. [CrossRef]
8. Madaeni, F.; Lhissou, R.; Chokmani, K.; Raymond, S.; Gauthier, Y. Ice jam formation, breakup and prediction methods based on hydroclimatic data using artificial intelligence: A review. *Cold Reg. Sci. Technol.* **2020**, *174*, 103032. [CrossRef]
9. Nafziger, J.; She, Y.; Hicks, F. Anchor ice formation and release in small regulated and unregulated streams. *J. Cold Reg. Sci. Technol.* **2017**, *141*, 66–77. [CrossRef]
10. Lindenschmidt, K.-E. RIVICE—A non-proprietary, open-source, one-dimensional river-ice and water-quality model. *Water* **2017**, *9*, 314. [CrossRef]
11. Graf, R.; Tomczyk, A.M. The Impact of cumulative negative air temperature degree-days on the appearance of ice cover on a river in relation to atmospheric circulation. *Atmosphere* **2018**, *9*, 204. [CrossRef]
12. Graf, R. Estimation of the dependence of ice phenomena trends on air and water temperature in river. *Water* **2020**, *12*, 3494. [CrossRef]
13. Kolerski, T. Modeling of ice phenomena in the mouth of the Vistula River. *Acta Geophys.* **2014**, *62*, 893–914. [CrossRef]
14. Shulyakovskii, L.G. *Manual of Forecasting Ice-Formation for Rivers and Inland Lakes. Manual of Hydrological Forecasting No. 4*; Translated from Russian, Israel Program for Scientific Translations; Central Forecasting Institute of USSR: Jerusalem, Israel, 1963.
15. Uzuner, M.S.; Kennedy, J.F. Theoretical model of river ice jams. *J. Hydraul. Eng. Div.* **1976**, *102*, 1365–1383. [CrossRef]
16. Yang, X.; Pavelsky, T.M.; Allen, G.H. The past and future of global river ice. *Nature* **2020**, *577*, 69–73. [CrossRef] [PubMed]
17. Weber, F.; Nixon, D.; Hurley, J. Semi-automated classification of river ice types on the Peace River using RADARSAT-1 synthetic aperture radar (SAR) imagery. *Can. J. Civil Eng.* **2003**, *30*, 11–27. [CrossRef]

18. Lindenschmidt, K.; Syrenne, G.; Harrison, R. Measuring ice thicknesses along the Red River in Canada using RADARSAT-2 satellite imagery. *J. Water Resour. Prot.* **2010**, *2*, 923–933. [CrossRef]
19. Kolerski, T. Mathematical modeling of ice dynamics as a decision support tool in river engineering. *Water* **2018**, *10*, 1241. [CrossRef]
20. Mahabir, C.; Hicks, F.; Fayek, A.R. Transferability of a neuro-fuzzy river ice jam flood forecasting model. *Cold Reg. Sci. Technol.* **2007**, *48*, 188–201. [CrossRef]
21. White, K.D.; Daly, S.F. Predicting Ice Jams with Discriminant Function Analysis. In Proceedings of the ASME 2002 21st International Conference on Offshore Mechanics and Arctic Engineering, Oslo, Norway, 23–28 June 2002; Volume 3, pp. 683–690.
22. Beltaos, S. Numerical modelling of ice-jam flooding on the Peace-Athabasca Delta. *Hydrol. Proc.* **2003**, *17*, 3685–3702. [CrossRef]
23. Shen, H.T. Mathematical modelling of river ice processes. *Cold Reg. Sci. Technol.* **2010**, *621*, 3–13. [CrossRef]
24. Wang, J.; Sui, J.; Guo, L.; Karney, B.W.; Jüpner, R. Forecast of water level and ice jam thickness using the back propagation neural network and support vector machine methods. *Int. J. Environ. Sci. Technol.* **2010**, *7*, 215–224. [CrossRef]
25. Beltaos, S. Progress in the study and management of river ice jams. *Cold Reg. Sci. Technol.* **2008**, *51*, 2–19. [CrossRef]
26. Lindenschmidt, K.-E.; Carstensen, D.; Fröhlich, W.; Hentschel, B.; Iwicki, S.; Kögel, K.; Kubicki, M.; Kundzewicz, Z.W.; Lauschke, C.; Łazarów, A.; et al. Development of an ice-jam flood forecasting system for the lower Oder River—Requirements for real-time predictions of water, ice and sediment transport. *Water* **2019**, *11*, 95. [CrossRef]
27. Sutyrina, E.N. Prediction of spring ice phenomena on lakes and reservoirs using teleconnection indices. *Limnol. Freshw. Biol.* **2020**, *4*, 946–947. [CrossRef]
28. Piotrowski, A.P.; Napiórkowski, M.J.; Napiórkowski, J.J.; Osuch, M. Comparing various artificial neural network types for water temperature prediction in rivers. *J. Hydrol.* **2015**, *529*, 302–315. [CrossRef]
29. Graf, R.; Zhu, S.; Sivakumar, B. Forecasting river water temperature time series using a wavelet–neural network hybrid modelling approach. *J. Hydrol.* **2019**, *578*, 124115. [CrossRef]
30. Zhu, S.; Heddam, S.; Nyarko, E.K.; Hadzima-Nyarko, M.; Piccolroaz, S.; Wu, S. Modeling daily water temperature for rivers: Comparison between adaptive neurofuzzy inference systems and artificial neural networks models. Environ. *Sci. Pollut. Res.* **2019**, *26*, 402–420. [CrossRef]
31. Morse, B.; Hessami, M.; Bourel, C. Mapping environmental conditions in the St. Lawrence River onto ice parameters using artificial neural networks to predict ice jams. *Can. J. Civ. Eng.* **2003**, *30*, 758–765. [CrossRef]
32. Wang, T.; Yang, K.L.; Guo, Y.X. Application of artificial neural networks to forecasting ice conditions of the Yellow River in the Inner Mongolia reach. *J. Hydrol. Eng.* **2008**, *13*, 811–816.
33. Massie, D.D. Neural-Network Fundamentals for Scientists and Engineers. In Proceedings of the Conference on Efficiency, Costs, Optimization, Simulation and Environmental Impact of Energy Systems (ECOS 01), Istanbul, Turkey, 4–6 July 2001.
34. Chokmani, K.; Khalil, B.; Ouarda, T.B.M.J.; Bourdages, R. Estimation of river ice thickness using artificial neural networks. In Proceedings of the 14th Workshop Hydraulics Ice Covered Rivers, Quebec, QC, Canada, 19–22 June 2007; CGU HS/CRIPE: Quebec, QC, Canada, 2007.
35. Hu, J.; Liu, L.; Huang, Z.; You, Y.; Rao, S. Ice Breakup Date Forecast with Hybrid Artificial Neural Networks. In Proceedings of the 4th International Conference on Natural Computation, ICNC, Jinan, China, 25–27 August 2007; Volume 2, pp. 414–418.
36. Chen, S.Y.; Ji, H.L. Fuzzy optimization neural network approach for ice forecast in the Inner Mongolia reach of the Yellow River. *Hydrol. Sci. J.* **2005**, *50*, 319–330.
37. Yan, Q.; Ding, M. Using Dynamic Fuzzy Neural Networks Approach to Predict Ice Formation. In Proceedings of the 2011 MSEC International Conference on Multimedia, Software Engineering and Computing, Wuhan, China, 26–27 November 2011; Jin, D., Lin, S., Eds.; Advances in Multimedia, Software Engineering and Computing Vol.1. Springer: Berlin/Heidelberg, Germany, 2011; Volume 128.
38. Liu, H.; Jiang, Q.; Ma, Y.; Yang, Q.; Shi, P.; Zhang, S.; Tan, Y.; Xi, J.; Zhang, Y.; Liu, B.; et al. Object-Based Multigrained Cascade Forest Method For Wetland Classification Using Sentinel-2 and radarsat-2 imagery. *Water* **2022**, *14*, 82. [CrossRef]
39. Shin, Y.; Kim, T.; Hong, S.; Lee, S.; Lee, E.; Hong, S.; Lee, C.; Kim, T.; Park, M.S.; Park, J.; et al. Prediction of chlorophyll-a concentrations in the Nakdong River using machine learning methods. *Water* **2020**, *12*, 1822. [CrossRef]
40. Flach, P.A. *Machine Learning: The Art and Science of Algorithms that Make Sense of Data*; Cambridge University Press: Cambridge, UK, 2012.
41. Prowse, T.D.; Bonsal, B.R. Historical trends in river-ice break-up: A review. *Nordic Hydrol.* **2004**, *35*, 281–293. [CrossRef]
42. Zounemat-Kermani, M.; Batelaan, O.; Fadaee, M.; Hinkelmann, R. Ensemble machine learning paradigms in hydrology: A review. *J. Hydrol.* **2021**, *598*, 126266. [CrossRef]
43. Graf, R.; Aghelpour, P. Daily river water temperature prediction: A comparison between neural network and stochastic techniques. *Atmosphere* **2021**, *12*, 1154. [CrossRef]
44. Woś, A. *The Climate of Poland in the Second Half of the 20th Century*; Scientific Publishing House UAM: Poznan, Poland, 2010; 490p. (In Polish)
45. Wrzesiński, D.; Perz, A. The features of the runoff regime in the basin of the Warta River. *Bad. Fizjogr. R. VII Ser. A Geogr. Fiz.* **2016**, *A67*, 289–304. (In Polish)
46. Graf, R.; Łukaszewicz, J.T.; Jawgiel, K. The analysis of the structure and duration of ice phenomena on the Warta River in relation to thermic conditions in the years 1991–2010. *Woda-Sr.-Obsz. Wiej.* **2018**, *18*, 5–28. (In Polish)
47. Pawłowski, B.; Gorączko, M.; Szczerbińska, A. Ice Phenomena on the Rivers of Poland. In *Hydrology of the Poland*; Jokiel, P., Marszelewski, W., Pociask-Karteczka, J., Eds.; PWN: Warsaw, Poland, 2017; pp. 195–200. (In Polish)

48. Graf, R.; Wrzesiński, D. Detecting patterns of changes in river water temperature in Poland. *Water* **2020**, *12*, 1327. [CrossRef]
49. Lan, T.; Lin, K.; Tan, X.; Xu, C.-Y.; Chen, X. Dynamics of hydrological model parameters: Calibration and Reliability. *Hydrol. Earth Syst. Sci.* **2020**, *24*, 1347–1366. [CrossRef]
50. Hintze, J.L.; Nelson, R.D. Violin plots: A box plot-density trace synergism. *Am. Stat.* **1998**, *52*, 181–184.
51. Brownlee, J. Train-Test Split for Evaluating Machine Learning Algorithms. 2020. Available online: https://machinelearningmastery.com/train-test-split-for-evaluating-machine-learning-algorithms (accessed on 15 September 2021).
52. Fritsch, S.; Guenther, F.; Wright, M.N. "Neuralnet: Training of Neural Networks." R package version 1.44.2. 2019. Available online: https://CRAN.R-project.org/package=neuralnet (accessed on 22 November 2021).
53. Riedmiller, M. Advanced supervised learning in multi-layer perceptrons—From backpropagation to adaptive learning algorithms. *Comput. Stand. Interfaces* **1994**, *16*, 265–278. [CrossRef]
54. Chen, T.; He, T.; Benesty, M.; Khotilovich, V.; Tang, Y.; Cho, H.; Chen, K.; Mitchell, R.; Cano, I.; Zhou, T.; et al. Xgboost: Extreme Gradient Boosting. 2020. Available online: https://cran.r-project.org/package=xgboost (accessed on 10 July 2021).
55. Loong, T.W. Understanding sensitivity and specificity with the right side of the brain. *BMJ* **2003**, *327*, 716–719. [CrossRef] [PubMed]
56. Parikh, R.; Mathai, A.; Parikh, S.; Chandra Sekhar, G.; Thomas, R. Understanding and using sensitivity, specificity and predictive values. *Indian J. Ophthalmol.* **2008**, *56*, 45–50. [CrossRef] [PubMed]
57. R Core Team, R. *A Language and Environment for Statistical Computing*; R Foundation for Statistical Computing: Vienna, Austria, 2019; Available online: http://www.R-project.org (accessed on 15 July 2021).
58. Pawłowski, B. *Course of Ice Phenomena on the Lower Vistula River in 1960–2014*; Nicholas Copernicus University Toruń: Torun, Poland, 2017. (In Polish)
59. Marszelewski, W.; Pawłowski, W. Long-term changes in the course of ice phenomena on the oder river along the Polish–German border. *Water Resour. Manag.* **2019**, *33*, 5107–5120. [CrossRef]
60. Ptak, M.; Choiński, A. Ice phenomena in rivers of the coastal zone (southern Baltic) in the years 1956–2015. Baltic Coastal Zone. *J. Ecol. Prot. Coastline* **2016**, *20*, 73–83.
61. Łukaszewicz, J.T.; Graf, R. The variability of ice phenomena on the rivers of the Baltic coastal zone in the Northern Poland. *J. Hydrol. Hydromech.* **2020**, *68*, 38–50. [CrossRef]
62. Bączyk, A.; Suchożebrski, J. Variability of ice phenomena on the Bug River (1903–2012). *Inżynieria Ekol.* **2018**, *49*, 136–142. (In Polish) [CrossRef]
63. Gorączko, M.; Pawłowski, B. Changing of ice phenomena on Warta River in vicinity of Uniejów. *Biul. Uniejowski* **2014**, *3*, 23–33. (In Polish)
64. Kornaś, M. Ice phenomena in the Warta River in Poznań in 1961–2010. *Quaest. Geogr.* **2014**, *33*, 51–59. [CrossRef]
65. Smith, E.H. Using extreme gradient boosting (XGBoost) to evaluate the importance of a suite of environmental variables and to predict recruitment of young-of-the-year spotted seatrout in Florida. *bioRxiv* **2019**, 543181. [CrossRef]
66. Wang, Z.; Li, C. River Ice Forecasting Based on Genetic Neural Network. In Proceedings of the International Conference on Information Engineering and Computer Science (ICIECS2009), Wuhan, China, 19–20 December 2009; pp. 1–4.
67. Li, S.; Qin, J.; He, M.; Paoli, R. Fast evaluation of aircraft icing severity using machine learning based on XGBoost. *Aerospace* **2020**, *7*, 36. [CrossRef]

resources

MDPI

Article

Changes in Terrestrial Evaporation across Poland over the Past Four Decades Dominated by Increases in Summer Months

Urszula Somorowska

Department of Hydrology, Faculty of Geography and Regional Studies, University of Warsaw, Krakowskie Przedmieście 30, 00-927 Warsaw, Poland; usomorow@uw.edu.pl

Abstract: Given the importance of terrestrial evaporation (ET) for the water cycle, a fundamental understanding of the water quantity involved in this process is required. As recent observations reveal a widespread ET intensification across the world, it is important to evaluate regional ET variability. The specific objectives of this study are the following: (1) to assess annual and monthly ET trends across Poland, and (2) to reveal seasons and regions with significant ET changes. This study uses the ET estimates acquired from the Global Land Evaporation Amsterdam Model (GLEAM) dataset allowing for multi-year analysis (1980–2020). The Mann–Kendall test and the Sen's slope were applied to estimate the significance and magnitude of the trends. The results show that a rising temperature, along with small precipitation increase, led to the accelerated ET of 1.36 mm/y. This was revealed by increased transpiration and interception loss not compensated by a decrease in bare soil evaporation and sublimation. The wide-spread higher water consumption especially occurred during the summer months of June, July, and August. Comparing the two subperiods of 1980–2020, it was found that in 2007–2020, the annual ET increased by 7% compared to the reference period of 1980–2006. These results can serve as an important reference for formulating a water resources management strategy in Poland.

Keywords: terrestrial evaporation; components; Global Land Evaporation Amsterdam Model (GLEAM); increasing trends; spatial–temporal pattern; evaporative ratio; Poland

Citation: Somorowska, U. Changes in Terrestrial Evaporation across Poland over the Past Four Decades Dominated by Increases in Summer Months. *Resources* **2022**, *11*, 6. https://doi.org/10.3390/resources11010006

Academic Editor: Diego Copetti

Received: 7 December 2021
Accepted: 9 January 2022
Published: 12 January 2022

1. Introduction

Terrestrial evaporation (ET), alternatively called land surface evaporation or evapotranspiration [1], is an important component of the global water cycle. It consists of biophysical (transpiration from vegetation) and physical (evaporation from the interception, bare soil, and open water) water fluxes. The contribution of these different fluxes to the total amount of evaporated water depends on both the climate controlling the atmospheric water demand, and the land surface features, especially vegetation characteristics, influencing the energy balance at the land surface and determining the volume of interception [2]. At a global scale, the largest share in terrestrial evaporation is transpiration (Et; 59%), followed by vegetation and floor interception (Ei; 31%), soil moisture evaporation (Es; 6%), and, lastly, open water evaporation (Ew; 4%) [3]. As changes in terrestrial evaporation can lead to either warming or cooling of the land surface [4], acquiring a better understanding of ET changes is of high priority for ongoing research. Additionally, knowledge of the temporal changes of ET is necessary for accurately quantifying global and regional water budgets and for a better understanding of the hydrological interactions between the land and the atmosphere [5]. Revealing the current ET trends might contribute to the prediction of the Earth's runoff changes, and to the introduction of water management strategies.

Globally, ET has shown a significant upward trend over the 32-year period of 1982–2013, mainly driven by vegetation greening and rising atmosphere moisture demand [6]. It has been mainly caused by increases in transpiration from vegetation and the vaporization of intercepted rainfall from vegetation [7]. An increase in global terrestrial ET was also reported for the period 1982–2011, as estimated by different remote sensing-based physical models, machine-learning algorithms, and land surface models [8]. Despite the overall increasing trend, decreasing ET trends were also observed in shorter periods of time. For example, it was reported that global annual evapotranspiration declined due to the limited moisture supply from 1998 to 2008 [9]. Additionally, a long-term (1950–2017) relative decline in evapotranspiration, accompanied by increasing runoff, was found in 27% of the global land areas, which was explained by a reduction in surface conductance [10]. Apart from long-term changes observed globally, the strongly regional and temporal differentiation of ET trends are evidenced, as proven for central Europe and central North America [11], and for the north-eastern United States [12]. For the period 1980–2010, large-scale increases in ET were observed in south-eastern China, while decreases in ET occurred over the north-east [13]. Other studies conducted over the 71-year period from 1948 to 2018 confirmed that ET exhibited an increasing trend over almost 90% of the territory of China [14].

In most of Europe, ET increased in response to land use (mainly large-scale reforestation and afforestation) and climate change, except in the Iberian Peninsula and some other parts of the Mediterranean where negative ET trends were found [15]. For example, Great Britain is becoming warmer and wetter, through which increases in precipitation and air temperature are driving increases in runoff and evapotranspiration, as proven for 1961–2015 [16]. In Germany, significant trends were observed in transpiration and evapotranspiration in the period of 1961–2019 [17]. Over European Russia, positive trends in the annual values of potential evapotranspiration were revealed for 1966–2017, the distribution of which has a strong, patchy character [18]. As projected for most of Europe, widespread and relatively uniform ET increases of around 75–125 mm are expected by the 2050s [19]. Overall, the trends in evapotranspiration exhibit wide regional differentiation, depending on varying climate and human impacts. The changes are linked to the main climatic drivers of ET that comprise precipitation, net solar radiation, air temperature, vapor pressure deficit, and wind speed. Transpiration, which is the main component of ET in forested landscapes [7,20], is dependent not only on the climate conditions, but also on the soil moisture regime, the physiological features of vegetation, and the duration of the growing season [21,22].

In Poland, only a few studies have focused on changes in ET from a long-term perspective. Based on data from 18 weather stations across Poland, an increase in the growing season reference evapotranspiration was detected in the period from 1971 to 2010 [23]. Moreover, a significant increase in the mean daily values of the summer reference ET in Poland in the same multiyear period was observed, with the potential to occur more frequently in the future [24]. Projections of changes of areal ET in the Wielkopolska Region (central-western part of Poland) indicate that the regional average increase in the annual ET is predicted to equal 45 mm, with the highest changes occurring during the winter, when comparing the control period of 1961–1990 to the projection horizon of 2061–2090 [25]. In the north-western part of Poland, significant increasing trends in the potential ET were observed during the period of 1952–2018, as investigated based on data acquired from six weather stations [26]. In this area, particularly high increases were detected during the spring and summer months, explained by a significant increase in air temperature and a decrease in relative humidity, marking the end of the 20th century and the start of the 21st century.

While the above-mentioned studies investigated the actual, potential, or reference ET trends from selected weather stations' data across Poland, none identified country-scale spatial–temporal patterns of ET derived from the satellite-based or reanalyzed data covering the whole territory of Poland. Hence, undertaking such research can bridge the gap and improve the understanding of the ongoing ET changes. This might provide insight into the hydrological implications that result from the ET variations helping to manage the water resources at the country and regional scales. This paper uses well-validated, satellite-based Global Land Evaporation Amsterdam Model (GLEAM) data that comprise spatially continuous estimates of terrestrial evaporation and its main components [27–29]. For the first time, ET is quantified throughout the recent 41-year period (1980–2020), and the spatial–temporal trends are estimated across the country. New insights into the ET variability and its changes are acquired by considering the grid cell and country scale patterns. Thus far, such an approach has not been given detailed consideration with respect to the territory of Poland. Apart from investigations focusing on specific regions and based on point estimates of ET, quantitative multi-year studies incorporating spatial–temporal characteristics have not been reported.

The key scientific questions that need to be answered in this study are the following:

(1) Is the ET across Poland increasing or decreasing at annual and monthly time scales, and, if this is the case, what is the magnitude of such temporal trends?
(2) Which components of ET are contributing to the country-scale inter-annual trend?
(3) What are the monthly spatial patterns of the ET trends and regions of significantly increasing/decreasing ET?

It is hypothesized that the annual and monthly ET has remarkably increased in the past four decades, and that extremely dry and warm years have led to a high evaporation ratio, defined as the ratio between the actual terrestrial evaporation and precipitation. It is also hypothesized that the ET trends are regionally differentiated, with the highest trend rates occurring in the summer months.

2. Materials and Methods

2.1. Basic Geographical Characteristics of Poland

Poland is located in the central part of Europe. The area of the country stretches from 54.83° N to 49.00° N and from 14.18° W to 24.15° W. The land area of the country covers an area of 311,888 km^2. While the average elevation is approximately 170 m a.s.l., the relief is characterized by significant diversity. The lowest point is located in Żuławy Wiślane (−1.8 m a.s.l.) at the Baltic coast in northern Poland, and the highest peaks are in the Tatras Mountains (Rysy 2499 m a.s.l.) in the south. Lowlands (0–300 m a.s.l.) dominate and cover 91% of the country's area. Highlands (300–500 m above sea level) occupy 6%, while mountains (above 500 m above sea level) occupy only 3%. As much as 99.7% of the country's area constitutes the drainage basin of the Baltic Sea, which consists of the Vistula River Basin (53.7%), Oder River Basin (33.9%), and the river basins draining directly to the Baltic Sea (12.1%). A large part of the country (from the 14th to approximately 20–22 meridians) is influenced by a humid, temperate (Cfb) climate [30]. The rest of the area to the east and the mountainous parts to the south are classified as a snowy, humid climate (Dfb) with a warm summer. The highest annual precipitation total of approximately 1600 mm/y occurs in the Tatra Mountains, while the lowest, below 500 mm/y, concern the central, lowland part of the country. The average air temperature in the months from April to October ranges from 9.3 °C in the mountains to 14.9 °C in the midwest lowland, with a country-wide average of 13.8 °C. The land use is dominated by agricultural land (60%) and forests (33%), while the rest of the territory is covered by artificial surfaces (6%) and other categories (1%) (Figure 1).

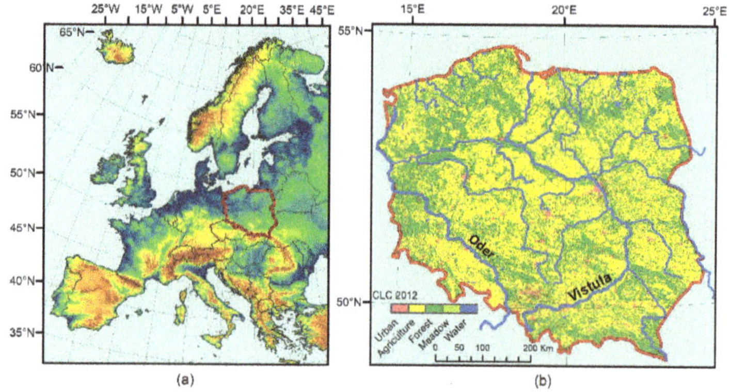

Figure 1. (**a**) Geographic location of the study area. (**b**) Land use according to CLC 2012 acquired from https://land.copernicus.eu, accessed on 7 June 2021.

2.2. Data

In this study, the satellite-based GLEAM (https://www.gleam.eu, accessed on 20 March 2021) dataset was used for detecting the magnitude and trends in terrestrial evaporation for 1980–2020. Driven by remote sensing data, GLEAM provides estimates of different components of land evaporation at a resolution of $0.25° \times 0.25°$, including transpiration (Et), vaporization of intercepted rainfall from vegetation (Ei), evaporation from the bare soil (Eb), snow sublimation (Es), and open-water evaporation (Ew). Version 3.5a is derived from the reanalysis of net radiation, air temperature, satellite and gauge-based precipitation, satellite-based vegetation optical depth (VOD), ESA-CCI (European Space Agency Climate Change Initiative) soil moisture, and snow water equivalent [27,28]. Changes in land cover are derived from the MEaSUREs Vegetation Continuous Fields dataset [31]. In this study, the monthly values of terrestrial evaporation were acquired and extracted for the territory of Poland. Moreover, the annual values of evaporation components (Et, Ei, Eb, Es, and Ew) were also acquired within the same spatial domain consisting 1056 grid cells. It is worth mentioning that spatial–temporal ET patterns might be captured by different regional and global ET products [32]. The GLEAM ET dataset belongs to the sophisticated land surface model group, which has an acceptable accuracy when compared to the benchmark ET values [33]. It has been recently used as a diagnostic dataset to investigate the global ET trends [34], to identify best-performing evaporation datasets [35], and to improve the structure of a simple conceptual rainfall–runoff model [36]. The key distinguishing features of these data are the multi-year observation period (1980–2020), the availability of five pieces of terrestrial evaporation component data, and the long-term satellite surface soil moisture and phenology observations assimilated into the GLEAM model.

For the analysis of the climate conditions, the air temperature and precipitation dataset was employed for 1980–2020 from version 23.1e of the station-based E-OBS gridded dataset (https://www.ecad.eu, accessed on 9 October 2021), available from the European Climate Assessment and Dataset Project [37]. It comprised daily values of precipitation (P) and air temperature (T) acquired from a regular latitude/longitude grid of $0.25° \times 0.25°$. From the daily gridded values, the country scale, annual air temperature, and precipitation estimates were calculated. Due to E-OBS uncertainties caused by gauge measurement errors, lack of wind corrections, and possible station relocations [38], the areal precipitation averages are considered to be underestimated [39,40]. To overcome the underestimation of the precipitation data available, a correction factor was applied to the annual precipitation totals derived from the E-OBS data. The correction factor results from the linear relationship between the E-OBS data ($P_{E\text{-}OBS}$) and the G2DC-PL+ data (P_{G2DC+}), a gridded 2 km daily climate dataset for the Polish territory, which is corrected for snowfall and rainfall under-catch and spans until 2019 [41,42]. The relationship is represented by the equation $P_{G2DC+} = 1.2 \times P_{E\text{-}OBS}$

established for the period 1980–2019. The value of the Pearson's correlation coefficient is 0.996, which confirms the strong linear relationship between the two variables. Thus, the $P_{E\text{-}OBS\text{-}corrected}$ value of the corrected E-OBS precipitation is estimated from the following equation: $P_{E\text{-}OBS\text{-}corrected} = 1.2 \times P_{E\text{-}OBS}$, where annual values of $P_{E\text{-}OBS}$ are acquired for the period 1980–2020. The corrected values of annual precipitation totals were used to calculate the values of the evaporation ratio, defined as a ratio of the actual evaporation over a given area to the precipitation falling on that area [43]. Then, the annual precipitation and air temperature time series were tested for the presence of trend, as explained in Section 2.3.

2.3. Methods

The non-parametric Mann–Kendall (MK) test [44,45] was applied for detecting trends in time series of terrestrial evaporation, precipitation, and air temperature data. It is widely applied test to analyze the hydro-meteorological time series [46]. Here, it was used to test the null hypothesis of no trend, H_0, against the alternative hypothesis, H_1, where there is an increasing or decreasing monotonic trend. The test statistic S is defined as

$$S = \sum_{k=1}^{n-1} sgn(x_j - x_k) \tag{1}$$

where x_j and x_k are the annual values in years j and k, $j > k$, respectively, n is the length of the time series, and

$$sgn(x_j - x_k) = \begin{cases} 1 & if\ (x_j - x_k) > 0 \\ 0 & if\ (x_j - x_k) = 0 \\ -1 & if\ (x_j - x_k) < 0 \end{cases} \tag{2}$$

The S statistic is approximately normally distributed when $n \geq 8$, with a mean of 0 and the variance of statistics S, σ^2, given as

$$\sigma^2 = [n(n-1)(2n+5)]/18 \tag{3}$$

The standardized test statistic Z is computed by

$$Z = \begin{cases} (S-1)/\sigma & if\ S > 0 \\ 0 & if\ S = 0 \\ (S+1)/\sigma & if\ S < 0 \end{cases} \tag{4}$$

For the country scale, annual time series of precipitation, air temperature, terrestrial evaporation, and its components, the trends were tested at a significance level of $\alpha = 0.05$. For $\alpha = 0.05$, the null hypothesis is accepted when $-1.960 \leq Z \leq 1.960$ (no significant trend), while it is rejected when $Z < -1.960$ (significant decreasing trend) or when $Z > 1.960$ (significant increasing trend). Thus, the trend is significant if the null hypothesis cannot be accepted. For the monthly terrestrial evaporation time series in a multi-year period 1980–2020, the significance level $\alpha = 0.1$ was assumed. For the assumed $\alpha = 0.1$, the null hypothesis is accepted when $-1.645 \leq Z \leq 1.645$ (no significant trend), while it is rejected when $Z < -1.645$ (significant decreasing trend) or when $Z > 1.645$ (significant increasing trend). For the computation of the magnitude (slope) of an existing trend (as change per year), the directional coefficient β expressed by the Theil–Sen estimator [47,48] is calculated by the formula:

$$\beta = Median\big((x_j - x_k)/(j-k)\big) \tag{5}$$

A positive value of β signals an increasing trend, and a negative value of β indicates a decreasing trend. In case the change is not statistically significant, but shows an inclination, it is called a tendency. The Climate Data Toolbox (CDT) for MATLAB [49] was used to calculate the MK test statistic (Z) and the p-values.

For the shift-type changes, the terrestrial evaporation time series were partitioned into two subseries by minimizing the sum of the residual (squared) error of each subset from its local mean, and finally returning the index, which, in this case, is the year in which the change occurs. The change point was identified for the country-scale time series of annual ET sums by using the MATLAB function "findchangepts". In the analysis performed here, a complete 41-element ET time series was divided into subperiods of differing length, consisting of 27 and 14 records covering the years 1980–2006 and 2007–2020, respectively. Finally, the Kruskal–Wallis test was applied to test for statistically significant differences between the subseries in two selected subperiods [50,51] by using HYDROSPECT software [52]. This test is a non-parametric test that compares mean ranks (i.e., medians). For this test, the null hypothesis is that the subseries medians are equal, versus the alternative that there is a difference between them. Under the null hypothesis of equal subperiod means, the statistic follows the Chi-square distribution. The Kruskal–Wallis test statistic is calculated as

$$H = \frac{12}{n(n+1) - T} \sum_{i=1}^{N} n_i (m_i - m)^2 \tag{6}$$

$$T = \frac{1}{n-1} \sum t_j^3 - t_j \tag{7}$$

where n is the series length, m the global mean of ranks, N denotes the number of subperiods, n_i is the number of values in the i-th subperiod, m_i is the mean rank for the i-th subperiod, T is the "tie correction", and t_j denotes the number of ties in subsequent tie groups [50]. The value of T for $t = 1$ is equal to 0. The Kruskall–Wallis test was also applied to the 41-element monthly ET time series partitioned into two 27- and 14-element subseries, as a split series of annual ET totals.

3. Results

3.1. *Country Scale, Inter-Annual Terrestrial Evaporation over the Period 1980–2020*

Figure 2 shows the course of the country scale of the annual sum of terrestrial evaporation comprised of its components in the analyzed period 1980–2020. It is within the range of 409–491 mm, and the multi-year average (1980–2020) equals 455 mm. The largest share in ET is Et (78%), and the second quantitatively important component is Ei (17%). The other contributions, Eb, Es, and Ew, have much smaller shares, i.e., 2%, 1%, and 2%, respectively. The annual ET shows an increasing trend of 1.36 mm/y in response to the slightly increasing tendency in precipitation (P) and the significant increase in air temperature (Figure 3a). The increasing ET is the result of the statistically significant increases in Et, Ei, and Ew, which outweigh the decreasing trend in Eb and the decreasing tendency in Es (Figure 3a,b). Noteworthy are the high values of Es in 1995 and 2006, which presumably were caused by particularly long-lasting snow cover, usually occurring from mid-December to the end of April. The multi-year rate of change of Et, Ei, Eb, Es, and Ew is equal to 0.84, 0.61, −0.04, −0.08, and 0.03 mm/y, respectively.

A shift-type temporal change of annual ET occurred in 2007, as detected by the change point analysis (Figure 3c). The Kruskal–Wallis test confirmed the significance of the difference between the median values of ET in the two subperiods. It shifted from 449 mm in the subperiod of 1980–2006 to 480 mm in the subperiod of 2007–2020. Additionally, the mean annual ET changed from 444 mm to 476 mm. There was also a significant shift-type change in the components of terrestrial evaporation, and this concerns transpiration, interception loss, and open-water evaporation. The average evaporation ratio (ET/P) in the period before the change was 65%, while after the change, it increased by 3% to a value of 68%. However, in particularly wet years, it was much lower, while in dry years, it reached very high values. For example, in the wet years of 2010 and 2017, it was equal to 50% and 56%, respectively, while in dry years, it was far above the long-term average, reaching 95% in 1982, 84% in 2015, and 87% in 2018.

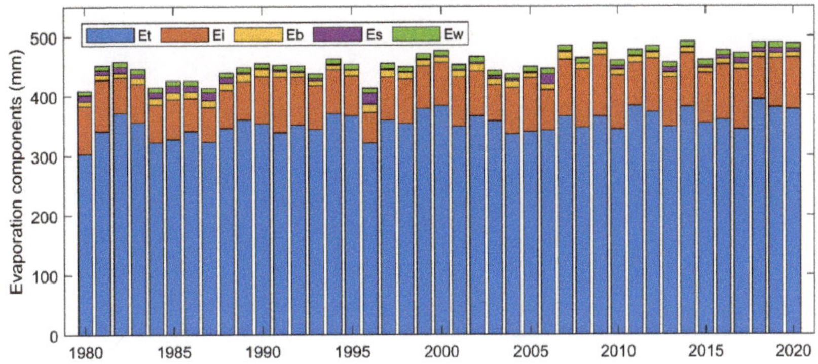

Figure 2. Annual terrestrial evaporation (ET) over Poland in 1980–2020, composed of transpiration (Et), interception loss (Ei), open-water evaporation (Ew), snow sublimation (Es), and bare soil evaporation (Eb).

Thus, atmospheric conditions, represented here by temperature and precipitation, control extreme ET values affecting the range of anomalies. In this work, precipitation and temperature are considered as selected ET drivers, although it should be noted that other factors, such as wind, solar radiation, and air humidity, can also play an important role. Except for climatological–meteorological conditions, other features such as hydrogeological, topographical, and physiological characteristics can also be critical.

3.2. Country-Scale, Monthly Terrestrial Evaporation over the Multi-Year Period 1980–2020

Figure 4 shows the course of the country-scale, monthly ET in the analyzed period 1980–2020. Statistically significant increasing trends occurred in summer and early fall, from June to October. In winter, an increase took place in January (0.08 mm/y). In the remaining months, the changes were not significant. The highest increases occurred in June (0.37 mm/y), July (0.30 mm/y), and August (0.21 mm/y). Much smaller changes concerned the months of September (0.13 mm/y) and October (0.05 mm/y).

An annual cycle is exhibited by ET, with the highest values in the summer season (June–August) and the lowest values in the winter season (November–February), as shown in Figure 5. Comparing the two subperiods, the second subperiod (2007–2020) is marked by a clear increase in ET in June (from 73 mm to 81 mm), July (from 77 mm to 84 mm), and August (from 63 mm to 68 mm) (Figure 5b). In these months, the increase in ET reached on average of 9%. The shift type changes in ET were statistically significant at $\mapsto = 0.05$ in January and in all months from June to September, as determined by the Kruskal–Wallis test. The same is true for ET totals in the months from April to October.

3.3. Monthly and Annual Spatial Patterns of Terrestrial Evaporation

Figure 6 shows the mean monthly spatial patterns of ET for 1980–2020. In December, January, and February, the values were the lowest. From March on, throughout April and May, they gradually increased, reaching the highest values in June, July, and August. Then, starting from September, they gradually decreased, reaching the lowest values in the winter months. Throughout the whole year, the highest monthly ET occurred in the south, and it concerned the mountainous, forested area. The area with relatively high ET values was also located in the western and northwestern part of Poland, where there were dense forest complexes. The lowest monthly ET were characteristic for the central zone, stretching from west to east.

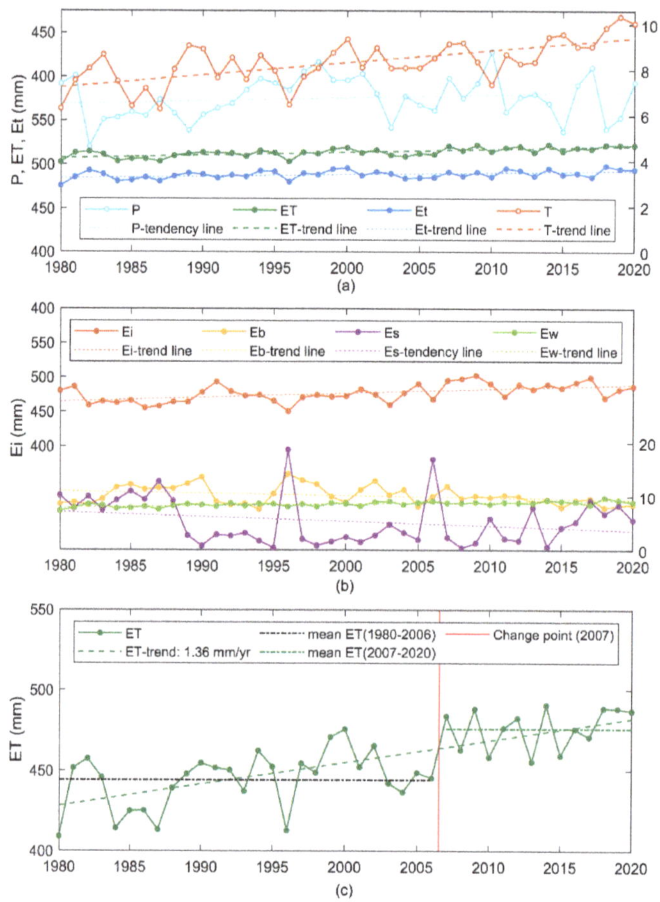

Figure 3. (**a**) Changes in annual precipitation (P), air temperature (T), terrestrial evaporation (ET), and transpiration (Et) in the multi-year period 1980–2020. (**b**) Changes in evaporation from interception (Ei), evaporation from bare soil (Eb), sublimation (Es), and evaporation from water (Ew) in the multi-year period 1980–2020. (**c**) Terrestrial evaporation (ET) in two subperiods (1980–2006 and 2007–2020), separated by a change point in 2007. The presence of a trend is determined at the significance level of $\alpha = 0.05$.

To check if there was a trend in monthly ET, 41-element time series were prepared for each grid cell for each month. Then, the Mann–Kendall test was applied. The results are presented in Figure 7 and Table 1. A statistically significant increasing trend occurred in many grid cells in June–September, covering 96% of the country's territory in June (Figure 7f), 89% in July (Figure 7g), 84% in August, and 75% in September. An increasing ET trend was also observed in 59% of the territory of Poland in January (Figure 7a).

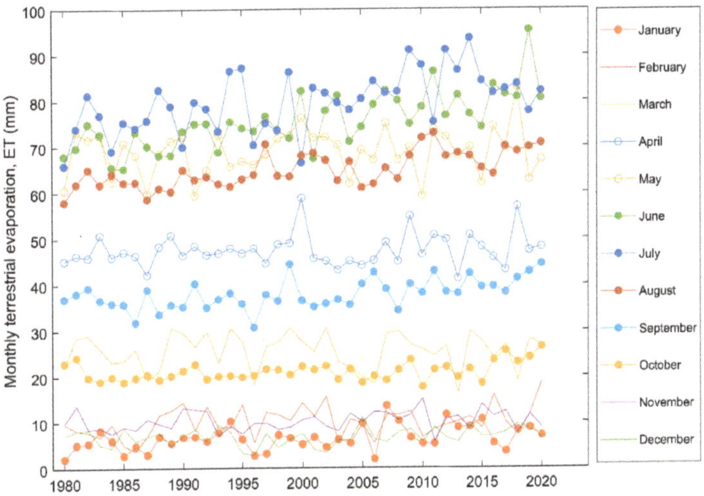

Figure 4. Monthly time series of terrestrial evaporation (ET) over Poland in 1980–2020. Statistically significant trends are indicated by filled markers. The presence of a trend is determined at a significance level of $\alpha = 0.05$.

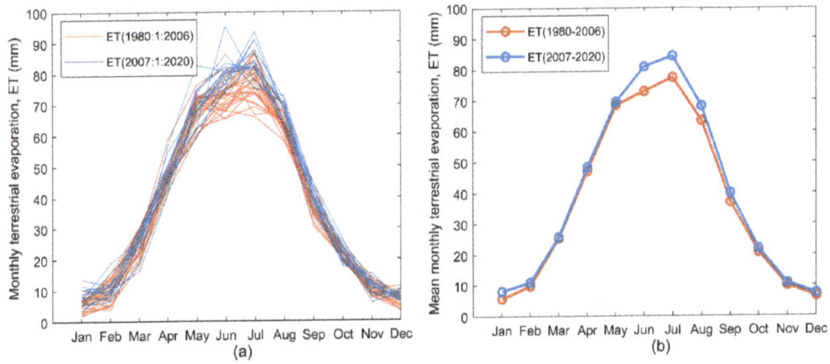

Figure 5. (**a**) Monthly terrestrial evaporation (ET). (**b**) Mean monthly terrestrial evaporation (ET) for subperiod 1 (1980–2006) and subperiod 2 (2007–2020). The ET values are averaged over the territory of Poland.

The trend rate of ET over the 41-year period was the highest in June, July, and August (Figure 8). In June, 63% of the territory experienced a trend rate of 0.20–0.39 mm/y, 29% had a trend rate of 0.40–0.59 mm/y, 2% was marked by a highest trend rate of 0.60–0.79 mm/y, and 2% was marked by a rate of 0.01–0.19 mm/y (Figure 8f). In July, the trend rate of 0.20–0.39 mm/y dominated as well, covering 57% of the territory, while the rate of 0.01–0.19 mm/y (either trend or tendency) was observed in 22% of the territory, a trend rate of 0.40–0.59 mm/y concerned 17% of the territory, and 0.60–0.79 mm/y concerned 2% of the territory. The remaining 2% of the territory experienced a slight decreasing tendency of −0.19–0 mm/y. The contribution of particular trend rate classes changed in August; 46% of the country had a trend or tendency rate of 0.01–0.19 mm/y, while the rates of 0.20–0.39 mm/y, 0.40–0.59 mm/y, and −0.19–0.00 mm/y concerned 43%, 9%, and 2% of the country, respectively. While the rising ET trend dominated in the summer, a slight downward trend or tendency was observed across the country during the spring months.

This applied to the months of March, April, and May when the changes were in the range of −0.19–0 mm/y, and they covered 68%, 48%, and 42% of the country territory, respectively.

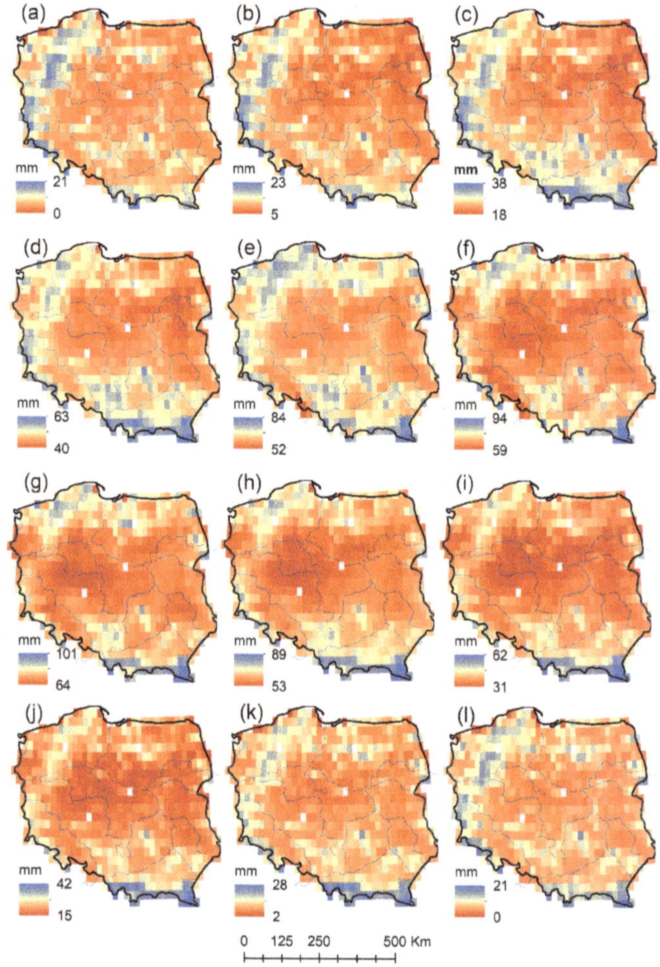

Figure 6. Spatial patterns of mean monthly ET during 1980–2020 for: (**a**) January, (**b**) February, (**c**) March, (**d**) April, (**e**) May, (**f**) June, (**g**) July, (**h**) August, (**i**) September, (**j**) October, (**k**) November, and (**l**) December.

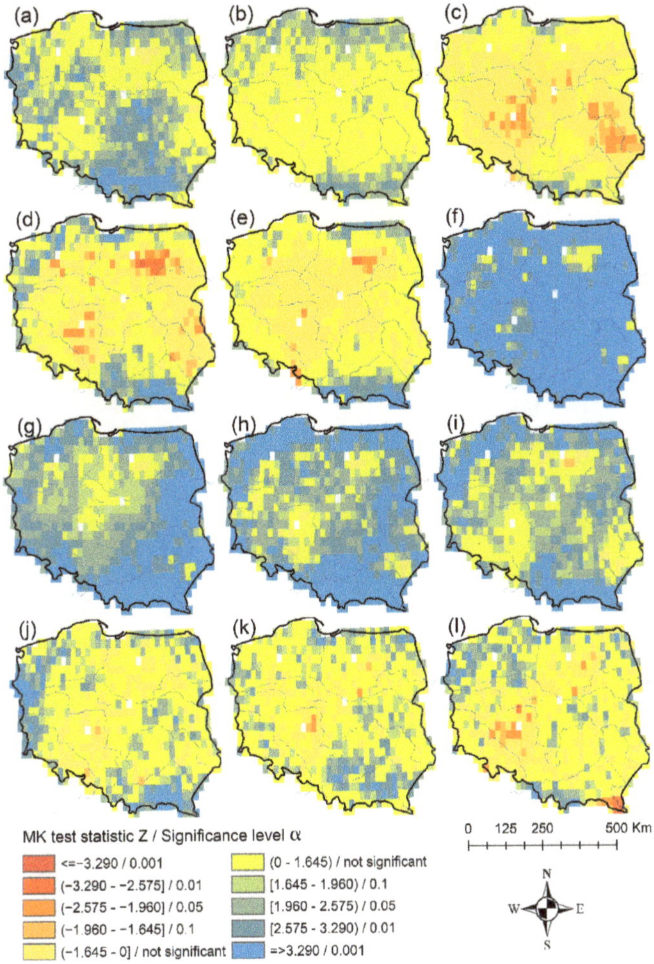

Figure 7. Mann–Kendall test results of ET trend detection over Poland at significance levels of 0.001, 0.01, 0.05, and 0.1 in the period 1980–2020 for: (**a**) January, (**b**) February, (**c**) March, (**d**) April, (**e**) May, (**f**) June, (**g**) July, (**h**) August, (**i**) September, (**j**) October, (**k**) November, and (**l**) December. Red and orange grid cells indicate a decreasing trend, light green and blue grid cells show an increasing trend, whereas yellow and yellowish green indicate no trend detected. Four grid cells with no data are shown in white.

Table 1. Percent of the country area with decreasing or increasing ET trends, or without changes in ET detected. The presence of a trend is determined at a significance level of $\alpha = 0.1$.

Month	Percent of Area with Increasing ET Trend %	Percent of Area with Decreasing ET Trend %	Percent of Area with No Changes in ET %
January	58	0	42
February	30	0	70
March	5	9	86
April	22	7	71
May	16	2	82

Table 1. *Cont.*

Month	Percent of Area with Increasing ET Trend %	Percent of Area with Decreasing ET Trend %	Percent of Area with No Changes in ET %
June	96	0	4
July	89	0	11
August	84	0	16
September	75	0	25
October	35	0	65
November	31	0	69
December	24	4	72

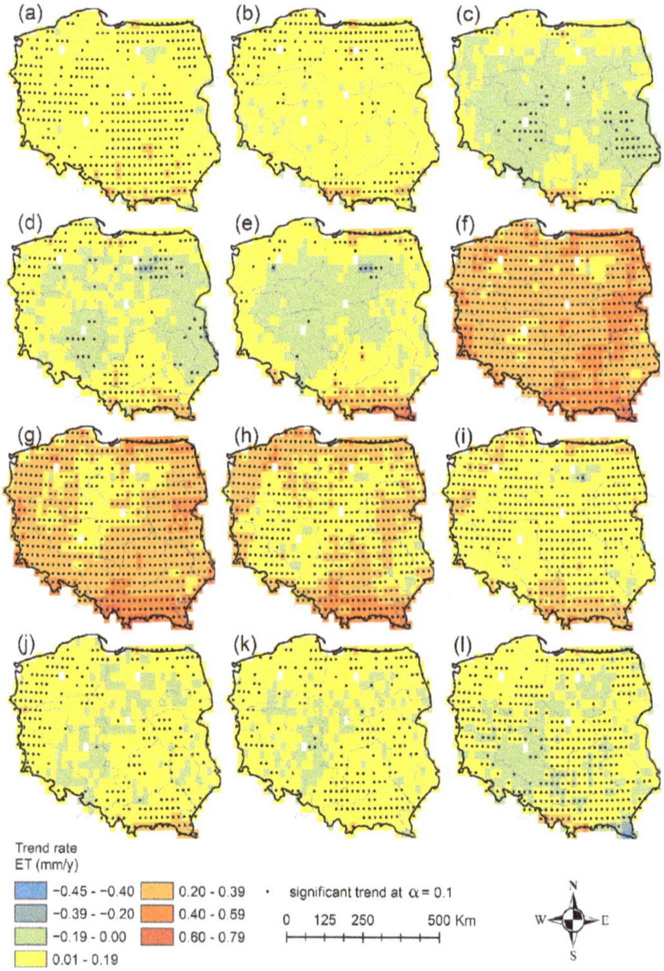

Figure 8. Trend rate of monthly terrestrial evaporation (ET) in the multi-year period 1980–2020 for: (**a**) January, (**b**) February, (**c**) March, (**d**) April, (**e**) May, (**f**) June, (**g**) July, (**h**) August, (**i**) September, (**j**) October, (**k**) November, and (**l**) December.

Finally, the spatial patterns of the annual ET were elaborated (Figure 9). Overall, the highest ET values occurred in the southern mountainous part of the country (Figure 9a), determined by the relatively high precipitation totals and the land use dominated by forests. The land strips in the north-west and west were also marked by relatively high ET values, associated with the presence of dense forest complexes in this area. Much lower ET occurred in the central part of the country, but also in the south-western, north-eastern, and eastern parts of the country, where agricultural fields have a large share in the land use. The comparison of ET in the two selected subperiods showed that subperiod 2 (2007–2020) had significantly higher ET values than subperiod 1 (1980–2006) (Figure 9b,c). The minimum ET values observed across the country were 35 mm higher, and the maximum values increased by 63 mm. The annual ET difference in the range of 0.1–20 mm/y was detected in the zone spreading from the south-west to the north-east, while much higher differences in the south reached, locally, up to 115 mm (Figure 9e). Only five grid cells were characterized by an ET decrease of up to 3% (Figure 9f), of which only one cell had a statistically significant decrease (Figure 9d). A total of 87% of the country experienced an increase of 1–10% ET, while a 11–22% increase was characteristic for the rest of the area (Figure 9f). Overall, the statistically significant increase in annual ET concerned 90% of the country's territory (Figure 9d).

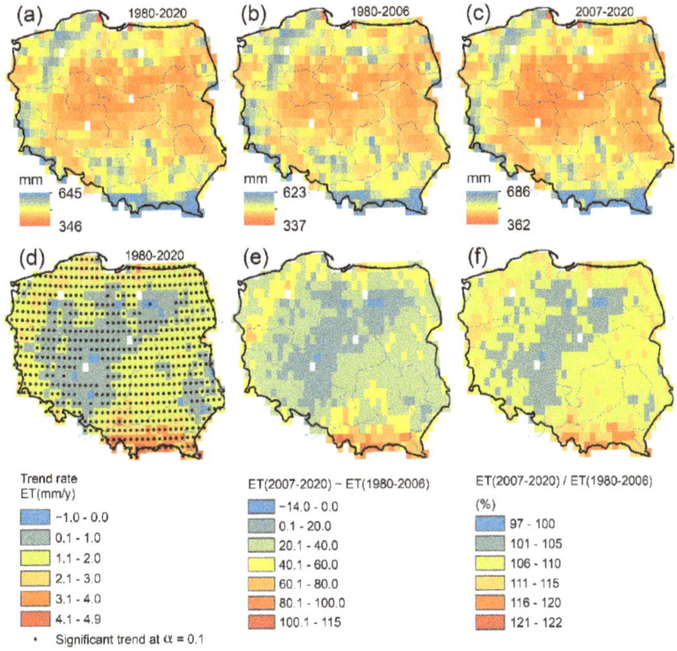

Figure 9. (**a**) Annual terrestrial evaporation (ET) in the multi-year period 1980–2020. (**b**) Annual terrestrial evaporation (ET) in the subperiod 1980–2006. (**c**) Annual terrestrial evaporation (ET) in the subperiod 2007–2020. (**d**) Trend rate of annual ET in the multi-year period 1980–2020. (**e**) Anomalies in annual ET detected as a difference between ET (2007–2020) and ET (1980-2006). (**f**) Anomalies in annual ET detected as a ratio of ET (2007–2020) to ET (1980–2006).

4. Discussion

This study demonstrated that terrestrial evaporation has increased both at the annual and monthly time scale. This finding supports the hypothesis that terrestrial evaporation has remarkably increased in the past four decades. The ET increase was caused by increases in transpiration, interception loss, and open-water evaporation, partially counteracted by bare soil evaporation and sublimation decreases. The increasing trends in transpiration (0.87 mm/y) and interception loss (0.68 mm/y) have the largest contribution to the ET changes. They are presumably driven by increases in the vegetation leaf area index, dominated by greening [6]. In Poland, greening has recently been confirmed by increasing NDVI trends, showing that 59% of the country has been greening continuously since the 1980s and that it has a widespread character [53]. This process affected over 44% of 1980s agricultural land and 87% of 1980s non-agricultural land. In total, this concerned nearly 60% of the country's area. Thus, the intensified greening all over the country coincides with the increase in terrestrial evaporation detected in this study; 81% of the territory is marked by an ET increase of 0.1–2 mm/y (trend or tendency), while 17% has a change of 2.1–4.9 mm/y. Only 2% of the territory has a decreasing tendency. To conclude, most of Poland has experienced an increasing ET trend in the last four decades, and the evaporation ratio was extraordinarily high in dry and warm years, such as 1982, 2015, and 2018.

The follow-up hypothesis is also confirmed; ET trends are regionally differentiated, with the highest multi-year trend rates occurring in June, July, and August, equaling, on average, 0.35 mm/y, 0.30 mm/y, and 0.22 mm/y, respectively. These are signals of an accelerated, more intense water cycle. The main implication of this finding is that increasing ET rates might have a great impact on the other components of the water cycle. If the observed ET trends continue, the associated changes in water budget components might present challenges for water managers at country and regional scales.

In this study, a well-validated satellite-based GLEAM dataset was used for analyzing ET dynamics for 1980–2020. To the best of the author's knowledge, this is the first study for the Polish territory that gives country-scale estimates of terrestrial evaporation over the last four decades. This enabled the analysis of total terrestrial evaporation and its components in the multi-year period. This study explained the relative contribution of each ET component, revealing the dominant role of transpiration and interception loss in the terrestrial evaporation. The mean annual ET of 455 mm/y is comparable to the ET of 442 mm/y, which was estimated for the whole of Europe [7]. However, the contribution of transpiration to the ET differs (69% for Europe and 78% for Poland). Additionally, clear differences concern the role of interception loss and bare soil evaporation. In this study, interception loss (77 mm/y) contributed in the amount of 17% to the ET (455 mm/y), which is almost double the amount estimated for Europe (9%). In turn, the bare soil evaporation contribution was found to be only 2%, while for Europe, it was found to be 21%. Such divergent results might be due to differences in model-dependent partitioning approaches, datasets forcing the ET models, the differences in climatic and land surface characteristics, and vegetation morphological attributes. The implication is that the relationships between particular components of ET should be further explored in a future study. It is worth noting that the multi-year ET average (455 mm) determined in this study is relatively consistent with earlier estimates for the territory of Poland; it is only slightly higher than estimates for the Oder River basin (454 mm) and for the Vistula River basin (432 mm) reported in 1950s [54]. The discrepancy is very low and may be due to the difference in source data (derived from the water budget) and a different period (1920–1940). Much higher ET values (553 mm) were signaled for 1961–1990 in the Wielkopolska region located in central and western Poland [25]. Such contrasting results are probably due to differences in applied methods and different sets of variables forcing the ET models from which areal ET values are derived. Moreover, this region cannot be treated as representative for the whole territory of Poland.

5. Conclusions

To conclude, this study evaluated the magnitude and multi-year trends in terrestrial evaporation and its components across Poland. Benefitting from the novel satellite observations assimilated to GLEAM, together with the reanalysis data used as the model forcing, it was found that most of Poland experienced significant terrestrial evaporation increases from 1980 to 2020. The warmer climate, along with a small precipitation increase, led to increased vegetation activity. This was revealed by increased transpiration and interception loss not compensated by a decrease in bare soil evaporation and sublimation. The enhanced vegetation activity was manifested by the wide-spread higher water consumption, especially during the summer months of June, July, and August. These monthly increases contributed to the annual changes in terrestrial evaporation of 1.36 mm/y. Comparing the two subperiods of 1980–2020, it was found that in the subperiod 2007–2020, the annual evaporation increased by 7% compared to the reference subperiod of 1980–2006. Further study on the quantification of terrestrial evaporation is required to explain its impacts on the changing water budget structure. This can serve as a reference for formulating a water resources management strategy in Poland.

Funding: This research was funded by the Faculty of Geography and Regional Studies, University of Warsaw, Poland, grant no. SWIB55/2021, and by the University of Warsaw, Poland, "Excellence Initiative–Research University" program, Action IV.3.1, grant no. BOB-661-519/2021.

Institutional Review Board Statement: Not applicable.

Informed Consent Statement: Not applicable.

Data Availability Statement: All the data used during this study are available at locations cited in the acknowledgements section.

Acknowledgments: The author thanks the editors and reviewers for their insightful comments and suggestions. The author acknowledges the GLEAM dataset (version 3.5a), which is available at https://www.gleam.eu (accessed on 20 March 2021). The author also acknowledges the E-OBS dataset from the ECA&D project (http://www.ecad.eu, accessed on 9 October 2021); version 23.1e of the dataset was downloaded from the Copernicus Climate Change Service (https://surfobs.climate.copernicus.eu/dataaccess/access_eobs.php) (accessed on 9 October 2021). The CORINE Land Cover was downloaded from the Copernicus Land Monitoring Service (https://land.copernicus.eu), European Union, European Environment Agency (EEA) (accessed on 7 June 2021). The author also acknowledges the G2DC-PL+ dataset openly available from the 4TU Centre for Research Data (https://doi.org/10.4121/uuid:a3bed3b8-e22a-4b68-8d75-7b87109c9feb, accessed on 16 September 2021).

Conflicts of Interest: The author declares no conflict of interest. The funder had no role in the design of the study; in the collection, analyses, or interpretation of data; in the writing of the manuscript; or in the decision to publish the results.

References

1. Miralles, D.G.; Brutsaert, W.; Dolman, A.J.; Gash, J.H. On the Use of the Term "Evapotranspiration". *Water Resour. Res.* **2020**, *56*, e2020WR028055. [CrossRef]
2. Paschalis, A.; Fatichi, S.; Pappas, C.; Or, D. Covariation of vegetation and climate constrains present and future T/ET variability. *Environ. Res. Lett.* **2018**, *13*, 104012. [CrossRef]
3. Wang-Erlandsson, L.; Van Der Ent, R.J.; Gordon, L.J.; Savenije, H.H.G. Contrasting roles of interception and transpiration in the hydrological cycle—Part 1: Temporal characteristics over land. *Earth Syst. Dyn.* **2014**, *5*, 441–469. [CrossRef]
4. Laguë, M.M.; Pietschnig, M.; Ragen, S.; Smith, T.A.; Battisti, D.S. Terrestrial Evaporation and Global Climate: Lessons from Northland, a Planet with a Hemispheric Continent. *J. Clim.* **2021**, *34*, 2253–2276. [CrossRef]
5. Zhan, S.; Song, C.; Wang, J.; Sheng, Y.; Quan, J. A Global Assessment of Terrestrial Evapotranspiration Increase Due to Surface Water Area Change. *Earth's Futur.* **2019**, *7*, 266–282. [CrossRef]
6. Zhang, K.; Kimball, J.S.; Nemani, R.R.; Running, S.W.; Hong, Y.; Gourley, J.J.; Yu, Z. Vegetation Greening and Climate Change Promote Multidecadal Rises of Global Land Evapotranspiration. *Sci. Rep.* **2015**, *5*, 15956. [CrossRef]
7. Zhang, Y.; Peña-Arancibia, J.L.; McVicar, T.R.; Chiew, F.H.S.; Vaze, J.; Liu, C.; Lu, X.; Zheng, H.; Wang, Y.; Liu, Y.Y.; et al. Multi-decadal trends in global terrestrial evapotranspiration and its components. *Sci. Rep.* **2016**, *6*, 19124. [CrossRef]
8. Pan, S.; Pan, N.; Tian, H.; Friedlingstein, P.; Sitch, S.; Shi, H.; Arora, V.K.; Haverd, V.; Jain, A.K.; Kato, E.; et al. Evaluation of global terrestrial evapotranspiration using state-of-the-art approaches in remote sensing, machine learning and land surface modeling. *Hydrol. Earth Syst. Sci.* **2020**, *24*, 1485–1509. [CrossRef]
9. Jung, M.; Reichstein, M.; Ciais, P.; Seneviratne, S.I.; Sheffield, J.; Goulden, M.L.; Bonan, G.; Cescatti, A.; Chen, J.; De Jeu, R.; et al. Recent decline in the global land evapotranspiration trend due to limited moisture supply. *Nature* **2010**, *467*, 951–954. [CrossRef]
10. Wang, R.; Gentine, P.; Yin, J.; Chen, L.; Chen, J.; Li, L. Long-term relative decline in evapotranspiration with increasing runoff on fractional land surfaces. *Hydrol. Earth Syst. Sci.* **2021**, *25*, 3805–3818. [CrossRef]
11. Teuling, A.J.; Hirschi, M.; Ohmura, A.; Wild, M.; Reichstein, M.; Ciais, P.; Buchmann, N.; Ammann, C.; Montagnani, L.; Richardson, A.D.; et al. A regional perspective on trends in continental evaporation. *Geophys. Res. Lett.* **2009**, *36*, L02404. [CrossRef]

12. Vadeboncoeur, M.A.; Green, M.B.; Asbjornsen, H.; Campbell, J.L.; Adams, M.B.; Boyer, E.W.; Burns, D.A.; Fernandez, I.J.; Mitchell, M.J.; Shanley, J.B. Systematic variation in evapotranspiration trends and drivers across the Northeastern United States. *Hydrol. Process.* **2018**, *32*, 3547–3560. [CrossRef]
13. Li, S.; Wang, G.; Sun, S.; Fiifi Tawia Hagan, D.; Chen, T.; Dolman, H.; Liu, Y. Long-term changes in evapotranspiration over China and attribution to climatic drivers during 1980–2010. *J. Hydrol.* **2021**, *595*, 126037. [CrossRef]
14. Zhang, F.; Geng, M.; Wu, Q.; Liang, Y. Study on the spatial-temporal variation in evapotranspiration in China from 1948 to 2018. *Sci. Rep.* **2020**, *10*, 17139. [CrossRef]
15. Teuling, A.J.; De Badts, E.A.G.; Jansen, F.A.; Fuchs, R.; Buitink, J.; Van Dijke, A.J.H.; Sterling, S.M. Climate change, reforestation/afforestation, and urbanization impacts on evapotranspiration and streamflow in Europe. *Hydrol. Earth Syst. Sci.* **2019**, *23*, 3631–3652. [CrossRef]
16. Blyth, E.M.; Martínez-de la Torre, A.; Robinson, E.L. Trends in evapotranspiration and its drivers in Great Britain: 1961 to 2015. *Prog. Phys. Geogr.* **2019**, *43*, 666–693. [CrossRef]
17. Ziche, D.; Riek, W.; Russ, A.; Hentschel, R.; Martin, J. Water budgets of managed forests in northeast germany under climate change—Results from a model study on forest monitoring sites. *Appl. Sci.* **2021**, *11*, 2403. [CrossRef]
18. Malinin, V.N.; Gordeeva, S.M.; Mitina, Y.V.; Kuleshova, A.V. Interannual variability of moistening in the European Russia. *Russ. J. Earth Sci.* **2021**, *21*, 1. [CrossRef]
19. Dezsi, Ş.; Mîndrescu, M.; Petrea, D.; Rai, P.K.; Hamann, A.; Nistor, M.M. High-resolution projections of evapotranspiration and water availability for Europe under climate change. *Int. J. Climatol.* **2018**, *38*, 3832–3841. [CrossRef]
20. Jasechko, S.; Sharp, Z.D.; Gibson, J.J.; Birks, S.J.; Yi, Y.; Fawcett, P.J. Terrestrial water fluxes dominated by transpiration. *Nature* **2013**, *496*, 347–350. [CrossRef] [PubMed]
21. Hursh, A.; Ballantyne, A.; Cooper, L.; Maneta, M.; Kimball, J.; Watts, J. The sensitivity of soil respiration to soil temperature, moisture, and carbon supply at the global scale. *Glob. Chang. Biol.* **2017**, *23*, 2090–2103. [CrossRef]
22. Liu, X.; Biondi, F. Transpiration drivers of high-elevation five-needle pines (*Pinus longaeva* and *Pinus flexilis*) in sky-island ecosystems of the North American Great Basin. *Sci. Total Environ.* **2020**, *739*, 139861. [CrossRef] [PubMed]
23. Łabędzki, L.; Bąk, B.; Smarzyńska, K. Spatio-temporal variability and trends of Penman-Monteith reference evapotranspiration (FAO-56) in 1971-2010 under climatic conditions of Poland. *Polish J. Environ. Stud.* **2014**, *23*, 2083–2091. [CrossRef]
24. Bogawski, P.; Bednorz, E. Atmospheric conditions controlling extreme summertime evapotranspiration in Poland (central Europe). *Nat. Hazards* **2016**, *81*, 55–69. [CrossRef]
25. Szwed, M. Projections of changes of areal evapotranspiration for different land-use units in the Wielkopolska Region (Poland). *Theor. Appl. Climatol.* **2017**, *130*, 291–304. [CrossRef]
26. Okoniewska, M.; Szuminska, D. Changes in potential evaporation in the years 1952–2018 in North-Western Poland in terms of the impact of climatic changes on hydrological and hydrochemical conditions. *Water* **2020**, *12*, 877. [CrossRef]
27. Martens, B.; Miralles, D.G.; Lievens, H.; Van Der Schalie, R.; De Jeu, R.A.M.; Fernández-Prieto, D.; Beck, H.E.; Dorigo, W.A.; Verhoest, N.E.C. GLEAM v3: Satellite-based land evaporation and root-zone soil moisture. *Geosci. Model Dev.* **2017**, *10*, 1903–1925. [CrossRef]
28. Miralles, D.G.; De Jeu, R.A.M.; Gash, J.H.; Holmes, T.R.H.; Dolman, A.J. Magnitude and variability of land evaporation and its components at the global scale. *Hydrol. Earth Syst. Sci.* **2011**, *15*, 967–981. [CrossRef]
29. Miralles, D.G.; Holmes, T.R.H.; De Jeu, R.A.M.; Gash, J.H.; Meesters, A.G.C.A.; Dolman, A.J. Global land-surface evaporation estimated from satellite-based observations. *Hydrol. Earth Syst. Sci.* **2011**, *15*, 453–469. [CrossRef]
30. Kottek, M.; Grieser, J.; Beck, C.; Rudolf, B.; Rubel, F. World map of the Köppen-Geiger climate classification updated. *Meteorol. Z.* **2006**, *15*, 259–263. [CrossRef]
31. Song, X.P.; Hansen, M.C.; Stehman, S.V.; Potapov, P.V.; Tyukavina, A.; Vermote, E.F.; Townshend, J.R. Global land change from 1982 to 2016. *Nature* **2018**, *560*, 639–643. [CrossRef]
32. Xu, T.; Guo, Z.; Xia, Y.; Ferreira, V.G.; Liu, S.; Wang, K.; Yao, Y.; Zhang, X.; Zhao, C. Evaluation of twelve evapotranspiration products from machine learning, remote sensing and land surface models over conterminous United States. *J. Hydrol.* **2019**, *578*, 124105. [CrossRef]
33. Chao, L.; Zhang, K.; Wang, J.; Feng, J.; Zhang, M. A comprehensive evaluation of five evapotranspiration datasets based on ground and grace satellite observations: Implications for improvement of evapotranspiration retrieval algorithm. *Remote Sens.* **2021**, *13*, 2414. [CrossRef]
34. Kim, S.; Anabalón, A.; Sharma, A. An assessment of concurrency in evapotranspiration trends across multiple global datasets. *J. Hydrometeorol.* **2021**, *22*, 231–244. [CrossRef]
35. Dembélé, M.; Ceperley, N.; Zwart, S.J.; Salvadore, E.; Mariethoz, G.; Schaefli, B. Potential of satellite and reanalysis evaporation datasets for hydrological modelling under various model calibration strategies. *Adv. Water Resour.* **2020**, *143*, 103667. [CrossRef]
36. Roy, T.; Gupta, H.V.; Serrat-Capdevila, A.; Valdes, J.B. Using satellite-based evapotranspiration estimates to improve the structure of a simple conceptual rainfall-runoff model. *Hydrol. Earth Syst. Sci.* **2017**, *21*, 879–896. [CrossRef]
37. Cornes, R.C.; van der Schrier, G.; van den Besselaar, E.J.M.; Jones, P.D. An Ensemble Version of the E-OBS Temperature and Precipitation Data Sets. *J. Geophys. Res. Atmos.* **2018**, *123*, 9391–9409. [CrossRef]
38. Lockhoff, M.; Zolina, O.; Simmer, C.; Schulz, J. Representation of precipitation characteristics and extremes in regional reanalyses and satellite-and gauge-based estimates over western and central Europe. *J. Hydrometeorol.* **2019**, *20*, 1123–1145. [CrossRef]

39. Hofstra, N.; Haylock, M.; New, M.; Jones, P.D. Testing E-OBS European high-resolution gridded data set of daily precipitation and surface temperature. *J. Geophys. Res. Atmos.* **2009**, *114*, D21101. [CrossRef]
40. Gampe, D.; Ludwig, R. Evaluation of gridded precipitation data products for hydrological applications in complex topography. *Hydrology* **2017**, *4*, 53. [CrossRef]
41. Berezowski, T.; Szczeniak, M.; Kardel, I.; Michalowski, R.; Okruszko, T.; Mezghani, A.; Piniewski, M. CPLFD-GDPT5: High-resolution gridded daily precipitation and temperature data set for two largest Polish river basins. *Earth Syst. Sci. Data* **2016**, *8*, 127–139. [CrossRef]
42. Piniewski, M.; Szcześniak, M.; Kardel, I.; Chattopadhyay, S.; Berezowski, T. G2DC-PL+: A gridded 2 km daily climate dataset for the union of the Polish territory and the Vistula and Odra basins. *Earth Syst. Sci. Data* **2021**, *13*, 1273–1288. [CrossRef]
43. McDonald, J.E. On the Ratio of Evaporation to Precipitation1. *Bull. Am. Meteorol. Soc.* **1961**, *42*, 185–189. [CrossRef]
44. Mann, H.B. Nonparametric Tests against Trend. *Econometrica* **1945**, *13*, 245–259. [CrossRef]
45. Kendall, M.G. A New Measure of Rank Correlation. *Biometrika* **1938**, *30*, 81–93. [CrossRef]
46. Theil, H. A rank-invariant method of linear and polynomial regression analysis. *R. Neth. Acad. Sci.* **1950**, *53*, 386–392, 521–525, 1397–1412.
47. Sen, P.K. Estimates of the Regression Coefficient Based on Kendall's Tau. *J. Am. Stat. Assoc.* **1968**, *63*, 1379–1389. [CrossRef]
48. Radziejewski, M.; Kundzewicz, Z.W. Detectability of changes in hydrological records. *Hydrol. Sci. J.* **2004**, *49*, 39–52. [CrossRef]
49. Greene, C.A.; Thirumalai, K.; Kearney, K.A.; Delgado, J.M.; Schwanghart, W.; Wolfenbarger, N.S.; Thyng, K.M.; Gwyther, D.E.; Gardner, A.S.; Blankenship, D.D. The Climate Data Toolbox for MATLAB. *Geochem. Geophys. Geosyst.* **2019**, *20*, 3774–3781. [CrossRef]
50. Kruskal, W.H.; Wallis, W.A. Use of Ranks in One-Criterion Variance Analysis. *J. Am. Stat. Assoc.* **1952**, *47*, 583–621. [CrossRef]
51. Kundzewicz, Z.W.; Robson, A.J. Change detection in hydrological records—A review of the methodology/Revue méthodologique de la détection de changements dans les chroniques hydrologiques. *Hydrol. Sci. J.* **2004**, *49*, 7–19. [CrossRef]
52. Radziejewski, M.; Kundzewicz, Z.W. *Development, Use and Application of the Hydrospect Data Analysis System for the Detection of Changes in Hydrological Time-Series for Use in WCP-Water and National Hydrological Services*; WCASP-65; WMO: Geneva, Switzerland, 2004; pp. 1–44.
53. Kolecka, N. Greening trends and their relationship with agricultural land abandonment across Poland. *Remote Sens. Environ.* **2021**, *257*, 112340. [CrossRef]
54. Dębski, K. *Hydrologia Kontynentalna. Cz. II. Fizyka Wody, Opady Atmosferyczne i Parowanie*; Wydawnictwa Komunikacyjne: Warszawa, Poland, 1959; pp. 1–547.

resources

MDPI

Article

Economic Indicators in Water and Wastewater Sector Contributing to a Circular Economy (CE)

Marzena Smol [1,*] and Renata Koneczna [1,2]

1 Mineral and Energy Economy Research Institute, Polish Academy of Sciences, Wybickiego 7A Str., 31-261 Cracow, Poland; rkoneczna@meeri.pl
2 ERBEKA Foundation, Kazimierza Morawskiego 5/222C Str., 30-102 Cracow, Poland
* Correspondence: smol@meeri.pl; Tel.: +48-12-617-16-60

Abstract: Protection and sustainable management of water was indicated as one of the strategic tasks in the process of transformation towards a circular economy (CE) in the European Union (EU), therefore, the water and wastewater sector plays an important role in this process. At the same time, the European Commission (EC) strongly underlined the importance of the possibility to assess the transformation process toward the CE, and developed a set of CE indicators that are available on the Eurostat website. However, these indicators have limited ability to assess the transformation progress in the water and wastewater sector. This paper presents a set of indicators for assessing the economic progress of transformation towards the CE in this sector. The proposed economic CE indicators were grouped into the following actions of the CE model in the water and wastewater sector: reduction, reclamation (removal), reuse, recycling, recovery and landfilling. The selection of specific indicators was based on a systematic review of the literature presenting economic indicators developed by international organisations and researchers (covering different thematic areas, scopes and potential applications). The selected economic CE indicators were assigned to three groups of the cash flow: income (revenues, expenses), costs, and investment financing. The proposed CE indicators can be used by water supply and sewage companies (i.e., supplying water to the public and wastewater treatment plants, and companies that use water in their production processes) to assess the level of the transformation toward the CE at a microeconomic level. An important aspect of future application and usage of the proposed set of CE economic indicators is the collection and processing of data needed for their reporting. The proposed set of CE economic indicators refers to information that are reported by the companies to prove its revenues, costs and investment outlays, and are collected by companies anyway. The proposed set of economic CE indicators is flexible, allowing the adaptation of indicators and areas of interest to maintain effectiveness throughout the transition period from linear to the CE model.

Keywords: circular economy (CE); monitoring; indicators; economic indicators; water; wastewater

Citation: Smol, M.; Koneczna, R. Economic Indicators in Water and Wastewater Sector Contributing to a Circular Economy (CE). *Resources* **2021**, *10*, 129. https://doi.org/10.3390/resources10120129

Academic Editor: Diego Copetti

Received: 2 November 2021
Accepted: 13 December 2021
Published: 16 December 2021

1. Introduction

A circular economy (CE) is defined as a regenerative system [1] where the value of materials, products and resources is maintained as long as possible in the economy and the production of waste is minimized [2]. The CE enables more efficient use of available resources, but also promotes a more sustainable management of waste. The integrated initiatives along the entire life cycle of raw materials [3], from extraction to the circular final processing are more and more often and successfully introduced in various industries [4]. It has to be underlined that the CE refers not only to raw materials (such as animal, vegetable or mineral), but also to water [5], which is an irreplaceable resource with life-giving property for nature, people and the economy [6].

Water resources are currently under unprecedented pressure in most countries [7]. The problem of water stress (reaching the level above 70%) occurs mainly in the regions of the world such as Northern Africa, Middle East, Western, Central and Southern Asia [8]. However, this problem also affects the European Union (EU), as the water scarcity was estimated to have affected at least 17% of the EU territory and at least 11% of the European population [9]. Moreover, next to water scarcity, an important issue in water management in Europe is water pollution, from industry and agriculture. To minimize the effects of anthropogenic use of water, both in agriculture and in industry, the European Commission (EC) announced further initiatives focused on water management in the second CE Action Plan [10]. This new Action Plan is one of the main blocks of the European Green Deal (EGD)-new agenda for sustainable growth of the EU. The main goal of the EGD is to achieve climate neutrality in Europe by 2050 [11] by turning climate and environmental challenges into new opportunities across all policy areas, and ensuring that the transition is fair and inclusive. In the coming years, the EC plans to facilitate water reuse and efficiency in both industrial processes and agriculture [10].

The protection and sustainable management of water and water-based waste (as wastewater, sewage sludge or sewage sludge ash) are indicated as one of the strategic tasks in the transformation process towards CE [12]. In the White Paper "Water and the Circular Economy" [13], common characteristics, ideas and approaches between the CE initiatives being implemented by organizations and Water System Management were identified. The three key dimensions of water use were grouped into the three themes of (i) water as service (consumptive use, production use, process use), (ii) water as source of energy (kinetic, thermal, bio-thermal), and (iii) water as carrier (nutrients, chemicals, minerals). The main areas of the transformation of water and wastewater sector to the CE model have been also indicated by the International Water Association (IWA) [14]. The IWA also proposed three pathways to support the water utility leaders in boosting their progress towards CE: (i) water pathways (upstream investments, rainwater harvesting, water recycling for non-potable use, water reuse for agriculture/aquaculture/industry), (ii) materials pathways (resource efficiency, used water sludge and products for agriculture, bioplastics, fertilizer and other materials), and (iii) energy pathways (energy saving, energy reduction and recovery, biosolids to energy production, renewable energy). In turn, the EC assumed that in applying the CE main principles—reduce, reuse and recycle—in the water and wastewater sector will accelerate the process of transition to the CE model in the EU. However, at the moment it is not possible to determine whether the subsequent actions are bringing the intended effects, whether environmental, social or economic. This is due to the lack of a dedicated CE monitoring framework for the water and wastewater sector, which would take into account indicators and measures allowing the assessment of the level of transformation towards CE in this sector. In 2018, the EC proposed the CE monitoring framework with ten CE indicators grouped in four thematic areas: production and consumption, waste management, secondary raw materials, competitiveness and innovation [15]. However, the potential application of these indicators in water and wastewater sector is limited, and does not evaluate all sector elements. Despite the fact that the EC underlined that the monitoring such important areas as production and consumption is essential for understanding progress towards the CE, the presented data does not take into account the water usage. Moreover, apart from the environmental benefits (resulting from the protection of water resources) and social benefits (securing drinking water supplies), the economic benefits of taking measures to implement CE in the water and wastewater sector should be also demonstrated [16]. In this area, the EC indicated that water savings in all sectors in the EU could lead up to 5% of reduced total primary energy consumption, which bring economic benefits for individual players [4]. To encourage companies to implement CE measures, which could generate greater value and commercial opportunity [13], economic indicators that allow for the assessment of the level of transformation towards the CE in different sectors should be identified. Therefore, the objective of the current paper was to present an inventory of economic indicators that can be used for the evaluation of the progress toward CE in the water and wastewater sector. In the previous years, several indicators have been proposed to access water-related aspects in the economy; however, their goals, scope and potential application for the assessment of level of transformation toward CE in water and wastewater sector must be analyzed and evaluated from the point of view of the possibility of their monitoring at the microeconomic and macroeconomic levels.

2. Materials and Methods

The current research includes three steps of collecting and processing data. The research framework is presented on Figure 1.

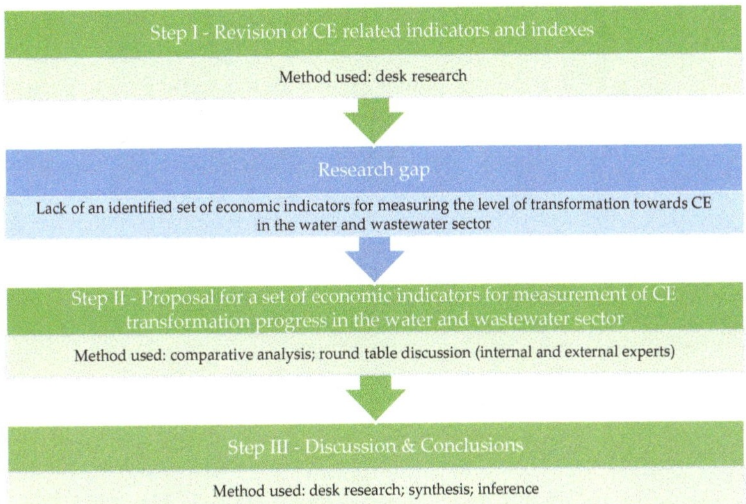

Figure 1. The research framework.

In the first step, a detailed analysis of the published research was conducted with the use of the desk research method. This state of the art analysis was based on the review approaches used in [17,18] to conduct searching and eligibility screening of available literature while retaining the procedural scope of analysis, and ensuring that that the review process is objective, repeatable and. The objective of this step of research was to review the indicators (economic, social, technological and environmental) from national and international organizations. The analyzed indicators regarded different aspects related to the CE and sustainability. Moreover, from the list of identified CE-related indicators, the specific indicators that can be used in the water and wastewater sector were also analyzed. The following data sources were analyzed: international and European official documents related to the water and wastewater sector and circular economy, published in EUR-Lex (eur-lex.europa.eu) and the official webpage of the EC (ec.europa.eu). The analysis also included the review of the statistical documents at the international and European levels (Organization for Economic Cooperation and Development—OECD, United Nations—UN, World Bank, European Investment Bank, Eurostat, European Environment Agency—EEA) and selected reviewed articles available in the scientific databases Elsevier Scopus, Elsevier Science Direct, Google Scholar and the Multidisciplinary Digital Publishing Institute (MDPI) database [19,20]. The selection of the articles was conducted based on the identified keywords "circular economy", "CE", "economic" "indicator", "index", "measurement", "assessment", "water", and "wastewater". The results of this step of research are presented in Section 3.1.

In the second step of the research, identified indicators, measures and indices have been analyzed and grouped according to the CE model for the water and wastewater sector, developed under the MonGOS project [14]. The objective of this step of research was to propose a set of economic indicators that could be used in water and wastewater management. At the beginning, the economic indicators were selected from the list o indicators analyzed in the first step of the research. Social and environmental indicator have been rejected as they are not the subject of the current research. Then, these indicator that are directly or indirectly related to the measurement of economic efficiency wer analyzed and grouped in the following actions of the CE model in the water and wastewate sector: *reduction*, *reclamation* (*removal*), *reuse*, *recycling*, *recovery* and *rethink* [14]. The concep of the CE model framework in the water and wastewater sector is presented in Table 1. Finally, new economic indicators have been proposed for each action of this mode The round table discussion which included the consultation in the group of internal anc external experts was used for this purpose. The research group consisted of 6 experts three representatives of enterprises operating in the water and wastewater sector, anc three specialists (scientists) dealing with economics and environmental technologies. Th criterion for the selection of experts from enterprises was to have at least a master's degre and a minimum of 10 years of experience in a managerial position in a company using wate or/and dealing with wastewater treatment. In turn, the criterion for selecting scientific experts was to have a doctoral degree in the economic and environmental sciences, taking into account experience in the water and wastewater sector. The indicators were analyzed and discussed during three meetings with these experts: (1) consultation online with industry experts, (2) consultation online with scientific experts; (3) consultation online during joint meeting of the MonGOS project. The final results of this step of the research was the list of economic indicators that can be used for the evaluation of the level of the transformation toward the CE in water and wastewater sector. The results of this step of research are presented in Section 3.2.

Table 1. CE model in the water and wastewater sector (own based on [14]).

Element of the CE Model	Description
Reduction (removal)	prevent wastewater generation in the first place by the reduction of water usage and pollution reduction at source
Removal	application of effective technologies for the prevention of inclusion of hazardous pollutants into wastewater and removal of pollutants from water and sewage
Reuse	reuse of wastewater as an alternative source of water supply (non-potable usage)
Recycling	recovery or reclamation of water from wastewater for potable usage
Recovery	recovery of resources as nutrients and energy from water-based waste
Lanfilling	lanfilling of waste
Rethink	rethinking how to use of resources to create a sustainable economy which is free of waste and emissions

The third step of the research includes the discussion of the possibility of the usage of the identified economic indicators at the microeconomic levels. The synthesis method, that is, formulating generalizations based on recognized partial theorems, was used to interpret the obtained results. Moreover, the desk research was used to compare obtained results from the perspective of previous studies and other authors. The findings and their implications, as well as future research directions, are discussed in Section 4.

3. Results

3.1. Inventory and Classification of CE Indicators

In recent years, several CE indicators have been developed by various organizations. The inventory of the groups of circularity indicators in the documents of international organizations is presented in Table 2. As a result, 10 documents proposing or discussing CE indicators at the macro (European and international monitoring framework) and micro (products, services and companies) levels were presented. There are also CE indicators proposed by individual authors. Therefore, to the listed documents, selected scientific papers presenting specific CE indicators were also analyzed (Table 3). In total, 742 indicators (provided by international organizations) and summary indicators and indexes (provided by other authors) were analyzed.

Table 2. Groups of circularity indicators in the documents of international organizations [19–28].

Programme	Thematic Area	Number of All Indicators	Number of Water-Related Indicators	Number of Economc Indicators	Source
EEA. Digest indicators	1. Climate state and impact indicators 2. Air pollutant emissions, 3. Energy indicators 4. Industrial pollution indicators 5. Land and soil indicators 6. Marine indicators 7. Outlook indicators 8. Sustainable consumption and production 9. Streamlining European biodiversity indicators 10. Transport and environment reporting mechanism 11. Water indicators 12. Water resource efficiency indicators 13. Waste indicators	122	37	6	[24]
EEA. Circular economy in Europe	1. Material input 2. Eco-design 3. Production 4. Consumption 5. Waste recycling	18	0	0	[29]
EC. Raw Materials Scoreboard	1. Raw materials in the global context 2. Competitiveness and Innovation 3. Framework conditions for mining 4. Circular economy and recycling 5. Environmental and social sustainability	27	1	10	[23]
EC. Resource Efficiency Scoreboard	1. Lead Indicator 2. Dashboard Indicators 3. Thematic indicator	32	2	7	[30]
EUROSTAT. Circular Economy indicators	1. Production and consumption 2. Waste management 3. Secondary raw materials 4. Competitiveness and innovation	10	0	2	[15]
OECD. Green Growth indicators	1. Environmental and resource productivity 2. Natural asset base 3. Environmental dimension of quality of life 4. Economic opportunities and policy responses	153	17	56	[31]
Global reporting initiative (GRI)	1. Environmental 2. Economic 3. Social	91	3	9	[21]

Table 2. *Cont.*

Programme	Thematic Area	Number of All Indicators	Number of Water-Related Indicators	Number of Economc Indicators	Source
The World Bank. Little Green data book	1. Economic 2. Agriculture 3. Forests and biodiversity 4. Oceans 5. Energy and emissions 6. Water and sanitation 7. Environmental and health 8. National accounting aggregates	43	8	9	[32]
UN. Sustainable Deve-lopment Goals	1. No poverty 2. Zero hunger 3. Good health and well-being 4. Quality education 5. Gender equality 6. Clean water and sanitation 7. Affordable and clean energy 8. Decent work and economic growth 9. Industry, innovation and infrastructure 10. Reduce inequalities 11. Sustainable cities and communites 12. Responsible consumption and production 13. Climate action 14. Life below water 15. Life on land 16. Peace, justice and strong institutions 17. Partnership for the gols	231	11	75	[22]
World Business Council for Sustainable Development (WBCSD)	1. Close the Loop 2. Optimize the Loop 3. Value the Loop	9	1	0	[33]

Table 3. Circularity indicators in scientific papers.

Indicator	Life Cycle Stages	Sustainability Pillars	Circular Level	Source
Circularity Measurement Toolkit (CMT)	Full life cycle	Environmental and Economic	Micro	[34–36]
Improved Water Circularity Index (WCI)	Make, Recover	Environmental and Economic	Micro, Macro	[37]
Material Circularity Indicator (MCI)	Full life cycle	Environmental	Micro, Macro	[38]
Eco-costs Value Ratio (EVR)	Full life cycle	Economic	Micro, Macro	[39]
Water Sustainable Development Index (WSDI)	Make	Environmental and Social	Micro, Macro	[40]
Value-Based Resource Efficiency (VRE)	Full life cycle	Environmental and Economic	Macro	[41]

There are various methods of grouping the indicators into specific thematic groups, for example, according to the perspective of sustainability—economic, social and environmental indicators proposed by the Global Reporting Initiative (GRI) [21] or due to individual 17 Sustainable Development Goals (SDGs) proposed by United Nations (UN) [22]. A holistic picture of the level of transformation towards a CE in European countries is indicated by the CE monitoring framework, developed by the EC in 2018 [15]. It proposes four groups of indicators, divided according to the key areas of CE implementation in the EU, such as production and consumption, waste management, secondary raw materials and competitiveness and innovation. Other organizations group indicators according to specific industries, environmental problems, individual elements of the life cycle [23], or components of the environment [24]. Depending on the potential application, there are different goals with different scope with regard to the proposed specific indicators.

As can be seen from Table 2, there are various classifications of indicators related to the features of a circular economy, focusing on the assessment, improvement, monitoring and communication on the results of the CE [25]. However, there are no official or recognized indicators, methods or tools for measuring a company's performance in the transition from a linear economic model to a more sustainable one, and there are no tools to support and track this transition [26]. Indeed, most of the CE indicators are in their early stages of development [27], and they cannot capture the overall performance of circular products and services [25]. However, many existing indicators can help measure performance in several areas (micro, meso, macro) that contribute directly or indirectly to the circular economy [28].

There are also interesting indicators developed by the individual authors. The summary of these indicators is presented in Table 3, taking into account the level of their measurement (micro, meso, macro). Most of these indicators can be grouped to the perspective of sustainability-economic, social and environmental. Most of the presented indicators focus on individual stages or all stages of life cycle, such as the Circularity Measurement Toolkit (CMT). They mainly measure the efficiency of materials use.

From the list of analyzed indicators, the CE-related indicators that can be used directly or indirectly for the circularity analysis in the water and wastewater management sector were selected. The conducted inventory shows that direct CE indicatos published by international organizations mainly focus on single indices for the water and wastewater sector not rotating the entire CE model. For example, the UN presented 231 indicators, including 11 water and 75 economic indicators, while the Organisation for Economic Cooperation and Development (OECD) published 153 indicators, including 17 water and 56 economic indicators.

CE-related indicators for the water and wastewater sector were published by the EC—water exploitation index, water productivity, price of water scarcit and water use (calculated as water abstraction minus returned water), the European Environment Agency (EEA)—water exploitation index, water productivity. Additionally, the World Business Council for Sustainable Development (WBCSD) presented the water circularity index (%), and the OECD presented a water productivity index. The European Investment Bank

(EIB) proposed circular value recovery models: reuse/recycling of wastewater. However, despite the proposed CE indicators, there are no statistical data that would allow them to be calculated and reported. The proposed indicators refer to both the micro and macro levels of water management. However, none of the organizations proposed a set of indicators that could assess the level of the CE transformation in the water and wastewater sector.

On the other hand, the indicators published in scientific papers only take into account individual aspects of the resources (incl. water) management, as cost analysis without analysis of possible revenue for CE. The economy aspect is combined with others, e.g., social or environmental. Only the Eco-costs Value Ratio (EVR) is an indicator of the economic dimension. The authors propose to calculate all environmental effects in monetary terms based on the costs that should be incurred to reduce environmental pollution and materials depletion to "no effect level" [39]. However, the EVR does not take into account the income that a company or household may gain as a result of involvement in CE.

The revision of the available CE-related indicators shows that there is a lack of a set of economic indicators that could be used for measurement of the level of transformation towards CE in the water and wastewater sector. Therefore, the next sections provide a proposal for a set of CE -related economic indicators in mentioned sector of the economy.

3.2. CE Economic Indicators in the Water and Wastewater Sector

As part of the MonGOS project, the assumptions for the CE model in water and wastewater management was proposed and published in [14]. The assumptions for this model were developed on the basis of the "xR" models in waste management, as well as the EU waste hierarchy [42]. The CE model in water and wastewater management has been classified into groups of activities that fit into the assumptions of the CE, i.e., *reduction, reclamation* (*removal*), *reuse, recycling, recovery* and *rethink* [14]. In the current section, the economic indicators have been proposed for each action of this model. The selected economic indicators were assigned to three groups of the cash flow (income - revenues, expenses – costs, and investment financing) and to the specific actions of the of the CE model in water and wastewater sector (*reclamation/removal*), *reuse, recycling, recovery)* and landfilling. The proposed indicators are dedicated to water supply and sewage companies, i.e., supplying water to the public and wastewater treatment plants (WWTPs) and companies that use water in their production processes. The proposed CE economic indicators in the water and wastewater sector are presented in Table 4.

The first element of the model (*reduce*) includes revenues from less water consumption. It creates two levels of value added-lower charges for water abstraction from the water supply, but also less pressure on the environment (lower water consumption and then less amount of generated wastewater that need to be treated). Investments are related to the cost of equipment for the optimization of water usage.

In the second element of the CE model (*reclamation/removal)*, the revenues come from the possible sale of water, the provision of wastewater collection and treatment services, to ensure the continuity of collective water supply of adequate quality and quantity, and collective wastewater disposal. The expenses are related to costs of water production and wastewater treatment services. The investments are related to equipment for water purification and wastewater treatment. In addition, there are also costs of water intake and abstraction, operation, maintenance and expansion of the water supply and sewage.

In the next level of the CE model (*reuse*), the revenues come from sales of non-drinking water, lower wastewater treatment services fees and reduced water abstraction from the waterworks. The expenses include the costs of non-drinking water production (costs of non-consumer water recovery). There are also costs of electricity production, water consumption, external services and employee remuneration, as well as investments in equipment for the wastewater treatment and water recovery. The economic added value of implementation of water reuse activities are lower annual water bills, an increase in additional business entities benefiting from improved wastewater treatment, and less waste landfilled. Analogous economic indicators are proposed for the next element of the CE model (*recycling*). However, in this case, due to the need for additional water purification (for human consumption), more energy and material costs will be incurred. On the other hand, higher revenues from sales of drinking water are expected.

Table 4. Proposed CE economic indicators in the water and wastewater sector.

Element of CE Model	Income (Revenues)	Expenses (Costs)	Investments (Investment Financing)
		Unit [Euro/Year]	
Reduce	Net revenues from less water consumption	-	Investments in equipment for the optimization of water usage
Reclamation (Removal)	Net revenues from sales of drinking water	Costs of water production	Investments in equipment for the water purification
	Net revenues from sales of wastewater treatment services	Costs of wastewater treatment services	Investments in equipment for the wastewater treatment
Reuse	Net revenues from sales of non-drinking water	Costs of non-drinking water production (costs of non-consumer water recovery)	Investments in equipment for the water reuse
	Net revenues for lower wastewater treatment services fees		
	Net revenues from the reduced water abstraction from the waterworks		
Recycling	Net revenues from sales of drinking water	Costs of drinking water production (costs of consumer water recovery)	Investments in equipment for the water recycling
	Net revenues for lower wastewater treatment services fees		
	Net revenues from the reduced water abstraction from the waterworks		
Recovery	Net revenues from sales of electricity	Costs of electricity production	Investments in equipment for the energy recovery
	Net revenues from the sale of fertilizers	Costs of fertilizers production	Investments in equipment for the nutrients recovery
	Net revenues for lower wastewater treatment services fees	Costs of materials production	Investments in equipment for the materials recovery
Landfilling	-	fees for waste landfilling	Investments in infrastructure for waste landfilling

In the next element of the CE model (*recovery*), the revenues are related not only to lower wastewater treatment services fees, but also the ability to sell electricity, fertilizers and other materials. Here, the expenses include costs of electricity, fertilizers and materials production, while investment are related to equipment needed for recovery of those resources.

It should be noted that the implementation of technological and organizational solutions in each of element of the presented CE levels can reduce the amount of landfilled waste and thus bring an economic benefit. In addition, emissions and environmental charges for landfilling are reduced. The proposed economic CE indicators selected can allow companies to identify both positive and negative trends in their activities. The measurement of these indicators can provide information about which actions are effective or created negative effects or regression at the CE economic level. The implementation of CE-related solutions is expected to increase in coming years, due to the tightening legal regulations on water and water-based waste, and EC recommendations for the transformation toward the CE model.

An important aspect of the future application of proposed indicators is the collection and processing of data needed for the reporting of these indicators. The proposed CE economic indicators are related to the information that is reported by the companies to demonstrate their revenues, costs and investment outlays. Therefore, the collecting of this economic data should not face significant barriers in the individual units.

4. Discussion

The transformation process towards the CE requires more rational use of resources and waste management practices in all sectors of the economy [43]. This also applies to the water and wastewater sector and its key elements, i.e., water, sewage, sewage sludge, other waste, and by-products arising from water purification and wastewater treatment [14]. In practice, the implementation of the CE assumptions in various sectors of the economy is often supported by the use of various methods of rational management of raw materials, products and resources, as well as sustainable waste management [44].

The process of transformation towards the circular economy in the water and wastewater sector [45] requires the involvement of all stakeholder groups, both experts working for innovative and pro-ecological solutions, and the society, which should reduce water waste in households. In addition, the implementation of circular economy principles in water and wastewater management is important for enterprises dependent on water (e.g., the cosmetics industry) and wastewater treatment plants, because their environmentally conscious decisions regarding the implementation of sustainable and circular solutions in the management of water and wastewater may accelerate the transformation process towards the CE. in the given country.

In the recent years, some significant progress has been made in the area of the assessment of circularity of products [46–48], companies [34,49,50], and regions [51,52]. The CE indicators are created to assess the progress of transformation towards CE at the micro, meso and macro levels. They are an important element of new business models for CE, which are systematically improved and introduced into the activities of enterprises, including those operating in the water and wastewater sector. The EC clearly indicates that the transition to the CE model brings economic benefits for those involved in the transformation process [10]. In order to assess the economic benefits of implementing CE solutions, specific financial data on the operation of the company and introduced changes must be reported. Therefore, the indicators developed in this research can be used by the enterprises to assess the level of transformation towards CE in the water and wastewater sector. The proposed indicators refer to the CE model for the sector developed in the MonGOS project [14], thus providing a broader perspective of the sector. In addition, they require reporting of information that is collected by the enterprise anyway, therefore its collection and processing should not pose a significant challenge to individual entities.

The added value of the presented economic indicators is the possibility of their application in various enterprises operating in the wastewater sector, i.e., supplying water to the public, WWTPs, and other companies (public and private) that use water in their production processes and create an innovations for the water and wastewater sector (e.g., Schwander [45], Veolia [53] or Outotec [54]). The proposed economic indicators could measure the CE-related activities in water and wastewater management, as minimization of water consumption, water and wastewater treatment, water reuse (for non-consumption purposes), water recycling (for consumption purposes) and the recovery of water, energy and raw materials produced in the water and wastewater treatment processes. The implementation of those activities may bring significant environmental benefits, resulting from the reduction of water consumption and the reduction of the impact of wastewater discharge on the quality of the aquatic environment. In the following years, further technological progress and new investments should be expected to reduce the consumption of water, raw materials and energy, in line with the CE model. Therefore, the proposed economic indicators can be widely applied in the sector and can complement new business models for CE.

Moreover, the usage of the proposed set of economic indicators at multiple levels would facilitate policy development, measuring economic performance, sector benchmarking, and improving business investment decisions. Such a framework should provide meaningful answers to decision-makers questions covering all relevant dimensions of the CE transition: resource consumption and material flows, economic parameters, financial flows and policy effectiveness. The presented set of economic indicators is flexible, allowing the adaptation of indicators and areas of interest to maintain effectiveness throughout the transition period. Further work on testing the developed indicators in the individual companies is undertaken as part of the MonGOS project.

5. Conclusions

Water and wastewater management is an important part of the CE model. The circular management of resources (as water, nutrients, energy) could generate financial benefits for companies operating in the water and wastewater sector. The financial benefits could be related to lower water consumption, the sale of drinking water, electricity, fertilizers, and lower fees for wastewater treatment services and taxes for waste landfilling. However, there are also unavoidable costs of the CE-related activities as investment outlays in new infrastructure. In the long run, there should be a reduction in average total costs that could bring further economic benefits.

The implementation of the CE solutions requires estimating the costs of such activities, both in terms of possible expenses and revenues. Therefore, the proposed set of economic CE indicators can be used by the companies not only to assess the transformation level but also to plan the future CE-related investments.

Author Contributions: Conceptualization, M.S. and R.K.; methodology, M.S.; investigation, M.S. and R.K.; resources, M.S. and R.K.; data curation, M.S. and R.K.; writing—original draft preparation, M.S. and R.K.; writing—review and editing, M.S. and R.K.; visualization, M.S. and R.K. All authors have read and agreed to the published version of the manuscript.

Funding: This work was funded by the Polish National Agency for Academic Exchange (NAWA) as the part of the project "Monitoring of water and sewage management in the context of the implementation of the circular economy assumptions" (MonGOS, mon-gos.eu).

MonGOS

NAWA

Institutional Review Board Statement: Not applicable.

Informed Consent Statement: Not applicable.

Data Availability Statement: The datasets generated during and/or analysed during the current study are available from the corresponding author on reasonable request.

Conflicts of Interest: The authors declare no conflict of interest. The funders had no role in the design of the study; in the collection, analyses, or interpretation of data; in the writing of the manuscript, or in the decision to publish the results.

References

1. European Commission. *Communication from the Commission—Towards a Circular Economy: A Zero Waste Programme for Europe*; COM 398; European Commission: Brussels, Belgium, 2014.
2. European Commission. *Communication from the Commission. Closing the Loop—An EU Action Plan for the Circular Economy*; COM 614; European Commission: Brussels, Belgium, 2015.
3. Rosenau-Tornow, D.; Buchholz, P.; Riemann, A.; Wagner, M. Assessing the long-term supply risks for mineral raw materials-a combined evaluation of past and future trends. *Resour. Policy* **2009**, *34*, 161–175. [CrossRef]
4. Voulvoulis, N. Water reuse from a circular economy perspective and potential risks from an unregulated approach. *Curr. Opin. Environ. Sci. Health* **2018**, *2*, 32–45. [CrossRef]
5. Jia, X.; Klemeš, J.J.; Alwi, S.R.W.; Varbanov, P.S. Regional Water Resources Assessment using Water Scarcity Pinch Analysis. *Resour. Conserv. Recycl.* **2020**, *157*, 104749. [CrossRef]
6. Al-Jawad, J.Y.; Alsaffar, H.M.; Bertram, D.; Kalin, R.M. A comprehensive optimum integrated water resources management approach for multidisciplinary water resources management problems. *J. Environ. Manag.* **2019**, *239*, 211–224. [CrossRef]

7. Lavrnić, S.; Zapater-Pereyra, M.; Mancini, M.L. Water Scarcity and Wastewater Reuse Standards in Southern Europe: Focus on Agriculture. *Water. Air. Soil Pollut.* **2017**, *228*, 251. [CrossRef]
8. European Commission. *Commission Staff Working Document—Leading the Way to a Global Circular Economy: State of Play and Outlook*; SWD 100; European Commission: Brussels, Belgium, 2020.
9. European Environment Agency. *Water Scarcity and Drought*; European Environment Agency: Copenhagen, Denmark, 2010.
10. European Commission. *Communication from the Commission. Circular Economy Action Plan for a Cleaner and More Competitive Europe*; COM 98; European Commission: Brussels, Belgium, 2020.
11. European Commission. *Communication from the Commission: The European Green Deal*; COM 640; European Commission: Brussels, Belgium, 2019.
12. Smol, M. Circular economy approach in the water and wastewater sector. In *Circular Economy and Sustainability*; Elsevier: Amsterdam, The Netherlands, 2021; pp. 1–19.
13. Tahir, S.; Steele, K.; Steichen, M.S.T.; Penning, P.; Martin, N. Water and Circular Economy: A White Paper. 2018. Available online: https://ceowatermandate.org/resources/water-and-circular-economy-2018/ (accessed on 10 October 2021).
14. Smol, M.; Adam, C.; Preisner, M. Circular economy model framework in the European water and wastewater sector. *J. Mater. Cycles Waste Manag.* **2020**, *22*, 682–697. [CrossRef]
15. European Commission. *A Monitoring Framework for the Circular Economy*; COM 29; European Commission: Brussels, Belgium, 2018.
16. Smol, M. Inventory and Comparison of Performance Indicators in Circular Economy Roadmaps of the European Countries. *Circ. Econ. Sustain.* **2021**. [CrossRef]
17. Smol, M.; Marcinek, P.; Duda, J.; Szołdrowska, D. Importance of sustainable mineral resource management in implementing the circular economy (CE) model and the european green deal strategy. *Resources* **2020**, *9*, 55. [CrossRef]
18. Barquet, K.; Järnberg, L.; Rosemarin, A.; Macura, B. Identifying barriers and opportunities for a circular phosphorus economy in the Baltic Sea region. *Water Res.* **2020**, *171*, 115433. [CrossRef] [PubMed]
19. Smol, M. Inventory of Wastes Generated in Polish Sewage Sludge Incineration Plants and Their Possible Circular Management Directions. *Resources* **2020**, *9*, 91. [CrossRef]
20. Smol, M.; Duda, J.; Czaplicka-Kotas, A.; Szołdrowska, D. Transformation towards circular economy (CE) in municipal waste management system: Model solutions for Poland. *Sustainability* **2020**, *12*, 4561. [CrossRef]
21. Global Reporting Initiative. *Performance Indicators*; GRI: Amsterdam, The Netherlands, 2016.
22. United Nations General Assembly. *United Nations Resolution Adopted by the General Assembly on 25 September 2015. Transforming Our World: The 2030 Agenda for Sustainable Development*; A/RES/70/1; United Nations: New York, NY, USA, 2015. [CrossRef]
23. European Commission. *Raw Materials Scoreboard*; European Commission: Brussels, Belgium, 2016; ISBN 9789279616990.
24. European Environment Agency (EEA). Digest of EEA indicators 2014. In *EEA Technical Report*; European Environment Agency; Publications Office of the European Union: Luxembourg, 2016.
25. Janik, A.; Ryszko, A. Circular economy in companies: An analysis of selected indicators from a managerial perspective. *Multidiscip. Asp. Prod. Eng.* **2019**, *2*, 523–535. [CrossRef]
26. McKinsey & Company. *Europe's Circular—Economy Opportunity*; Mckinsey & Company: New York, NY, USA, 2015; pp. 1–7.
27. Latawiec, A.; Agol, D. *Sustainability Indicators in Practice*; De Gruyter Open: Berlin, Germany; Warsaw, Poland, 2015; ISBN 9783110450491.
28. Van Hoof, V.; Maarten, C.; Vercalsteren, A. Indicators for a Circular Economy. 2018. Available online: https://ce-center.vlaanderen-circulair.be/en/publications/publication/1-indicators-for-a-circular-economy (accessed on 14 October 2021).
29. European Environment Agency (EEA). *Circular Economy in Europe—Developing the Knowledge Base*; European Environment Agency; Publications Office of the European Union: Luxembourg, 2016.
30. European Commission. *EU Resource Efficiency Scoreboard 2015*; European Commission: Brussels, Belgium, 2016; ISBN 15525260.
31. Organization for Economic Co-operation and Development (OECD). *Green Growth Indicators 2014*; OECD: Paris, France, 2014; ISBN 9789264202030.
32. World Bank. *The Little Green Data Book*; World Bank: Washington, DC, USA, 2017; ISBN 9781464810343.
33. WBCSD. Circular Transition Indicators V2.0. Metrics for Business, by Business. 2021. Available online: https://www.wbcsd.org/Programs/Circular-Economy/Factor-10/Metrics-Measurement/Resources/Circular-Transition-Indicators-v2.0-Metrics-for-business-by-business (accessed on 14 October 2021).
34. Garza-Reyes, J.A.; Salomé Valls, A.; Peter Nadeem, S.; Anosike, A.; Kumar, V. A circularity measurement toolkit for manufacturing SMEs. *Int. J. Prod. Res.* **2019**, *57*, 7319–7343. [CrossRef]
35. Santibanez Gonzalez, E.D.R.; Koh, L.; Leung, J. Towards a Circular Economy Production System: Trends and Challenges for Operations Management. *Int. J. Prod. Res.* **2019**, *57*, 7209–7218. [CrossRef]
36. Gustavsson, J.; Svaren, S. Generating Design Improvements Using Circularity Indicators—A Case Study on How to Measure, Increase and Evaluate the Circularity of a Fully-Automated Coffee Machine. Master's Thesis, Linköping University, Linköping, Sweden, 2021.
37. Sartal, A.; Ozcelik, N.; Rodriguez, M. Bringing the circular economy closer to small and medium enterprises: Improving water circularity without damaging plant productivity. *J. Clean. Prod.* **2020**, *256*, 120363. [CrossRef]

38. Franconi, E.; Bridgeland, B.; Webster, K. *A New Dynamic 2—Effective Systems in a Circular Economy*; Ellen MacArthur Foundation Publishing: Isle of Wight, UK, 2016; ISBN 978-0-9927784-4-6.
39. Scheepens, A.E.; Vogtlander, J.G.; Brezet, J.C. Two life cycle assessment (LCA) based methods to analyse and design complex (regional) circular economy systems. Case: Making water tourism more sustainable. *J. Clean. Prod.* **2016**, *114*, 257–268. [CrossRef]
40. Morris, J. Developing and exploring indicators of water sustainable development. *Heliyon* **2019**, *5*, e01778. [CrossRef]
41. Di Maio, F.; Rem, P.C.; Baldé, K.; Polder, M. Measuring resource efficiency and circular economy: A market value approach. *Resour. Conserv. Recycl.* **2017**, *122*, 163–171. [CrossRef]
42. European Union. Directive 2018/851 amending Directive 2008/98/EC on waste Framework. *Off. J. Eur. Union* **2018**, *L-150*, 109–140.
43. Marcinek, P.; Smol, M. Bioeconomy as one of the key areas of implementing a circular economy (CE) in Poland. *Environ. Res. Eng. Manag.* **2020**, *76*, 20–31. [CrossRef]
44. Bianchini, A.; Rossi, J. Design, implementation and assessment of a more sustainable model to manage plastic waste at sport events. *J. Clean. Prod.* **2021**, *281*, 125345. [CrossRef]
45. Ramm, K. Considerations Related to the Application of the EU Water Reuse Regulation to the Production of Snow from Reclaimed Water. *Circ. Econ. Sustain.* **2021**. [CrossRef]
46. Angioletti, C.M.; Despeisse, M.; Rocca, R. Product Circularity Assessment Methodology. *IFIP Int. Fed. Inf. Process.* **2017**, *514*, 411–418. [CrossRef]
47. Niero, M.; Kalbar, P.P. Coupling material circularity indicators and life cycle based indicators: A proposal to advance the assessment of circular economy strategies at the product level. *Resour. Conserv. Recycl.* **2019**, *140*, 305–312. [CrossRef]
48. Harris, S.; Martin, M.; Diener, D. Circularity for circularity's sake? Scoping review of assessment methods for environmental performance in the circular economy. *Sustain. Prod. Consum.* **2021**, *26*, 172–186. [CrossRef]
49. Otero, J.C. Circularity Assessment for Companies: Elements for a General Framework—Challenge Lab 2015: Sustainable urban development. Master's Thesis, Chalmers University of Technology, Gothenburg, Sweden, 2015.
50. Pauer, E.; Wohner, B.; Heinrich, V.; Tacker, M. Assessing the environmental sustainability of food packaging: An extended life cycle assessment including packaging-related food losses and waste and circularity assessment. *Sustainability* **2019**, *11*, 925. [CrossRef]
51. Towa, E.; Zeller, V.; Achten, M.J.W. Assessing the circularity of regions: Stakes of trade of waste for treatment. *J. Ind. Ecol.* **2021**, *25*, 834–847. [CrossRef]
52. Virtanen, M.; Manskinen, K.; Uusitalo, V.; Syvänne, J.; Cura, K. Regional material flow tools to promote circular economy. *J. Clean. Prod.* **2019**, *235*, 1020–1025. [CrossRef]
53. Silano, V.; Barat Baviera, J.M.; Bolognesi, C.; Chesson, A.; Cocconcelli, P.S.; Crebelli, R.; Gott, D.M.; Grob, K.; Lambré, C.; Mengelers, M.; et al. Safety assessment of the process Veolia URRC used to recycle post-consumer PET into food contact materials. *EFSA J.* **2020**, *18*, e06125.
54. Hermann, L.; Schaaf, T. Outotec (AshDec®) process for P fertilizers from sludge ash. *Phosphorus Recover. Recycl.* **2018**, *15*, 221–233. [CrossRef]

resources

MDPI

Article

Toxicity of Antiretrovirals on the Sea Urchin *Echinometra lucunter* and Its Predicted Environmental Concentration in Seawater from Santos Bay (Brazilian Coastal Zone)

Renato Sakai Cid [1,2], Vinicius Roveri [3], Diogo Guedes Vidal [4], Maria Alzira Pimenta Dinis [4], Fernando Sanzi Cortez [2], Flávia Rigos Salgueiro [1], Walber Toma [1], Augusto Cesar [2,5] and Luciana Lopes Guimarães [1,*]

1 Natural Products Research Laboratory, Santa Cecília University (UNISANTA), Rua Cesário Mota 8, F83A, Santos 11045-040, SP, Brazil; renatosakai@hotmail.com (R.S.C.); flaviarigos@gmail.com (F.R.S.); walbertoma@unisanta.br (W.T.)
2 Ecotoxicology Laboratory, Santa Cecília University (UNISANTA), Rua Oswaldo Cruz 266, Santos 11045-907, SP, Brazil; cortezfs@hotmail.com (F.S.C.); acesar@unifesp.br (A.C.)
3 Education Department, Metropolitan University of Santos (UNIMES), Avenida Conselheiro Nébias, 536, Encruzilhada, Santos 11045-002, SP, Brazil; viniciusroveri@bol.com.br
4 UFP Energy, Environment and Health Research Unit (FP-ENAS), Faculty of Science and Technology, University Fernando Pessoa (UFP), 4249-004 Porto, Portugal; diogovidal@ufp.edu.pt (D.G.V.); madinis@ufp.edu.pt (M.A.P.D.)
5 Department of Marine Sciences, Federal University of São Paulo, Rua Maria Máximo, 168, Santos 11030-100, SP, Brazil
* Correspondence: lucianafarm@unisanta.br

Citation: Cid, R.S.; Roveri, V.; Vidal, D.G.; Dinis, M.A.P.; Cortez, F.S.; Salgueiro, F.R.; Toma, W.; Cesar, A.; Guimarães, L.L. Toxicity of Antiretrovirals on the Sea Urchin *Echinometra lucunter* and Its Predicted Environmental Concentration in Seawater from Santos Bay (Brazilian Coastal Zone). *Resources* **2021**, *10*, 114. https://doi.org/10.3390/resources10110114

Academic Editor: Diego Copetti

Received: 22 September 2021
Accepted: 5 November 2021
Published: 9 November 2021

Abstract: Antiretrovirals (ARVs) have been detected in aquatic ecosystems throughout the world; however, studies focused on assessing their ecotoxicological effects on marine aquatic organisms are still rare. In the present study, the predicted environmental concentration (*PEC*) of 13 ARVs was estimated for surface seawater from Santos Bay, Brazil, according to the European Medicines Agency (EMEA) guidelines. The results indicated that all ARVs need to be assessed for their ecological effects, considering that they all exceeded the EMEA guideline limits ($PEC > 0.01$ μg L^{-1}). In this sense, three ARVs (namely atazanavir, nevirapine and efavirenz) were selected for the acute and chronic tests with sea urchin (*Echinometra lucunter*). Furthermore, the Environmental Risk Assessment (ERA) for these three ARVs was also performed by calculating the risk quotient. The acute and chronic toxicity results showed inhibitory concentrations (*IC*) for the fertilization (IC_{50}; 1 h; range: 11.46–84.61 mg L^{-1}) and for the embryo–larval development (IC_{50}; 42 h; range: 0.52–0.97 mg L^{-1}) of the sea urchin, respectively. Moreover, the ERA showed that these three ARVs are potentially hazardous for aquatic life in Santos Bay, raising concerns about the continuous introduction of ARVs in aquatic ecosystems. The data presented may contribute to the provision of subsidies for the development of monitoring public policies that aim to reduce the introduction of ARVs into the aquatic environment.

Keywords: subtropical coastal zone; waste treatment; ocean dumping; antiretrovirals; predicted environmental concentration; ecotoxicology; risk assessment; pollution effects

1. Introduction

Currently, 40% of the world's population, estimated at 7.2 billion people, live in about 2100 coastal cities [1–3]. However, the high population concentration in the world's coastal cities raises some concerns because there are many people living in this small area of the world's land surface, estimated at between 4 and 8%, thus causing intensive anthropic use of the narrow coastal areas [2,4]. Consequently, this high concentration of people exposes coastal ecosystems to different anthropogenic pressures, such as the disposal of municipal wastewater in the marine environment [5–7]. This sewage can contain thousands of chemical substances, such as pharmaceuticals and personal care products (PPCPs), that constitute a vast group of emerging environmental contaminants, from different therapeutic classes, including antiretroviral (ARV) drugs [8–10].

ARVs emerged in the 20th century when acquired immunodeficiency syndrome (AIDS) quickly spread across the five continents [11]. Nowadays, therapies with ARVs aim to reduce viral load, improving the host immune system once HIV mainly attacks the CD4+T cells, a crucial component in the body's immune system [12–14]. Based on their molecular mechanism of action, three classes of ARVs are widely used: (i) nucleoside/nucleotide reverse transcriptase inhibitors (NRTIs), namely abacavir, didanosine, lamivudine, tenofovir, and zidovudine; (ii) non-nucleoside reverse transcriptase inhibitors (NNRTIs), namely efavirenz, etravirine, and nevirapine; and (iii) protease inhibitors (PI), namely atazanavir, darunavir, lopinavir, ritonavir, and tipranavir. Each of these classes, represented by 13 ARVs, target a phase of the HIV virus life cycle [12–14]. Advances in ARVs treatment, essentially after the development of Highly Active Antiretroviral Therapy (HAART), a treatment regimen comprising a combination of three or more ARVs, made it possible to transform a syndrome, that was previously perceived to be a diagnostic to an announced death, into a disease with chronicity prospects [12–14].

The Joint United Nations Program on HIV/AIDS [11] estimates that 37.9 million people worldwide were living with HIV/AIDS in 2018. Consequently, after the high worldwide consumption of ARVs, these drugs have constantly been introduced into aquatic ecosystems through their main routes of human excretion, urine and faeces, a process that is allied to the absence or inefficacy of wastewater treatment plants (WWTPs) [8–10]. Several studies conducted in Norway [8], Kenya [15], Greece [9] and South Africa [10] reported high concentrations of ARVs, generally at $\mu g\ L^{-1}$ levels, in influents and effluents of WWTPs of the secondary treatment level (e.g., atazanavir, nevirapine and efavirenz). Moreover, the presence of ARVs in the aquatic environment was previously reported in different environmental matrices such as river water samples in South Africa (e.g., nevirapine, efavirenz and emtricitabine) and in sediments and waters of an estuary in France (namely abacavir, lamivudine, nelfinavir, nevirapine, ritonavir and saquinavir) [16,17]. However, despite the proven occurrence of ARVs in aquatic environments, there is a great scarcity of studies concerning the biological effects of ARVs in aquatic organisms, such as algae, crustaceans, fish and echinoderms, and especially in tropical marine organisms [18].

Specifically in Brazil, the fifth largest country in the world, where approximately 50 million people live in 463 coastal municipalities along 8500 km of coastline [19], there seems to exist no previous studies about the occurrence and the potential ecological risk of ARVs in different environmental matrices (e.g., seawater or sediment), despite the high incidence of the disease and the existence of a government program that distributes ARVs to those affected by HIV/AIDS. For instance, in 2018, 43,941 cases of HIV infection were reported in Brazil [11]. Currently, the Brazilian HIV/AIDS program, the budget of which is approximately US$ 408 million/year, recommends an immediate start of ARV therapy for all people living with HIV, regardless of their clinical and/or immunological stage, and indicates that the initial therapy should always include combinations of three different ARVs, with two NRTIs associated with another class of ARVs. These ARVs are distributed for the 463 Brazilian coastal municipalities, among them Santos, the city that is the focus of this study, through a logistics management system—SILCOM/Ministry of Health [20–22]. The municipal sewage of Santos, in the State of São Paulo, is treated through a WWTP with a preliminary treatment [23–25]. This WWTP performs only a mechanical treatment, i.e., railing and screening for the removal of solids, that is followed by chlorination [23–25]. The final destination of the preconditioned sewage is a submarine outfall, which is 4500 m long and 10 m deep, that, on a daily basis, disposes of sewage into Santos Bay, South Atlantic Ocean, a semi-closed and low-energy coastal system [23–25]. Consequently, these ARVs (in parental, metabolized or conjugated forms in human excreta) can be released indiscriminately into the receiving waters of Santos Bay, because this WWTP is not efficient in removing these emerging pollutants [8,9,15].

Considering this lack of data in the Brazilian coastal zone, the information about the prescription and/or consumption has been demonstrated to be very valuable in terms of estimating the occurrence of PPCPs in aquatic ecosystems [26]. According to the approach suggested by the European Medicines Agency (EMEA), the calculation of the predicted environmental concentration (*PEC*), based on consumption data, excretion, elimination in the WWTP, and dilution in receiving waters, constitutes extremely useful information to prioritize compounds for further monitoring, to establish the potential incidence of PPCPs in a specific area, and even to assess their risk according to toxicological data [26]. Further, the EMEA recommended the assessment of risk when *PEC* values in surface water were equal or above the threshold value of 0.01 $\mu g\ L^{-1}$ [26].

In this scenario of high consumption of ARVs in Brazil, combined with the lack of data regarding the biological effects of ARVs in marine organisms, this study estimated the concentration and potential ecological risk of 13 ARVs in the surface waters of Santos Bay, São Paulo, Brazil, through the *PEC* values. Moreover, the study employed ecotoxicological assays using sea urchins (*Echinometra lucunter*) as a test organism, with three selected ARVs: atazanavir, efavirenz and nevirapine. The data presented may contribute to the provision of subsidies for the development of monitoring government policies with the aim of reducing the introduction of ARVs into the aquatic environment and promoting good practices in the development and implementation of indicator systems regarding this urban issue.

2. Materials and Methods

2.1. Predicted Environmental Concentration Calculation—PEC

The *PEC* calculation of surface water of Santos Bay was performed according to the EMEA guideline [26], using the following Equation (1)

$$PEC\ Surface\ Waters = \frac{Maximum\ daily\ dose\ consumed\ per\ inhabitant \times Mpen}{Amount\ liquid\ waste\ \left(\frac{\frac{L}{inhabitant}}{day}\right) \times dilution\ factor} \quad (1)$$

considering:

Mpen = market penetration factor, Mpen [26], was calculated according to Equation (2)

$$Mpen(\%) = \frac{Consumption\ (mg/year) \times 100}{DDD\ (mg/inhabitant/day) \times Pop \times 365\ days} \quad (2)$$

considering:

DDD = defined daily dose, according to the World Health Organization (WHO)

Pop = number of the city inhabitants

Considering that the WWTP of Santos receives discharges from the cities of Santos and São Vicente, data were obtained from both cities [23–25]. According to the Brazilian Institute of Geography and Statistics data [19], the Santos population in 2018 was 432,957 inhabitants, and the São Vicente population was 363,173 inhabitants, totalling 796,130 inhabitants. According to the Basic Sanitation Company of São Paulo State [25], the amount of liquid waste/inhabitant/day is 575 L. The dilution factor used is 10, and thus, the following values of *Mpen* and *PEC* were obtained and are included in Table 1.

Information regarding the number of distributed ARVs was obtained through access to the Federal Government website [27], where, through the Law on Access to Information (LAI) No. 12527/2011, annual withdrawal data were requested for the antiretroviral drugs dispensed in Santos and São Vicente municipalities, from July 2017 to July 2018. Liquid waste generation data were collected with the publications available online from SABESP and Environmental Protection Agency of the State of São Paulo [24].

Table 1. *Mpen* results for the antiretrovirals based on *DDD*.

Antiretroviral	*DDD* (mg)	Annual Consumed Amount (g)	*Mpen* (%)
Abacavir	600	14,625	0.008388525
Atazanavir	400	106,723	0.091816852
Darunavir	600	158,805	0.091082736
Didanosine	400	24	0.0000206478
Efavirenz	600	65,635	0.037645122
Etravirine	400	4560	0.003923088
Lamivudine	300	94,802	0.10874833
Lopinavir	1066	19,114	0.006170718
Nevirapine	400	17,940	0.015434252
Ritonavir	1200	64,061	0.01837132
Tenofovir	300	9634	0.011051751
Tipranavir	1000	540	0.00018583
Zidovudine	600	169,750	0.09736025

Mpen = market penetration factor; *DDD* = defined daily dose, according to the World Health Organization (WHO).

2.2. Ecotoxicological Assessment

An EMEA [26] document establishes that if the *PEC* value is equal to or above 0.01 μg L^{-1}, then the environmental fate should be investigated and effect assessment carried out. In this sense, toxicity tests were performed to assess the acute (fertilization rate) and chronic (embryo–larval development) effects using sea urchins (*Echinometra lucunter*), with three selected ARVs that presented *PEC* values above 0.01 μg L^{-1}: atazanavir (Nortec® Lab, Duque de Caxias, Rio de Janeiro, Brazil), nevirapine and efavirenz (Farmanguinhos, Fio Cruz® Lab, Manguinhos, Rio de Janeiro, Brazil). The organisms were collected through free diving, at Ilha das Palmas, in the municipality of Guarujá, São Paulo state, stored in a thermal box, and covered with the algae genus *Ulva sp*. The temperature conditions were maintained until transport to the laboratory, where they were kept in a tank, under strong aeration and ideal conditions, until the time of testing. The water used in the tanks was natural, originally collected on Ilha das Palmas. For these organism's maintenance, their physical–chemical parameters, such as temperature, salinity, pH and dissolved oxygen were observed daily, obeying the ideal conditions, according to the Brazilian NBR 15,350 standard [28]. For the test substances' dilution, gamete handling and control preparation, reconstituted water was used, from the mixture CORAL PRO SALT brand (RED SEA®, São Paulo, Brazil), composed of commercial salt in processed water, kept under agitation for total solubilization and preservation of the characteristics found in the organisms' natural environment. The solution was filtered, with a filtration support aid and 0.45 μm Millipore® cellulose membrane. The water was maintained at a physical–chemical standard value of 15.350 (pH between 7.8 and 8.4 and salinity between 30 and 37 (g L^{-1}) [28]. A seawater control and a solvent (DMSO) control were set in parallel with the ARVs assays. There were no statistically significant differences between the control and the highest concentration of the DMSO solvent.

2.2.1. Acute Toxicity Test (Fertilization Assay)

The procedures were based on the USEPA protocol [29], adapted for the *Echinometra lucunter* species. Sea urchin sperm were exposed to different ARVs concentrations (3.12, 6.25, 12.5, 25, 50 and 100 mg L^{-1}) during the 1-h period. After this period, a solution containing eggs was added to the test flasks. Twenty minutes after the addition of the eggs, the test was ended with the 0.5 mL borax-buffered formaldehyde addition in all replicates. After the exposure period, the test was ended with the addition of buffered formaldehyde. Afterwards, the reading was performed, and the effect concentration was estimated. At the end of the test, the larvae were divided into two groups, according to their morphological aspects, to identify normal and abnormal larvae. The test reading was performed by counting the first 100 organisms according to the development stage. For these tests, the results are expressed as IC_{50} values (mean inhibitory concentrations) [28].

2.2.2. Chronic Toxicity Tests (Embryo–Larval Development Assay)

Newly fertilized sea urchin embryos were exposed to different ARV concentrations (0.195, 0.39, 0.78, 1.56 and 3.12 mg L^{-1}) during the embryo–larval development period, that is, from 36 h to 42 h for *Echinometra lucunter*, according to the technical standard ABNT/NBR 15350 [28]. At the end of the test, the larvae were divided into 2 groups, according to their morphological aspects, to identify normal and abnormal larvae. The test reading was performed by counting the first 100 organisms according to the development stage. In these tests, the results are expressed as IC_{50} (medium inhibitory concentration), *NOEC* (no observed effect on the concentration of the test organism) and *LOEC* (lowest observed that causes a statistically significant effect on the test organisms) [28].

2.2.3. Environmental Risk Assessment (ERA)

The Environmental Risk Assessment (ERA) for atazanavir, efavirenz and nevirapine to aquatic organisms was performed by calculating the risk quotient (*RQ*) for 4 different aquatic organisms, algae, crustaceans, fish and echinoderms, following Equation (3)

$$RQ = \frac{PEC}{PNEC} \tag{3}$$

considering:

RQ = Risk Quotient;
PEC = Predicted Environmental Concentration;
PNEC = Predicted No-Effect Concentration.

The *PEC* and *PNEC* values were predicted, and both were expressed in μg L^{-1}. *PNEC* values were obtained from reliable base-set ecotoxicity data that were available for the aquatic compartment regarding short-term (Lethal Concentration 50 (LC_{50}) or median Effective Concentration (EC_{50})) and long-term (No Observed Effect Concentration (*NOEC*)) toxicological endpoints. According to the existent studies and current marine risk assessment practices, a reasonable correlation exists between the ecotoxicological responses of freshwater and saltwater biota, at least for the usual aquatic taxa (i.e., acute and chronic toxicity to algae, crustacean and fish) [26,30,31]. In this context, an attempt was made to specifically compile *PNEC* data for marine and coastal species. When these data were not available, data from freshwater communities were used. In order to collect the available ecotoxicity test endpoints, an extensive search was carried out in the Ecotoxicology Database (ECOTOX) from the United States Environmental Protection Agency [32], as well as in other literature sources using the PubMed database. When the experimentally-derived data from the ecotoxicity laboratory were not available short [$L\ (E)C_{50}$] and long toxicological endpoints [*Chv*, geometric mean of *NOEC* and *LOEC*, *ChV* = 10^ ([log (*NOEC*x*LOEC*)]/2)] were estimated using the Ecological Structure Activity Relationships Program (ECOSAR, v 2.0) [33]. The derived *PNEC* values for the acute and chronic toxicity data were thereafter calculated by dividing each toxicological endpoint by an assessment factor (*AF*). For saltwater environments, an *AF* of 10,000 and 100 should be considered in short and long-term data sets. For further details, see the European Chemical Bureau [34] and the European Chemicals Agency [35] guidelines. Finally, the risk was categorized into four levels: no ($RQ < 0.01$), low ($0.01 \leq RQ < 0.1$), moderate ($0.1 \leq RQ < 1.0$) and high ecological risk ($RQ \geq 1.0$) to aquatic organisms [36].

2.3. Statistical Analysis

The linear interpolation method was used to calculate the set of inhibitory concentrations (IC_{50}) (1 h) for the fertilization assays and the set of IC_{50} (42 h) for the embryo–larval development assays, using the ICPin program. For each embryo–larval development assay, ANOVA followed by the Dunnett's test were used to identify the concentrations that were significantly different from the control (*NOEC* and *LOEC*). For all analyses, significant differences were determined when $p < 0.05$. Statistical analysis was performed by employing TOXSTAT 3.5.

3. Results and Discussion

3.1. Predicted Environmental Concentration Calculation—PEC

Predictive models have been used as an approach for investigating the presence of pharmaceuticals in the environment, especially in hospitals and WWTP effluents [37,38]. *PEC* is an estimation of the concentration of a substance in the environment, considering the initial amount released into the environment in terms of its fate, transformation and removal, either by artificial or natural means [13,26]. In the present study, it was possible to estimate ARV *PEC* values for Santos Bay surface waters considering that ARV dispensation and stocks in Brazil are monitored by the logistics management system—SILCOM/Ministry of Health [20–22]. The *PEC* values are shown in Table 2. Of the 13 drugs assessed, didanosine displayed a *PEC* value equal the EMEA document limit (0.01 μg L^{-1}) and the other 12 ARVs' *PEC* values were above the limit, indicating that all ARVs included in this study need to be assessed in terms of the destination and the environmental effect of this compound [26].

Table 2. Predicted environmental concentration (*PEC* surface waters) values of antiretrovirals (ARVs) in seawater from Santos Bay (Brazil).

Therapeutic Classes	ARV	*PEC* (μg L^{-1})
NRTI	Abacavir	0.875
	Didanosine	0.001
	Lamivudine	5.673
	Tenofovir	0.576
	Zidovudine	10.159
NNRTI	Efavirenz	3.928
	Etravirine	0.272
	Nevirapine	1.073
PI	Atazanavir	6.387
	Darunavir	9.504
	Lopinavir	1.143
	Ritonavir	3.834
	Tipranavir	0.032

Nucleoside reverse transcriptase inhibitor (NRTI); non-nucleoside reverse transcriptase inhibitor (NNRTI); protease inhibitor (PI).

For the NRTI class, zidovudine and lamivudine presented the highest *PEC* values (10.159 and 5.673 μg L^{-1}, respectively). The combination of zidovudine and lamivudine is one of the most studied in randomized clinical trials and is usually well tolerated for HIV/AIDS treatment, and it can also be used in schemes for postexposure prophylaxis (PEP) in situations involving potential exposure to HIV. It is available in co-formulation, which contributes to greater dose comfort [22]. Since 1999, Brazil have offered PEP as a strategy against HIV transmission [20,39]. The presence of both zidovudine and lamivudine and other ARVs was reported in WWTP effluents and river water in France [16]. The presence of zidovudine and carboxy-lamivudine (a stable carboxy-transformation product of lamivudine formed in WWTPs) was reported in surface waters in Germany [40].

Regarding the NNRTI class, the highest *PEC* value was observed for efavirenz (3.928 μg L^{-1}) followed by nevirapine (1.073 μg L^{-1}). According to the Brazilian Clinical protocol and guidelines, therapeutics for the management of HIV infections in adult women of childbearing age at the start of treatment should preferably involve the use of regimens containing efavirenz and the performance of pre-treatment genotyping [22]. Both efavirenz and nevirapine were detected in dam and ground waters in South Africa [17] and they were also detected in WWTP effluents in Kenya [15].

Among the protease inhibitors (PI) included in this study, darunavir and atazanavir presented the highest *PEC* values (9.504 and 6.387 μg L^{-1}, respectively). Atazanavir is the first option in the IP class among the ARVs of choice for first failure of HIV treatment rescue, and darunavir is used as an alternative for contraindication, intolerance or proven toxicity to atazanavir, according to the Brazilian treatment guidelines [22]. The occurrence of atazanavir was previously reported in WWTP effluents in Norway [8] and darunavir was detected in water samples (tap and river) in Poland [41]. Furthermore, Ncube et al. [13] mentioned that some ARVs, such as nevirapine and atazanavir, are persistent enough to by-pass most conventional wastewater treatment processes and remain in surface waters. ARVs can also reach water sources and enter food chains.

In the coastal areas of Brazil, there are 20 conventional WWTPs, such as Santos, that, on a daily basis, dispose of their sewer contents into the Atlantic Ocean [23–25]. These WWTPs are not efficient in removing these 13 ARVs. Consequently, these chemical stressors (in parental, metabolized or conjugated forms in human excreta) can be released indiscriminately into the receiving waters of the Brazilian coastline [8,9,15]. It is worth mentioning that there are advanced wastewater technologies that potentially remove ARVs residues (e.g., ozonation and activated carbon) but they are usually expensive to adopt in developing countries such as Brazil [13].

In the present study, eight ARVs displayed high *PEC* values, exceeding 100× the EMEA *PEC* values limits. Three of them were selected to perform acute and chronic toxicity tests with *Echinometra lucunter*: atazanavir, efavirenz and nevirapine.

3.2. Acute and Chronic Toxicity Tests

This study assessed the biological effects of atazanavir, efavirenz and nevirapine in different life stages of the sea urchin *Echinometra lucunter* through fertilization and embryo–larval development assays (acute and chronic toxicity tests, respectively). The mean concentrations that inhibited the fertilization of *Echinometra lucunter* (IC_{50}; 1 h) were: 73.04 mg L^{-1} for atazanavir; 11.46 mg L^{-1} for efavirenz; and 84.61 mg L^{-1} for nevirapine. These results are presented in Table 3. It is noteworthy to mention that of these three compounds, efavirenz demonstrated the highest toxicity in acute tests, and also displayed the highest octanol-water partitioning coefficient (*Kow*), with a log *Kow* value of 4.7 for efavirenz, as compared 4.54 for atazanavir and 2.5 for nevirapine. Normally, log *Kow* ≥ 3 indicates that ARVs could bioaccumulate and/or exert toxicity [26,32,33].

Table 3. Results of the fertilization assays (n = 4) of antiretrovirals (ARVs) (IC_{50} and confidence limits) on *Echinometra lucunter*.

ARV	IC_{50} (mg L^{-1})	Lower-Upper Confidence Intervals
Atazanavir	73.04	71.98–73.90
Efavirenz	11.46	11.18–11.81
Nevirapine	84.61	80.79–89.35

IC_{50} = average inhibitory concentration.

In the embryo–larval development assay, the mean concentrations of ARVs that inhibited the normal embryo–larval development of *Echinometra lucunter* (IC_{50}; 42 h) were: 0.63 mg L^{-1} for atazanavir; 0.52 mg L^{-1} for efavirenz; and 0.195 mg L^{-1} for nevirapine (Table 4).

Table 4. Results of the embryo–larval assays (n = 4) of antiretrovirals (ARVs) (*NOEC*, *LOEC*, and IC_{50}) on *Echinometra lucunter*.

ARV	*LOEC* (mg L^{-1})	*NOEC* (mg L^{-1})	IC_{50} (mg L^{-1})
Atazanavir	0.78	0.39	0.63 (0.62–0.64)
Efavirenz	0.195	0.0975	0.52 (0.50–0.53)
Nevirapine	0.39	0.195	0.97 (0.95–0.98)

LOEC = lowest observed effect concentration; *NOEC*: no observed effect concentration.

Even when present in low concentrations, PPCPs have the ability to persist in the environment, which increases the possibility of the occurrence of chronic effects, since many aquatic species are continuously exposed to these pollutants for long time periods or throughout the life cycle [26,33,42]. According to Daouk et. al. [43], ARVs can be seen as pseudo-pollutants that are persistent in the environment because of their continuous release. It should be noted that studies involving toxicity assays with ARVs are rare, especially in marine organisms. Robson et al. [44] have demonstrated that acute (96 h) exposure of *Oreochromis mossambicus* to 20.6 ng L^{-1} of efavirenz resulted in liver damage and an overall decline in fish health, when compared to control fish. Ngumba et al. [45] calculated the *RQ* values of nevirapine for aquatic life and observed that the chemical presented potential ecotoxicological effects on algae, daphnia and fish. Considering this scenario, in the present study, ERA was performed with atazanavir, efavirenz and nevirapine.

3.3. *Environmental Risk Assessment (ERA)*

The ERA of pharmaceuticals released in the aquatic environment is very important to protect the environmental and public health. Hence, considering the worst-case scenario in accordance with the Technical Guidance Document on Risk Assessment of the European Union [46], an environmental risk assessment screening was conducted for atazanavir, efavirenz and nevirapine.

For Echinoderm, the *PNEC* was estimated from data obtained in the present study, specifically the acute and chronic toxicity assay with *Echinometra lucunter*. On the other hand, for algae, crustacean and fish, 100% of the acute and chronic *PNEC* were estimated using the ECOSAR program, as described in Table 5. It is noteworthy to mention that the lack of data regarding toxicity with marine organisms with ARVs led the authors to use toxicity data from freshwater species to calculate the *PNEC*, highlighting the importance of ecotoxicological studies with ARVs, especially with tropical marine organisms.

Table 5. Results from the Environmental Risk Assessment tests regarding the antiretroviral drugs Atazanavir, Efavirenz and Nevirapine.

Compound	*PEC* (µg L^{-1})	Trophic Level		Toxicity Data: Organisms/ Species	Toxicity Data: Endpoint	Concentrations (µg L^{-1})	*AF*	*PNEC* (µg L^{-1})	Reference	*RQ*
Atazanavir	6.4	Acute	Algae	Green algae (1)	96 h EC_{50}	1410.0	10,000	0.14	ECOSAR	45.39
			Crustacea	mysid (2)	96 h LC_{50}	3680.0		0.37	ECOSAR	17.39
			Fish	Fish (2)	96 h LC_{50}	44,100.0		4.41	ECOSAR	1.45
			Echinoderm	*Echinometra lucunter*	IC_{50}	73,040		7.30	This study	0.88
		Chronic	Algae	Green algae (1)	10^([log (*LOEC* × *NOEC*)]/2)	3370.0	100	33.70	ECOSAR	0.19
			Crustacea	mysid (2)		188.0		1.88	ECOSAR	3.40
			Fish	Fish (2)		8830.0		88.30	ECOSAR	0.07
			Echinoderm	*Echinometra lucunter*	*NOEC*	390.0		3.90	This study	1.64
Efavirenz	3.9	Acute	Algae	Green algae (1)	96 h EC_{50}	1510.0	10,000	0.15	ECOSAR	25.83
			Crustacea	mysid (2)	96 h LC_{50}	153.0		0.02	ECOSAR	254.90
			Fish	Fish (2)	96 h LC_{50}	1270.0		0.13	ECOSAR	30.71
			Echinoderm	*Echinometra lucunter*	IC_{50}	11,460.0		1.15	This study	3.40
		Chronic	Algae	Green algae (1)	10^([log (*LOEC* × *NOEC*)]/2)	686.0	100	6.86	ECOSAR	0.57
			Crustacea	mysid (2)		6.0		0.06	ECOSAR	65.00
			Fish	Fish (2)		715.0		7.15	ECOSAR	0.55
			Echinoderm	*Echinometra lucunter*	*NOEC*	97.5		0.98	This study	4.00

Table 5. *Cont.*

Compound	*PEC* (µg L^{-1})	Trophic Level		Toxicity Data: Organisms/ Species	Toxicity Data: Endpoint	Toxicity Data: Concentrations (µg L^{-1})	*AF*	*PNEC* (µg L^{-1})	Reference	*RQ*
Nevirapine	1.1	Acute	Algae	Green algae (1)	96 h EC_{50}	600.0	10,000	0.06	ECOSAR	18.33
			Crustacea	mysid (2)	96 h LC_{50}	365.0		0.04	ECOSAR	30.14
			Fish	Fish (2)	96 h LC_{50}	3170.0		0.32	ECOSAR	3.47
			Echinoderm	*Echinometra lucunter*	IC_{50}	84,610.0		8.46	This study	0.13
		Chronic	Algae	Green algae (1)	10^([log	884.0	100	8.84	ECOSAR	0.12
			Crustacea	Mysid (2)	(*LOEC* ×	7.5.0		0.08	ECOSAR	14.67
			Fish	Fish (1)	*NOEC*)]/2)	74.0		0.74	ECOSAR	1.49
			Echinoderm	*Echinometra lucunter*	*NOEC*	195.0		0.95	This study	0.56

Notes: *PEC* (predicted environmental concentration) (in µg L^{-1}); acute and chronic toxicity data: ((trophic level; freshwater or seawater organisms/species; endpoint and concentrations (µg L^{-1}); *AF*: assessment factor; *PNEC*: predicted no-effect concentration (µg L^{-1})) obtained from the ECOSAR program [33]. In the last column, risk quotients (*RQ*) for the acute and chronic tests (i.e., without risk, signalled in white; low risk, signalled in **green**; medium risk, signalled in **yellow**; and high risk, signalled in **red**) (for more details, see Section 2.2.3); freshwater (1); seawater (2); EC_{50}: 50% effective concentration; LC_{50}: 50% lethal concentration; *NOEC*: no observed effect concentration; and *LOEC*: lowest observed effect concentration; IC_{50} = average inhibitory concentration.

The results obtained revealed that atazanavir showed high risks of acute toxicity for green algae, mysid and fish, and moderate risk for *Echinometra lucunter* sea urchin. In the chronic assessment, atazanavir displayed low risk for fish and moderate/high risks for the other organisms included in this assessment. Efavirenz displayed high risks in acute toxicity for all organisms and medium to high risks in the chronic risk assessment. For nevirapine, a high risk was observed for almost all organisms in the acute assessment, and a high/moderate risk was observed in the chronic assessment.

The data presented here suggest that atazanavir, efavirenz and nevirapine are potentially dangerous in Santos Bay waters. Despite this, the hypothesis that the mixtures of these compounds produce synergistic effects, which may potentiate their individual effects, in the environment, cannot be ruled out, as the toxicity tests were performed with isolated compounds. Furthermore, human metabolites must also be considered for further ecotoxicological studies to assess the acute and chronic toxicity of these compounds. In fact, although most drugs can undergo biotransformation, they can still be excreted in the active form, and can accumulate in other organisms or in the environment [26,33,47].

4. Conclusions

In Brazil, the use of ARVs is widely disseminated among the HIV/AIDS population through the unified, universal and free-of-charge public health system (SUS). ARVs are also used as prevention tool against HIV infection through PEP. It is worth mentioning that the modern and efficient decentralized HIV testing capability of the Brazilian HIV/AIDS program results in an increasing number of HIV+ patients in Brazil's health system. Considering the high consumption of ARVs in Brazil and the main routes of excretion of these pharmaceuticals (urine and faeces), the *PEC* values of 13 ARVs in the surface waters of Santos Bay, Brazil, were estimated. The results revealed that all of the ARVs included in this study need to be assessed for their environmental fate and an analysis of their effects on aquatic organisms needs to be conducted, according to the limits established by the EMEA guidelines. In the present study, acute and chronic toxicity tests were performed with atazanavir, efavirenz and nevirapine to gametes and embryos of the sea urchin *Echinometra lucunter*, constituting one of the few studies carried out with ARVs in aquatic (marine) organisms. The ERA, which was also performed here, showed that atazanavir, efavirenz and nevirapine are potentially hazardous for aquatic life in Santos Bay waters, which raises concerns about the continuous introduction of ARVs in aquatic ecosystems. The data presented here may contribute to the provision of subsidies for the development of regular government programs to monitor ARVs levels in Effluent Treatment Stations, as well as to the development of solutions to reduce and/or eliminate the releasing of ARVs into the aquatic environment. Future research will focus on the development of new experiments, using different marine media such as algae, with a combination of the compounds included in this study as well as using the real concentrations found in the sea, in order to assess their effects on different sea organisms.

Author Contributions: Conceptualization, R.S.C.; Data curation, R.S.C.; Formal analysis, R.S.C. and V.R.; Investigation, R.S.C. and V.R.; Methodology, R.S.C.; Resources, R.S.C.; Software, R.S.C.; Supervision, L.L.G.; Validation, V.R., D.G.V., M.A.P.D., F.S.C., F.R.S., W.T., A.C. and L.L.G.; Visualization, V.R., D.G.V., M.A.P.D., F.S.C., F.R.S., W.T., A.C. and L.L.G.; Writing—original draft, R.S.C.; Writing—review and editing, V.R., D.G.V., M.A.P.D., F.S.C., F.R.S., W.T., A.C. and L.L.G. All authors have read and agreed to the published version of the manuscript.

Funding: A. Cesar thanks the National Council, known as CNPq (Conselho Nacional de Desenvolvimento Científico e Tecnológico, MEC—Brazil), for his research productivity fellowship (PQ#305869/2013-2). R. S. Cid thanks CAPES (Coordenação de Aperfeiçoamento de Pessoal de Nível Superior) for his research fellowship.

Institutional Review Board Statement: Not applicable.

Informed Consent Statement: Not applicable.

Data Availability Statement: The data used in this study are available in the manuscript.

Acknowledgments: The authors are grateful for the important comments and suggestions raised by the anonymous reviewers.

Conflicts of Interest: The authors declare no conflict of interest.

References

1. Blackburn, S.; Pelling, M.; Marques, C. Megacities and the Coast: Global Context and Scope for Transformation. In *Coasts and Estuaries the Future*; Elsevier: Amsterdam, The Netherlands, 2019; pp. 661–669.
2. United Nations—Department of Economic and Social Affairs—Population Division. *World Urbanization Prospects: The 2018 Revision (ST/ESA/SER.A/420)*; United Nations: New York, NY, USA, 2019; Volume 12, ISBN 9789211483192.
3. Oliveira, G.M.; Vidal, D.G.; Ferraz, M.P. Urban Lifestyles and Consumption Patterns. In *Sustainable Cities and Communities. Encyclopedia of the UN Sustainable Development Goals*; Filho, W.L., Azul, A.M., Brandli, L., Özuyar, P.G., Wall, T., Eds.; Springer Nature Switzerland AG: Cham, Switzerland, 2020; pp. 851–860, ISBN 978-3-319-71061-7.
4. Barragán, J.M.; de Andrés, M. Analysis and trends of the world's coastal cities and agglomerations. *Ocean Coast. Manag.* **2015**, *114*, 11–20. [CrossRef]
5. Dafouz, R.; Cáceres, N.; Rodríguez-Gil, J.L.; Mastroianni, N.; de Alda, M.L.; Barceló, D.; de Miguel, Á.G.; Valcárcel, Y. Does the presence of caffeine in the marine environment represent an environmental risk? A regional and global study. *Sci. Total Environ.* **2018**, *615*, 632–642. [CrossRef]
6. Desbiolles, F.; Malleret, L.; Tiliacos, C.; Wong-Wah-Chung, P.; Laffont-Schwob, I. Occurrence and ecotoxicological assessment of pharmaceuticals: Is there a risk for the Mediterranean aquatic environment? *Sci. Total Environ.* **2018**, *639*, 1334–1348. [CrossRef] [PubMed]
7. Ojemaye, C.Y.; Petrik, L. Pharmaceuticals in the marine environment: A review. *Environ. Rev.* **2019**, *27*, 151–165. [CrossRef]
8. Ferrando-Climent, L.; Reid, M.J.; Rodriguez-Mozaz, S.; Barceló, D.; Thomas, K.V. Identification of markers of cancer in urban sewage through the use of a suspect screening approach. *J. Pharm. Biomed. Anal.* **2016**, *129*, 571–580. [CrossRef] [PubMed]
9. Ibañez, M.; Borova, V.; Boix, C.; Aalizadeh, R.; Bade, R.; Thomaidis, N.; Hernández, F. UHPLC-QTOF MS screening of pharmaceuticals and their metabolites in treated wastewater samples from Athens. *J. Hazard. Mater.* **2017**, *323*, 26–35. [CrossRef]
10. Schoeman, C.; Dlamini, M.; Okonkwo, O. The impact of a Wastewater Treatment Works in Southern Gauteng, South Africa on efavirenz and nevirapine discharges into the aquatic environment. *Emerg. Contam.* **2017**, *3*, 95–106. [CrossRef]
11. UNAIDS Data e Trends: Global Statistics. Available online: https://www.hiv.gov/hiv-basics/overview/data-and-trends/global-statistics (accessed on 4 March 2021).
12. Domingues, C.-S.B.; Waldman, E.A. Causes of Death among People Living with AIDS in the Pre- and Post-HAART Eras in the City of São Paulo, Brazil. *PLoS ONE* **2014**, *9*, e114661. [CrossRef]
13. Ncube, S.; Madikizela, L.M.; Chimuka, L.; Nindi, M.M. Environmental fate and ecotoxicological effects of antiretrovirals: A current global status and future perspectives. *Water Res.* **2018**, *145*, 231–247. [CrossRef]
14. Ginat, D.T.; Schaefer, P.W. *Highly Active Antiretroviral Therapy (HAART)*; Ginat, D., Small, J., Schaefer, P., Eds.; Neuroimaging Pharmacopoeia; Springer: Cham, Switzerland, 2015.
15. K'Oreje, K.; Vergeynst, L.; Ombaka, D.; De Wispelaere, P.; Okoth, M.; Van Langenhove, H.; Demeestere, K. Occurrence patterns of pharmaceutical residues in wastewater, surface water and groundwater of Nairobi and Kisumu city, Kenya. *Chemosphere* **2016**, *149*, 238–244. [CrossRef]
16. Aminot, Y.; Litrico, X.; Chambolle, M.; Arnaud, C.; Pardon, P.; Budzinski, H. Erratum to: Development and application of a multi-residue method for the determination of 53 pharmaceuticals in water, sediment, and suspended solids using liquid chromatography-tandem mass spectrometry. *Anal. Bioanal. Chem.* **2015**, *407*, 8623. [CrossRef]

17. Rimayi, C.; Odusanya, D.; Weiss, J.; de Boer, J.; Chimuka, L. Contaminants of emerging concern in the Hartbeespoort Dam catchment and the uMngeni River estuary 2016 pollution incident, South Africa. *Sci. Total Environ.* **2018**, *627*, 1008–1017. [CrossRef] [PubMed]
18. Nannou, C.; Ofrydopoulou, A.; Evgenidou, E.; Heath, D.; Heath, E.; Lambropoulou, D. Antiviral drugs in aquatic environment and wastewater treatment plants: A review on occurrence, fate, removal and ecotoxicity. *Sci. Total Environ.* **2020**, *699*, 134322. [CrossRef] [PubMed]
19. Instituto Brasileiro de Geografia e Estatística (IBGE). Available online: https://cidades.ibge.gov.br/ (accessed on 30 August 2021).
20. Benzaken, A.S.; Pereira, G.F.M.; Costa, L.; Tanuri, A.; Santos, A.F.; Soares, M.A. Antiretroviral treatment, government policy and economy of HIV/AIDS in Brazil: Is it time for HIV cure in the country? *AIDS Res. Ther.* **2019**, *16*, 1–7. [CrossRef] [PubMed]
21. Brazil Lei, No. 313, de 13 de Novembro de 1996, Dispõe Sobre a Distribuição Gratuita de Medicamentos aos Portadores do HIV e Doentes de AIDS [Law No. 9313, of November 13, 1996, Provides for the Free Distribution of Medication to HIV Carriers and AIDS Patients]. Available online: https://www.planalto.gov.br/ccivil_03/leis/l9313.htm (accessed on 4 March 2021).
22. Brazil. Ministério da Saúde Clinical Protocol and Guidelines Therapeutics for Management of Infection by HIV in Adults. Available online: http://www.aids.gov.br/pt-br/profissionais-de-saude/hiv/protocolos-clinicos-e-manuais (accessed on 13 April 2021).
23. Abessa, D.M.D.S.; Rachid, B.R.D.F.; Moser, G.A.D.O.; Oliveira, A.J.F.C. Environmental effects of sewage oceanic disposal by submarine outfalls: A review. *Mundo O Da* **2012**, *36*, 643–661. [CrossRef]
24. Cetesb-Companhia Estadual de Tecnologia e Saneamento ambiental. *Relatório de Qualidade das Águas Costeiras no Estado de São Paulo 2018 [Report on the Quality of Coastal Waters in the State of São Paulo 2018]*; Cetesb: São Paulo, Brazil, 2018.
25. SABESP. Companhia de Saneamento Básico do Estado de São Paulo [Basic Sanitation Company of the State of São Paulo]. Available online: http://site.sabesp.com.br/ (accessed on 4 March 2021).
26. EMEA-European Medicines Agency, C. for M.P. for H. Use (CHMP). *Guideline on the Environmental Risk Assessment of Medicinal Products for Human Use: Doc. Ref.: EMEA/CHMP/SWP/4447/00 corr 1*; EMEA: London, UK, 2006.
27. Governo Federal Federal Government Website. Available online: http://www.acessoainformacao.gov.br (accessed on 13 March 2021).
28. ABNT Ecotoxicologia. *Aquática—Toxicidade Crônica de Curta Duração—Método de Ensaio com Ouriço-do-mar (Echinodermata: Echinoidea) [Aquatic Ecotoxicology—Short Term Chronic Toxicity—Sea Urchin Test Method]-NBR 15350*; ABNT: Rio de Janeiro, Brazil, 2012.
29. USEPA-United States Environmental Protection Agency Sperm Cell Toxicity Tests Using the Sea Urchin (*Arbacia punctulata*) EPA 833-C-09-001. Available online: https://www.epa.gov/sites/production/files/2015-09/documents/seaurchintesting.pdf (accessed on 1 March 2021).
30. Li, Y.-X.; Zhang, X.-L.; Li, W.; Lu, X.-F.; Liu, B.; Wang, J. The residues and environmental risks of multiple veterinary antibiotics in animal faeces. *Environ. Monit. Assess.* **2013**, *185*, 2211–2220. [CrossRef]
31. Thomaidi, V.S.; Stasinakis, A.S.; Borova, V.L.; Thomaidis, N. Is there a risk for the aquatic environment due to the existence of emerging organic contaminants in treated domestic wastewater? Greece as a case-study. *J. Hazard. Mater.* **2015**, *283*, 740–747. [CrossRef]
32. USEPA. *United States Environmental Protection Agency ECOTOX User Guide: Ecotoxicology Database System*; USEPA: Washington, DC, USA, 2019.
33. USEPA. *United States Environmental Protection Agency Ecological Structure-Activity Relationship Model (ECOSAR) Class Program*; USEPA: Washington, DC, USA, 2017.
34. De Bruijn, J.; Hansen, B.; Johansson, S.; Luotamo, M.; Munn, S.; Musset, C.; Olsen, S.; OlssoN, H.; Paya-Perez, A.; Pedersen, F.; et al. *Technical Guidance Document on risk Assessment*; Part 1. Part 2; European Commission: Brussels, Belgium, 2002.
35. European Chemicals Agency. *Guidance on Information Requirements and Chemical Safety Assessment Chapter R.10: Characterisation of Dose [Concentration]-Response for Environment*; Dictus Publishing: Chisinau, Moldova, 2008.
36. Hernando, M.D.; Mezcua, M.; Alba, A.R.F.; Barceló, D. Environmental risk assessment of pharmaceutical residues in wastewater effluents, surface waters and sediments. *Talanta* **2006**, *69*, 334–342. [CrossRef]
37. Escher, B.I.; Baumgartner, R.; Koller, M.; Treyer, K.; Lienert, J.; McArdell, C.S. Environmental toxicology and risk assessment of pharmaceuticals from hospital wastewater. *Water Res.* **2011**, *45*, 75–92. [CrossRef]
38. Minguez, L.; Pedelucq, J.; Farcy, E.; Ballandonne, C.; Budzinski, H.; Halm-Lemeille, M.-P. Toxicities of 48 pharmaceuticals and their freshwater and marine environmental assessment in northwestern France. *Environ. Sci. Pollut. Res.* **2016**, *23*, 4992–5001. [CrossRef]
39. Brazil. Ministério da Saúde Clinical Protocol and Therapeutic Guidelines for Post-Exposure Prophylaxis (PEP) for the Risk of HIV, STIs and Viral Hepatitis Infections. Available online: http://www.aids.gov.br/pt-br/pub/2015/protocolo-clinico-e-diretrizes-terapeuticas-para-profilaxia-pos-exposicao-pep-de-risco (accessed on 4 March 2021).
40. Funke, J.; Prasse, C.; Ternes, T.A. Identification of transformation products of antiviral drugs formed during biological wastewater treatment and their occurrence in the urban water cycle. *Water Res.* **2016**, *98*, 75–83. [CrossRef] [PubMed]
41. Giebułtowicz, J.; Tyski, S.; Wolinowska, R.; Grzybowska, W.; Zaręba, T.; Drobniewska, A.; Wroczyński, P.; Nałęcz-Jawecki, G. Occurrence of antimicrobial agents, drug-resistant bacteria, and genes in the sewage-impacted Vistula River (Poland). *Environ. Sci. Pollut. Res.* **2018**, *25*, 5788–5807. [CrossRef] [PubMed]

42. Jones, O.A.H.; Voulvoulis, N.; Lester, J.N. Ecotoxicity of pharmaceuticals. *Compr. Anal. Chem.* **2007**, *50*, 387–424.
43. Daouk, S.; Chèvre, N.; Vernaz, N.; Bonnabry, P.; Dayer, P.; Daali, Y.; Fleury-Souverain, S. Prioritization methodology for the monitoring of active pharmaceutical ingredients in hospital effluents. *J. Environ. Manag.* **2015**, *160*, 324–332. [CrossRef] [PubMed]
44. Robson, L.; Barnhoorn, I.; Wagenaar, I. The potential effects of efavirenz on Oreochromis mossambicus after acute exposure. *Environ. Toxicol. Pharmacol.* **2017**, *56*, 225–232. [CrossRef]
45. Ngumba, E.; Gachanja, A.; Tuhkanen, T. Occurrence of selected antibiotics and antiretroviral drugs in Nairobi River Basin, Kenya. *Sci. Total Environ.* **2016**, *539*, 206–213. [CrossRef]
46. European Commission. *Technical Guidance Document on Risk Assessment for Existing Substances, Part II*; European Commission: Brussels, Belgium, 2003.
47. Mompelat, S.; Le Bot, B.; Thomas, O. Occurrence and fate of pharmaceutical products and by-products, from resource to drinking water. *Environ. Int.* **2009**, *35*, 803–814. [CrossRef]

resources

MDPI

Article

Influence of Hydrologic Alteration on Sediment, Dissolved Load and Nutrient Downstream Transfer Continuity in a River: Example Lower Brda River Cascade Dams (Poland)

Dawid Szatten *, Michał Habel and Zygmunt Babiński

Institute of Geography, Kazimierz Wielki University, 85-064 Bydgoszcz, Poland; hydro.habel@ukw.edu.pl (M.H.); zygmunt.babinski@ukw.edu.pl (Z.B.)

* Correspondence: szatten@ukw.edu.pl; Tel.: +48-52-349-62-50

Abstract: Hydrologic alternation of river systems is an essential factor of human activity. Cascade-dammed waters are characterized by the disturbed outflow of material from the catchment. Changes in sediment, dissolved load and nutrient balance are among the base indicators of water resource monitoring. This research was based on the use of hydrological and water quality data (1984–2017) and the Indicators of Hydrologic Alteration (IHA) method to determine the influence of river regime changes on downstream transfer continuity of sediments and nutrients in the example of the Lower Brda river cascade dams (Poland). Two types of regimes were used: hydropeaking (1984–2000) and run–of–river (2001–2017). Using the IHA method and water quality data, a qualitative and quantitative relationship were demonstrated between changes of regime operation and sediment and nutrient balance. The use of sites above and below the cascade made it possible to determine sediment, dissolved load, and nutrient trapping and removing processes. Studies have shown that changes in operation regime influenced the supply chain and continuity of sediment and nutrient transport in cascade-dammed rivers. The conducted research showed that sustainable management of sediment and nutrient in the alternated catchment helps achieve good ecological status of the water.

Keywords: Indicators of Hydrologic Alteration (IHA); cascade dams; river regime; suspended sediment transport; dissolved load transport; nutrient transport; Brda river

Citation: Szatten, D.; Habel, M.; Babiński, Z. Influence of Hydrologic Alteration on Sediment, Dissolved Load and Nutrient Downstream Transfer Continuity in a River: Example Lower Brda River Cascade Dams (Poland). *Resources* **2021**, *10*, 70. https://doi.org/10.3390/resources10070070

Academic Editors: Diego Copetti and Manolis G. Grillakis

Received: 6 May 2021
Accepted: 30 June 2021
Published: 1 July 2021

1. Introduction

The hydrological cycle is limited by natural and anthropogenic factors. The first group includes, e.g., atmospheric circulation, precipitation, geological structure and soil erosion among others; the second group includes modifying river systems for flood control, water supply, irrigation, land cover changes and electricity production, among others. The factors listed in both groups are directly or indirectly influenced by human activities, which additionally modify the hydrological cycle. Water flow results from the phenomena and processes taking place in the catchment area and is responsible for the nature of fluvial processes, sediment dynamics, and bedforms forming in the channel [1]. Understanding the river regime is key to defining climate change as a human impact [2,3].

The regime of inland waters on a regional and global scale regulates the transport of sediments and nutrients from land to the ocean [4]. Climate change is reflected in the global balance of sediment supply to the world's oceans [5]. Nutrients are one of the main components of soil, water, sediments and organisms [6], especially phosphorus, which is considered a key factor of eutrophication [7]. Research by Allan [8] indicated that anthropogenic eutrophication processes are related to agricultural area development, especially with respect to phosphorus supply to reservoirs [9]. Water pollution can be expressed by the processes of sorption and desorption of nutrients from sediments [10]. Revenga et al. [11] showed that 60% of the world's rivers are fragmented by hydrological alteration and have lost their hydromorphological and ecological continuity [12]. Belletti et al. [13] pointed out there are at least 1.2 million instream barriers in 36 European countries. However, the appearance of large dams and their reservoirs increased sediment retention on the terrestrial part of the hydrosphere, with estimated retention of suspended particles as 50% of the load that would have been transported to the oceans [4,14]. The number of large dams in the world has been estimated at 58,700 [15]. The impact of dams is reflected in many aspects: hydrological, morphological and ecological, among others. The change concerns not only the commission of the reservoir itself, i.e., the increase of the water surface in the catchment area, but also has impacts above and below, often over many kilometers. Dams influence changes in the timing and frequency of high and low flows [16]. In many cases, this is due to hydropower plant regimes, which should preserve ecological base flow (environmental flow). In this respect, participation is a significant renewable energy source (RES) in overall energy consumption. Poland should reach 15% RES by 2020 [17], and in 2016 the share of RES in total energy consumption reached 13.5%, including 2% from hydropower plants [18]. Transformation of river regimes below dams has been studies by Wiliams and Wolman [19], Vörösmarty et al. [20] and Obodovskyi et al. [21]. The aggradational or degradational state of a river system influences channel morphology and substrate textures of the bottom channel below dams [22]. Research on the dynamics of sediments transport includes that by Van Rijn [23], Kondolf [24] and Vörösmarty et al. [4], also taking into account research in our study area, the Brda River catchment [25]. Dams reduce the natural connectivity of the fluvial system [26], and the natural regime of rivers is changed [27]. This reflects on the sediment budget, where accumulation processes are dominant. For example, a 90% sediment load reduction was observed after the Akosombo Dam construction on the Volta River in West Africa [28], a 45% reduction of sediments in the Sulejów Reservoir [29], sediment accumulation estimated at 66% on the Danube River in Serbia [30], and accumulation of suspended sediments by the Wloclawek dam (Vistula River in Poland) estimated by Babiński [31] at 42%. Changes in river regime caused by reservoir function is reflected by nutrient trapping [32] or altering the original nutrient biogeochemical cycle [33].

On the other hand, environmental flow is guaranteed only for a small number of world rivers [34]. The Water Framework Directive (WFD) [35] includes a hydromorphological assessment. Achieving the requirement of WFD good status of the water is possible for hydromorphological, biological and chemical elements.

The amount and components of nutrients transported by rivers play an essential role in maintaining the ecosystem of downstream rivers [36]. The river regime is reflected in the ecological health of a river [37] and influences the biodiversity of the river system [38]. Disturbed by barriers of various types, sediment and nutrient transport has severe consequences for downstream ecosystems, including sensitive wetlands and delta systems [39]. According to Harrison et al. [40], much nitrogen removal takes place in reservoirs, and according to Bosch [41], reservoir sediments are effective at trapping phosphorus. Despite these general trends, single dams may temporarily serve as sources of nutrients [42]. Thus, the functioning of dams affects the condition of the habitats of aquatic organisms (e.g., living in a river or a reservoir) and organisms dependent on the water (e.g., living on a floodplain or in oxbow lakes). On the other hand, river flow regulation favors some fish species that spawn during a specific period [43]. This state is widely used by humans, introducing fishing on artificial reservoirs and rivers under their influence.

The seasonal characteristics of streamflow at the global scale were recognized by Dettinger and Diaz [44]. Many studies have used IHA to investigate the impact of commissioned dams or cascades of reservoirs on river regime [14,45–53], including the hydrological impact of the dam operation [54,55]. IHA was also used to predict future conditions in the river regime during climate change [56,57] and to detect urbanization impacts [58].

The main aim of this research was to demonstrate the impact of changes in the operating regime of the Lower Brda Cascade dams (LBC) to quantify both the trapping and removal of suspended sediment, dissolved load, and nutrients. Data included (a) daily discharge, concentration, and load of (b) suspended sediment, (c) dissolved load and (d) nutrients (nitrogen and phosphorus), measured at the positions above and below reservoir cascade. The main research objectives related to: (i) the presence of dams changing the regime above and below the dam, conditioned by the dam operating system; (ii) equalization of water outflow from the cascade of reservoirs affecting the dynamics of suspended sediment, dissolved load, and nutrients; (iii) disruptions of the continuum of sediment and nutrient transport below the water stages resulting in increased sediment accumulation in LBC reservoirs and limitation of nutrients responsible for primary production. The IHA method of hydrological data (1984–2017) was used for quantitative analysis of sediment and nutrient concentration, and calculated loads of the analyzed catchments. The conducted research is essential not only from the point of view of the future water management in the cascade of hydropower plants, but also with respect to river regime, sediment and nutrient data for long-term observation of environmental pressures.

2. Materials and Methods

2.1. Study Area

The Brda River is 245 km long and is a left-bank tributary of the Vistula River—the longest river in Poland, forming the largest basin in the Baltic Sea catchment area (Figure 1). The Brda catchment area amounts to 4661 km^2 [59]. The equal river regime of the catchment area above the LBC is influenced by the river's functioning in its river-lake system. The Brda flows through 16. lakes with a capacity from 0.7 million m^3 to 134.5 million m^3 (the largest is Lake Charzykowskie) [25]. The Brda valley's geomorphological evolution is related to the accumulation of outwash sands in the younger phases of the Pomeranian stage of the last glaciation [60,61]. The moraine uplands surrounding the Brda River valley are built of postglacial clays separated by different-grained sands. In the study area, several postglacial forms occur, e.g., terminal moraines, kames and subglacial channels [25].

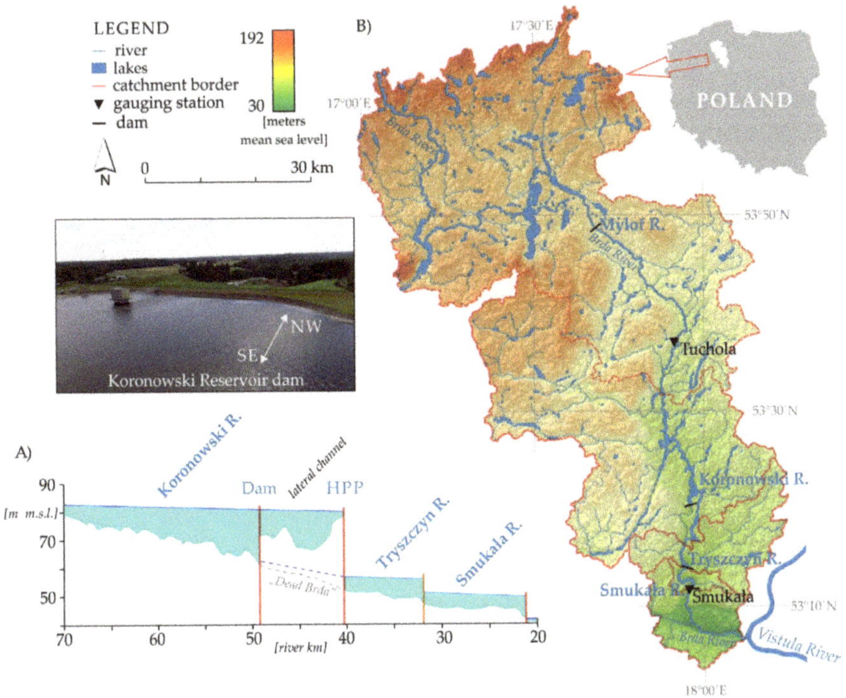

Figure 1. Sketch of the study area. (A) Longitudinal profile of LBC (B) on the background of the Brda catchment area.

The observation sites for hydrological conditions for the period 1984–2017 were water gauge stations of the Institute of Meteorology and Water Management National Research Institute in Tuchola (85.9 km) and Smukała (20.1 km) (Figure 1). The average annual flow in the first of these is 18.85 m^3s^{-1}, and in Smukała, it is 26.19 m^3s^{-1}. The Brda River is characterized by the lowest variability of flows among all Polish rivers [62]. The mean annual flow irregularity coefficients (C.V.) for the observation sites are 0.27 and 0.35, respectively. In the annual course, the regularity of high water flows of the Brda River in the winter period (spring snowmelt supply combined with ground supply), and low water flows in the summer period, are noticeable. This course is conditioned by the existence of a significant underground supply, the share of which in the total outflow from the Brda catchment area is estimated at 85% [63]. The functioning of the ecosystems of the Koronowski, Tryszczyn, and Smukała reservoirs is determined primarily by the size and variability of the flows of the Brda River, and less by the direct tributaries: Kamionka, Sępolna, Krówka, Kręgiel, Kotomierzyca (approximately 15% of the total inflow).

The climatic conditions of the study area were characterized for the period 1971–2000. The average annual sum of precipitation for the study area is 550 mm [64], the average annual temperature is 7.5 °C, of which the average annual temperature for the winter is 0.5 °C, and for summer is 14.0 °C [65].

The dominant land cover in the Brda catchment area, defined by Szatten and Habel [66] based on the Corine Land Cover (CLC) database, is forest and intensively used arable land covering 46.9% and 36.6% of the area, respectively. Cover types of arable (9.1%), anthropogenic areas (4.2%) and surface waters (3.2%) occupy a much smaller area. A characteristic feature of the Brda catchment area is a change in the dominant type of land cover with an increase in the catchment area, i.e., a decrease in the forest area and increase of intensively used arable land and the anthropogenic regions [66].

The Brda River is strongly modified by hydrotechnical structures. In 1848, in the upper part of the catchment, the first dam with a reservoir in Mylof (km 138.0) (Figure 1) [67] and weirs in the estuary section of the river were built [68]. In the 1960s and 1970s, a cascade was built in the lower part of the Brda River (LBC) with the Koronowski, Tryszczyn, and Smukała reservoirs. The basic characteristics of the LBC reservoirs are presented in Figure 1 and Table 1.

Table 1. Characteristics of the reservoirs included in the LBC.

Reservoir	Dam Localization [River km]	Commission Year	Water Damming [m]	Area [km^2]	Capacity [Million m^3]	Hydraulic Resistance Time [$Year^{-1}$]
Koronowski	49.1 * 40.1 **	1960	24.7	14.357	81.0	8.07
Tryszczyn	31.5	1962	4.6	0.87	1.8	0.16
Smukała	22.3	1951	7.5	0.94	2.2	0.16

Explanations: * dam, ** hydropower plant.

LBC performs three primary functions: energy production, flood prevention and recreation. The turbines in the hydropower plants allow for a total average annual electricity production of 60.32 GWh. In the case of the Tryszczyn and Smukała reservoirs, the hydropower plants are located on the dam. In the Koronowski Reservoir, the hydropower plant is located about 10 km below the dam, to which the waters are led through a lateral canal running at the bottom of the flooded glacial gutter of the Lipkusz—Białe lakes (Figure 1). This fact affects the functioning of the Brda section between the dam and the hydropower plant of the Koronowski Reservoir, called the "Dead Brda", due to the limitation of water flow from the average annual flow rate from before the dam was built, at a level of about 21.6 m^3s^{-1}, to a resulting level of about 2.0 m^3s^{-1} [69].

The LBC operating regime has changed twice (Figure 2) since its start. During regime I, called the hydropeaking operation from January 1951 (when the first dam was commissioned) to 2000, the powerplant was put into operation only in the highest electricity demand periods. This operation of the cascade of dams caused an impact on the temporary increase in flow and water stage both in the reservoirs and below the hydropower plant (Figure 2). The maximum amplitudes of the flows reached 50 $m^3 s^{-1}$ [70]. These fluctuations were recorded especially on weekdays (Monday–Friday). This intensity decreased during days off (Saturday–Sunday) when reservoir retention dominated over energy production outflow. The highest amplitudes of the water flows occurred during average flows. During this time, water was flushed from the reservoirs twice a day. The hydropower plants worked in the run-of-river operation for the remainder of the day, maintaining the environmental flow at 15 $m^3 s^{-1}$. In this way, the LBC hydropower plants operated for about 70% of days in a year. The second regime implemented in 2001, called run–of–river operation, was characterized by a different system of dam operation [71]. Its introduction was related to the implementation of proecological provisions of the Water Law Act [72]. Under this operating regime, hydropower plants were operated in run–of–river, taking into account the environmental flow of 20 $m^3 s^{-1}$. The operation scheme in this regime consisted of "working up" the water flowing into the LBC through the first hydropower plant at the Koronowski Reservoir dam. Next, the hydropower plant's turbines on the Tryszczyn dam, and then in Smukała, were put into operation,. This way, water level fluctuations in these reservoirs were eliminated, which, according to the research by Szatten [73], oscillated in the range of up to 0.3 m. For comparison, during the operation of the hydropower plant in the hydropeaking regime (before 2000), water fluctuations in the reservoirs were as high as 0.5 m (Figure 2).

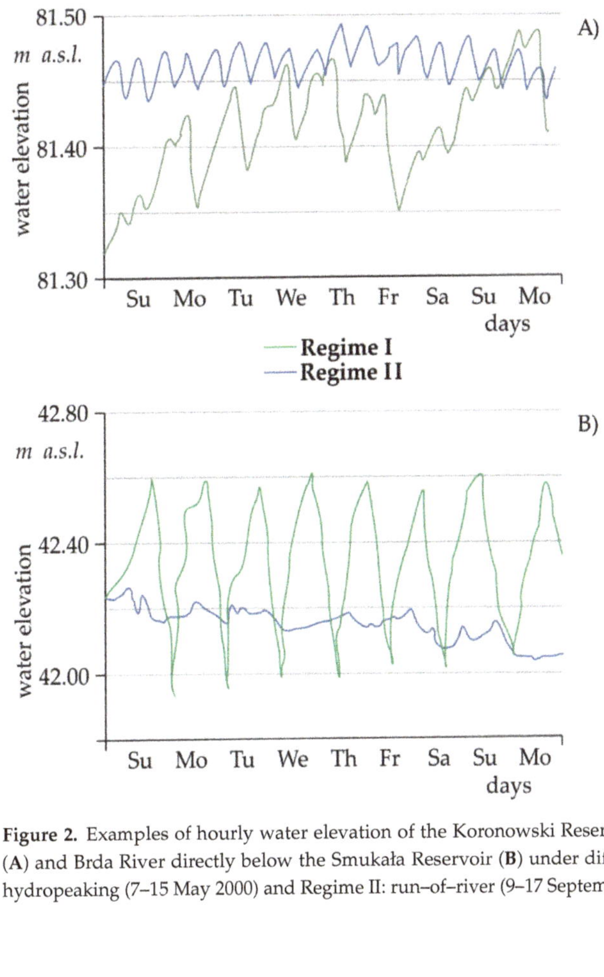

Figure 2. Examples of hourly water elevation of the Koronowski Reservoir on lacustrine section (**A**) and Brda River directly below the Smukała Reservoir (**B**) under different regimes. Regime I: hydropeaking (7–15 May 2000) and Regime II: run–of–river (9–17 September 2015).

2.2. *Input Data*

2.2.1. Hydrological Data

Hydrological data was the daily discharge (in $m^3 s^{-1}$), obtained from the public database of the Institute of Meteorology and Water Management National Research Institute in Warsaw (IMWM—NRI) for the years 1984–2017 for the measurement stations located above and below the LBC, i.e., Tuchola (km 85.9) and Smukała (km 20.1). The location of the measurement stations is shown in Figure 1. The hydrological data for IHA analysis were used covering the range from the beginning of observations (the 1980s) at the Tuchola and Smukała sites to 2017. An equal period (16-years) of the operation of the LBC under regime I (1984–2000) and regime II (2001–2017) was specified.

2.2.2. Indicators of Hydrologic Alteration

To determine changes in the hydrological regime resulting from LBC functioning, the Indicators of Hydrologic Alteration (IHA) method proposed by Richter et al. [74] was used. The characteristics of the Brda river flow from hydrological years 1984–2000 were compared in the gauging station located upstream of the reservoirs in Tuchola with the flow characteristics in the Smukała hydrological stations situated directly downstream of the dam. The assessment of the impact of changing the hydroelectric power plant's operation method from the hydropeaking regime to the run–of–river system was analyzed on a series of mean daily water flow values from the hydrological years 2001–2017 in Tuchola and Smukała.

Calculation were conducted with IHA software developed by The Nature Conservancy (Arlington, VA, U.S.) (version 7.1.0) and according to commonly used methodology: (i) defining data series described in Section 2.2.1; (ii) calculating values of hydrologic attributes for the hydropeaking operation data series (1984–2000) and the run–of–river operation data series (2001–2017); (iii) computing interannual statistics; and (iv) calculating values for IHA between two operation data series. The result is a list of 33 hydrologic parameters with their characteristics, in five statistics groups: (1) the magnitude of monthly water conditions, (2) magnitude and duration of annual extreme water conditions, (3) timing of annual extreme water conditions, (4) frequency and duration of high and low pulses and (5) rate and frequency of water condition changes, described in detail by Richter et al. [74].

The Range Variability Approach (RVA) method was used to determine the variability of IHA parameters [75]. For nonparametric data, the RVA breakpoints were set accordingly: low category $\leq$33rd percentile, <33rd middle category $\leq$67th, and >67th high category. Hydrologic alternation (HA) for analyzed parameters was calculated using Equation (1). HA is equal to zero when the observed frequency (OF) of post impact annual values falling with the RVA target range equals the expected frequency (EF). A positive deviation indicates that annual parameter values fell inside the RVA target window more often than expected. Negative values indicate that annual values fell within the RVA target window less often than expected [76].

$$HA = \frac{(OF - EF)}{EF} \times 100 \tag{1}$$

The transformation coefficient parameter in the range of | 0.0–0.33 | means no change or a low discharge transformation; the range | 0.34–0.67 | means a moderate transformation; and the range | 0.68–1.0 | a large alternation [76].

2.2.3. Suspended Sediment, Dissolved Load and Nutrient Data

The analysis of the transformation of the concentration of river sediment and nutrients was made possible by monthly measurements of suspended sediment concentration (SSC in mg L^{-1}), dissolved load concentration (DLC in mg L^{-1}), total nitrogen (TN in mgN L^{-1}), and total phosphorus (TP in mgP L^{-1}) at the sites of Tuchola and Smukała (Figure 1) in the period 1991–2012. The measurements were carried out as part of the State Environmental Monitoring Program (SEM). The data analysis period used for qualitative analyzes differed from the hydrological data period due to the change in the SEM. After considering WFD in Poland, the number of measurement stations and the frequency of sampling was limited. In research, a shorter, uniform data quality period was used. The suspended sediment concentration was determined following the standard in [77] using the traditional filtration method, filters with a porosity of 0.45 μm and a drying procedure. Sampling was done with a slowly filling bathometer according to commonly used methods [78]. Determination of the dissolved load concentration was carried out by the weight method, according to the procedure described by Hermanowicz et al. [79]. The concentration of nutrients was determined using generally used methods, including chemiluminescent detection for total nitrogen [80] and a spectrometry method for total phosphorus [81]. Using the Redfield relationship [82], the dominant nutrient primary production source was determined.

Balance (β) of suspended sediment load (SSL), dissolved load (DL), and nutrient loads (TNL, TPL) were calculated according to Equation (2) as material delivered to (D, in tons) and flushed off (F, in tons) the cascade reservoirs [83]:

$$\beta = \frac{D}{F}. \tag{2}$$

3. Results

3.1. Impact of LBC Regime Change on the Hydrologic Alternation

The IHA model results indicate that the mean annual flow in Tuchola increased from 18.57 $m^3\ s^{-1}$ (1984–2000) to 19.14 $m^3\ s^{-1}$ (2001–2017). At the Smukała site for the same periods, the mean annual flows decreased from 26.46 $m^3\ s^{-1}$ to 25.93 $m^3\ s^{-1}$, respectively. Annual C.V. for the above gauge stations was 0.25 and 0.29 (Tuchola) and 0.38 and 0.31 (Smukała), respectively. The calculated flood-free season had higher values for the Tuchola water gauge station (47 days for the period 1984–2000 and 30 days for the period 2001–2017) than for the Smukała site, where the number of days was 5 and 23, respectively.

Hydrologic alternation (HA) for indicators from the first group for Tuchola, assume values indicating low discharge transformation in most of the months. They oscillated in the range from 0.188 (November) to 0.313 (June) (Figure 3). Only for the months of July–August, HA values indicated moderate transformation. The situation was different for Smukała, where for the period up to 2000, the transformation of the flow reached moderate values in almost all months. On the other hand, for the period after 2001, for most of the year (except for summer months), the flow transformation was classified as low.

The HA parameters from the second group mostly assume a range of low regime transformation for both sites (Figure 3). The Base Flow Index (BFI) had relatively low values (0.56–0.61); however, there was a noticeable decrease in BFI values on the Smukała site for the second analyzed period (Table 2). There were no zero-flow days in the period 1984–2017. For the Tuchola station, the values of minimum flows (Table 2) in the first analyzed period fluctuated in the range from 11.00 $m^3\ s^{-1}$ (1-day) to 13.92 $m^3\ s^{-1}$ (90-days). In the second period, they slightly increased and assumed values ranging from 12.00 $m^3\ s^{-1}$ (1-day) to 14.46 $m^3\ s^{-1}$ (90-day). The same upward trend was characteristic for maximum flows (Table 2). On the other hand, for the Smukała position, for the first analyzed period, the minimum flows: 1-day, 3-day, and 7-day were higher than the second analyzed period. One-day and three-day maximum flows followed the same trend. Long-term flow parameters (30-day, 90-day minimum, and 90-day maximum) assumed higher values for the second research period (Table 2).

In the case of HA parameters belonging to the third group, the acceleration of the date of minimum and date of maximum discharges was noticeable. For the Tuchola station, the occurrence of minimum and maximum flows in the second period was accelerated by about 25 days, indicating the end of April and the end of November, respectively. The same tendency between the analyzed periods was characteristic of the Smukała station. However, the date of minimum discharge acceleration was 19 days (beginning of May), while the date of maximum discharge was 47 days (end of December) (Table 2).

The fourth group's parameters were characterized by the highest values of the transformation of flows from all analyzed indicators (Figure 3). In the case of the first analyzed period, for the Tuchola site, the number of low and high pulses were determined at seven and eight in the hydrological year, respectively. At the position below the LBC, their numbers significantly increased, amounting to 33 and 24. In the second studied period, the difference between the number and duration of low and high pulses was much lower (Table 2). However, the classification of HA parameters still indicated a significant transformation of the river regime (Figure 3).

The last fifth group of the parameters was characterized by internal differentiation, indicating a low transformation of the regime for the position of Tuchola and a high transformation for the position of Smukała (Figure 3). For the first of them, the annual discharge number of reversals was 69 for both analyzed periods, indicating an aligned rise and fall rate of discharges. On the other hand, for the site in Smukała, this parameter for the two tested periods had values of 166 and 129, respectively, with less equal values of rising and falling flow rates (Table 2).

Table 2. Calculated values for 33 indicators of Hydrologic Alteration (IHA) for LBC under a hydropeaking (1984–2000) and run-of-river (2001–2017) regime periods.

	Parameter	**Tuchola**		**Smukała**	
		1984–2000	**2001–2017**	**1984–2000**	**2001–2017**
1.1	November mean flow	20.0	20.0	25.0	25.0
1.2	December mean flow	21.0	21.0	26.0	26.0
1.3	January mean flow	21.0	21.0	29.0	30.0
1.4	February mean flow	21.0	22.0	30.0	30.0
1.5	March mean flow	21.0	22.0	31.0	31.0
1.6	April mean flow	23.0	21.0	32.0	30.0
1.7	May mean flow	18.0	19.0	26.0	25.0
1.8	June mean flow	16.0	15.0	21.5	19.0
1.9	July mean flow	13.0	15.0	18.0	21.0
1.10	August mean flow	13.0	14.0	19.0	19.0
1.11	September mean flow	16.5	16.0	21.5	20.5
1.12	October mean flow	19.0	17.0	24.0	23.0
2.1	1-day minimum	11.0	12.0	14.0	14.0
2.2	3-day minimum	11.0	12.0	15.3	14.3
2.3	7-day minimum	11.1	12.3	16.0	15.0
2.4	30-day minimum	12.0	13.5	17.0	17.1
2.5	90-day minimum	13.9	14.5	18.9	19.0
2.6	1-day maximum	27.0	30.0	48.0	44.0
2.7	3-day maximum	27.0	29.3	46.0	41.7
2.8	7-day maximum	26.6	28.0	38.9	39.4
2.9	30-day maximum	25.3	25.7	36.5	36.4
2.10	90-day maximum	22.8	23.4	32.3	32.9
2.11	Number of zero days	0	0	0	0
2.12	Base flow index	0.609	0.616	0.605	0.562
3.1	Date of minimum	198	172	207	188
3.2	Date of maximum	53	30	96	49
4.1	Low pulse count	7.0	5.0	33.0	9.0
4.2	Low pulse duration	4.0	3.5	2.0	2.0
4.3	High pulse count	8.0	6.0	24.0	10.0
4.4	High pulse duration	3.0	6.0	2.5	3.0
5.1	Rise rate	1.0	1.0	3.0	1.0
5.2	Fall rate	−1.0	−1.0	−2.5	−1.0
5.3	Number of reversals	69.0	69.0	166.0	129.0

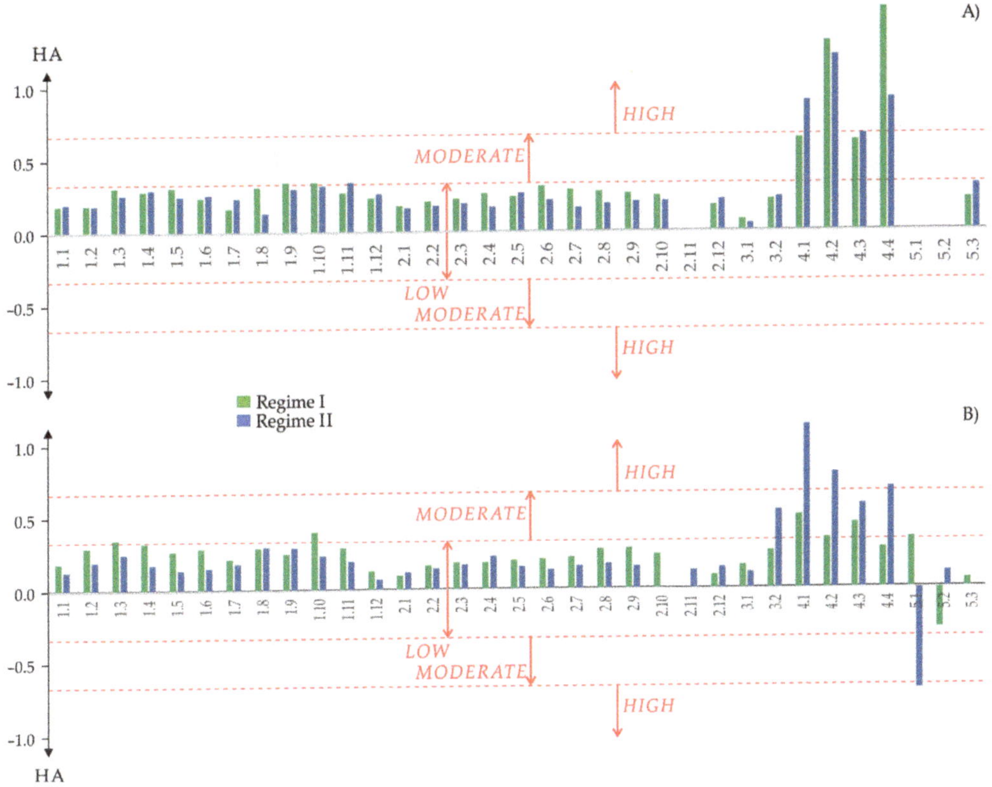

Figure 3. Comparison of changes in the value of hydrologic alteration (HA) for 33 Indicators of Hydrologic Alteration (IHA) for Tuchola (**A**) and Smukała (**B**) stations using RVA target range for a hydropeaking (1984–2000) and run-of-river (2001–2017) regime periods. Explanations for parameters 1.1–5.3 in Table 2.

3.2. Link between Sediment, Dissolved Load, Nutrient Transport and Flow Regime on LBC

One of the simplest analyses of the influence of damming structures on the trapping of suspended sediment is the comparison of inflow and outflow loads. In our work, we additionally analyzed the impact of two different regimes of hydropower operation on the regulation of sediment and nutrient transport downstream. The average value of suspended sediment concentration (SSC) for the period 1991–2000 (regime I), compared to Tuchola 5.39 mg L^{-1} and Smukała 4.56 mg L^{-1}, showed a reduction of 15.4%. In the second period (2001–2012), when the cascade was run-of-river (regime II) and the average concentration of suspended sediments in Tuchola was 3.66 mg L^{-1} and 2.21 mg L^{-1} in Smukała, we observed a reduction of SSC caused by LBC of about 39.6%. Comparing the average DSC value for the period 1991–2000 (regime I), the value in Tuchola was 210.62 mg L^{-1}, and in Smukała was 208.55 mg L^{-1}, which is only about 1% reduction in the concentration of dissolved load in the water. In the second period (2001–2012), when the cascade operated in run-of-river (regime II), the average DSC value in Tuchola was 234.99 mg L^{-1}, and that in Smukała 232.07 mg L^{-1}. We observed a slightly higher concentration reduction caused by LBC of about 1.25%. From the comparison of the averaged TN value for the period 1991–2000 (regime I), the value in Tuchola was 0.92 mgN L^{-1}, and in Smukała 1.03 mgN L^{-1}, i.e., a nitrogen enrichment by about 12%. In the second period (2001–2012), when the cascade operated in run-of-river (regime II) and the average TN value in Tuchola was 1.03 mgN L^{-1}, and in Smukała 1.20 mgN L^{-1}, we observed even more significant nitrogen enrichment of waters than in the previous period, i.e., a concentration reduction caused by LBC of about 16.5%. For the second analyzed nutrient, i.e., TP for the period 1991–2000 (regime I), the value in Tuchola was 0.17 mgP L^{-1}, and that in Smukała 0.14 mgP L^{-1}, which reduced phosphorus by about 18%. In the second period (2001–2012), when the cascade operated in run–of–river (regime II) and the average TN value in Tuchola was 0.14 mgP L^{-1}, and in Smukała 0.10 mgP L^{-1}, we observed an even more evident reduction of phosphorus in the water than in the previous period, the concentration reduction caused by LBC being about 28.5%. It should be mentioned that the general tendency to reduce SSC, DSC, and TP in the Brda waters passing through the cascade reservoirs in both regimes (I and II) could have been even more remarkable because the main reservoir is supplied by water directly from the catchment area of several small tributaries, i.e., Kamionka, Sępolna and Krówka. The inverse relationship should be taken into account in the case of enrichment of Brda waters in TN.

Another relationship that distinguishes the LBC river regimes was noted in the case of the occurrence of maximum SSC values in individual years at the site in Smukała (Figure 4). In the case of the cascade operation in regime I, slightly higher SSC maxima were achieved (average of the annual maximum was 9.3 mg L^{-1}), compared to the run–of–river operation (regime II) (average of the annual maximum was 7.78 mg L^{-1}). This relationship is analogous for the DSC indicator, i.e., in regime I, a DSC of 280.6 mg L^{-1}, and in regime II a DSC of 277 mg L^{-1}. In the case of nutrients, the change of the hydropeaking to run–of–river regime created better conditions for enriching the waters with nitrogen. In the periods of its maximum concentration, more favorable conditions for increasing nitrogen in the waters occurred, i.e., in regime I a TN 1.68 mgN L^{-1}; in regime II a TN 2.04 mgN L^{-1}. Additionally, if we compared the mean maximum TN concentrations at the station above the cascade with those below the cascade, it follows that during the period of regime II operation, the increase in mean values calculated from the maximum TN concentrations was higher by 4%. In the case of phosphorus, with the transition from regime I to regime II, the conditions for its reduction slightly improved in that TP decreased from 0.24 mgP L^{-1} to 0.21 mgP L^{-1}.

To demonstrate the change in the LBC's operating regime of regulating the suspended sediment concentration (SSC), dissolved load concentration (DLC), total nitrogen (TN) and total phosphorus (TP), the results of the indicators mentioned above in the form of monthly values in the years 1991–2000 (regime I) and 2001–2012 (regime II) were compared to the value of water flows (Figure 4). Concerning the SSC, a trend line change after the transition from hydropeaking to the run–of–river regime appeared to be evident from all four analyzed indicators on the site in Smukała (Figure 4A); a decrease in the suspended sediment concentration was observed as the flow rate increases for the run–of–river regime. The opposite trend occurred when the dams operated in a hydropeaking regime. For comparison, at the site located above the LBC, beyond the range of the reservoirs' impact, an increase in flows was recorded, and a decrease in the suspended sediment concentration was recorded, both in the first and the second period. In the case of DSC, during both regime I and regime II, a similar relation with the water flow was observed, i.e., with the increase of the flow rate, the dissolved load concentration increased (Figure 4B). The course of TN concentrations indicated that in both observed sites there was an increase in the average concentration of the indicator between the analyzed periods. In the case of TN, both during the operation of regime I and regime II, a similar relation to water flow was observed, i.e., with the increase of the flow rate, there was an increase in total nitrogen concentration in the Brda river waters (Figure 4C). In the relation of TP to flows, we observed an opposite relationship to that in the case of TN, and this decrease was even more pronounced during the operation of regime II (Figure 4D).

The change in the LBC operating regime to regulate the transport of SSC, DSC, TN, and TP were determined, as well as the periods of the occurrence of the extreme monthly mean values of the indicators mentioned above on the base of average flow rates for stations above and below the reservoir cascade. The results showed eight pattern loops (Figure 5).

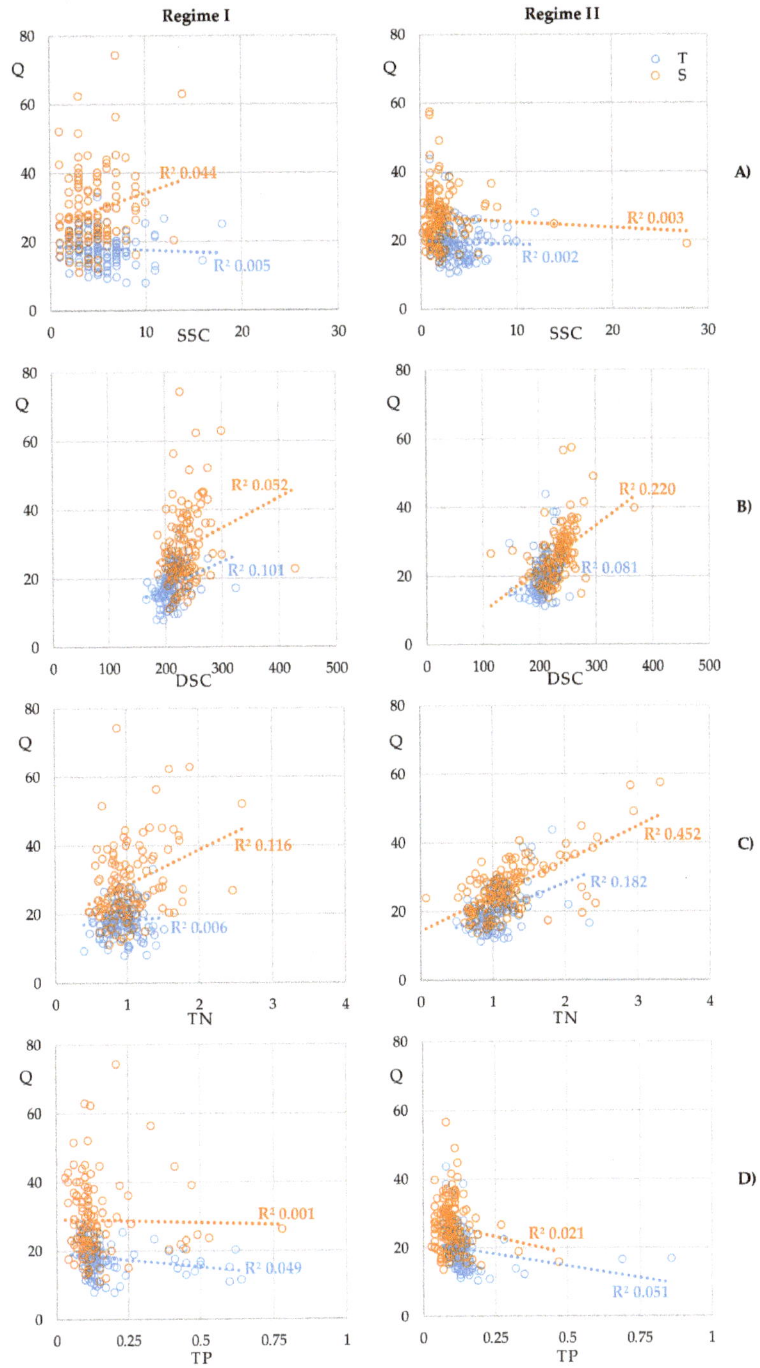

Figure 4. The relationship between water discharge (Q, in $m^3\ s^{-1}$) with: (**A**) suspended sediment concentration (SSC, in mg L^{-1}), (**B**) dissolved load concentration (DLC, in mg L^{-1}), (**C**) total nitrogen (mgN L^{-1}), (**D**) total phosphorus (mgP L^{-1}) at the stations Tuchola (T) and Smukała (S) during Regime I (hydropeaking) and Regime II (run–of–river).

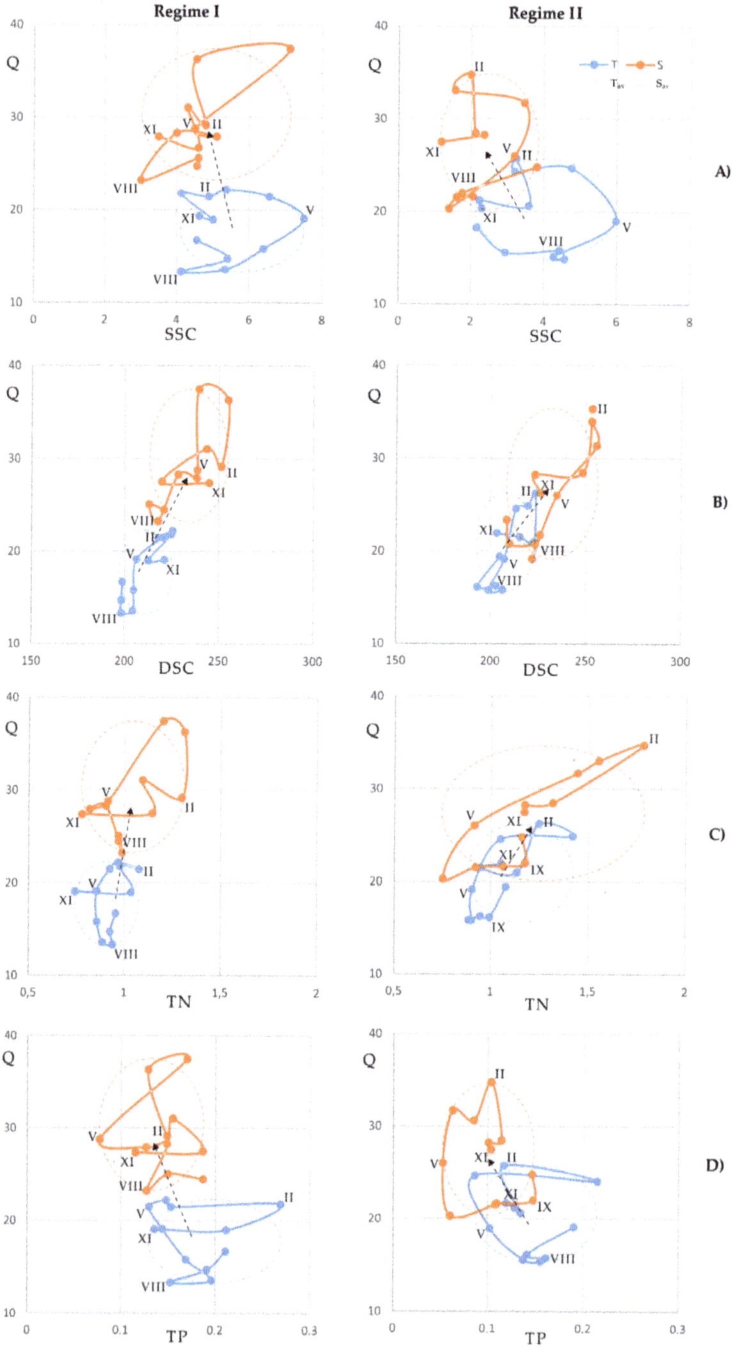

Figure 5. The pattern loops for seasonal relationship for water discharge (Q, in $m^3\ s^{-1}$) for stations Tuchola (T) and Smukała (S) with: (**A**) suspended sediment concentration (SSC, in mg L^{-1}), (**B**) dissolved load concentration (DLC, in mg L^{-1}), (**C**) total nitrogen (mgN L^{-1}) and (**D**) total phosphorus (mgP L^{-1}), during Regime I (hydropeaking) and Regime II (run–of–river).

The SSC loop at the Tuchola site for the first analyzed period 1991–2000 (regime I) shows that the spring's culmination of flows preceded the maximum suspended sediment concentration in the waters. Then, as the water flow decreased, the SSC decreased too (June–August). The loop for the second analyzed period 2001–2012 (regime II) takes a similar course. However, in this case, the summer ahead of maximum suspended sediment concentration was even earlier and occurred in July (Figure 5A). In the case of the Smukała station, for both regimes I and II, the SSC loop follows a similar course for the spring months. However, the difference in the course of the loop was noticeable in the case of summer and autumn months because for regime I, there were periods of maximum SSC ahead of the flow peak for the months June and September. In contrast, for regime II they were characteristic for the later months, i.e., August, October (Figure 5A).

The Tuchola station loop for the indicator reflecting the DSC for regime I (1991–2000) shows that the spring period of maximum flows preceded the maximum concentrations of DSC. Then, as the flow decreased, the dissolved loads' concentration decreased, which occurred from July to August. In the case of the second analyzed operating regime of the hydropower plant, the spring period was similar. At the same time, in July, the maximum DSC concentration was observed with the peak of the flow intensity. For the Smukała station in regime I (1992–2000), a different course of the loop was observed compared to the Tuchola station. For both the spring and summer flow culmination, they were exceeded by the maximum DSC concentrations. In the case of regime II (2001–2012), the course of the loop also has a different course compared to the Tuchola station because the flow rate ahead of the maximum DSC concentrations was observed both in spring and summer (Figure 5B).

In the case of nutrients, the loop at the Tuchola site assumes a similar course (Figure 5C,D). For regime I (1991–2000), in the spring period, the maximum concentration of total nitrogen (February) and total phosphorus (January) was ahead of the flow rate (March). A similar situation was observed in summer and autumn (Figure 5C). For the second analyzed regime (2001–2012), there was an opposite tendency during spring. The maximum flow rate (February) was ahead of the maximum concentration of nutrients in the river water (March). However, in summer and autumn, the maximum concentration of total nitrogen (September) and total phosphorus (August) occurred ahead of the peak flow rate. At the Smukała station, located below the cascade, a different course of the loop is observed than at the station above the reservoirs' cascade. During spring floods, for both analyzed operating regimes of hydropower plants (1991–2000 and 2001–2012), the concentration of total nitrogen in the river water (February) was ahead of the culmination of the flow rate (April), while for total phosphorus the opposite tendency was observed (Figure 5D). In summer and autumn, for total nitrogen in the first analyzed period, the culmination of the flow rate in the river was ahead of the concentration of the indicator mentioned above, while the course was reversed for the second period (Figure 5C). The advance of the maximum total phosphorus concentrations in the Brda waters before the maximum flow rate was also characteristic for both analyzed periods in the case of the Smukała site (Figure 5D).

All three reservoirs, Koronowski, Tryszczyn and Smukała, operating in the cascade system, affect the continuity of sediment transport and the distribution of nutrients along the Brda River. The course of the β coefficient for LBC, concerning the two, analyzed operating regimes of dams in the years 1991–2012, showed differences in the balance of suspended sediment load (β_{SSL}), dissolved load (β_{DSL}) and nutrient load (β_{TNL} and β_{TPL}).

In the first analyzed period, when the cascade was operating in the hydropeaking regime, in the case of the β_{SSL} coefficient, the removal of the suspended sediment from the LBC reservoirs was dominant. The annual mean value for the period 1991–2000 was −44.0%, ranging from −0.3% (1995) to −113.0% (1991). In the second analyzed period (2001–2012), the annual mean value of β_{SSL} was 18.2%, demonstrating SSL accumulation in LBC reservoirs in almost all years (Figure 6A). In the case of β_{DSL}, removing the dissolved load out of the cascade dominated in both periods, but with less intensity in the second, run–of–river regime (Figure 6B). The annual mean value for the period 1991–2000 (regime I) was −85.2%, while for the period 2001–2012 (regime II), it was −40.2%. The first of the analyzed nutrients, total nitrogen, was also characterized by the dominance of removal from the LBC reservoirs (Figure 6C). The annual mean value of β_{TNL} for the first period was −88.5%, while for the second period was −70.9%. The balance for total phosphorus also showed the dominance of the removal process in the first analyzed period (β_{TPL} −32.5%). However, in the second period, there were times of accumulation (mean value of β_{TNL} −1.9%) (Figure 6D). For all the analyzed indicators, there was a clear trend of changing the nature of the balance from the removal/transit of sediment and nutrients towards the accumulation/retention of matter in LBC reservoirs.

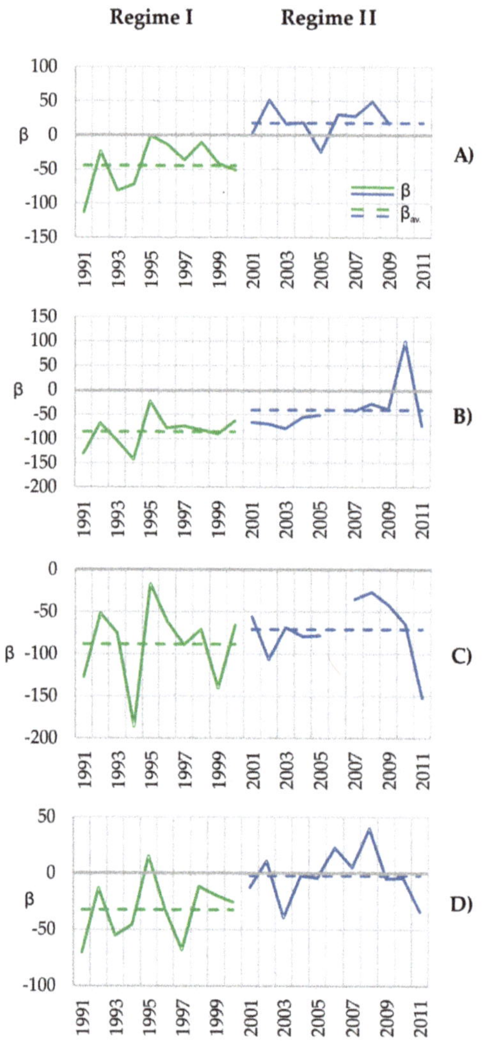

Figure 6. The ability of yearly (β) and average ($\beta_{av.}$) retention of river load and nutrients for LBC in the period 1991–2011. (**A**) suspended sediment load (SSL, in tones), (**B**) dissolved load (DL, in tones), (**C**) total nitrogen load (TNL, in tones) and (**D**) total phosphorus load (TPL, in tones). The green color line is for Regime I (hydropeaking), and the blue color line is for Regime II (run–of–river).

4. Discussion

Artificial regulation of sediment and nutrient transport is closely related to the hydrological situation, determining the energy carrier for the circulation of matter. Mean annual flow at the station in Tuchola, located beyond the range of influence of LBC reservoirs, indicated a slight increase in water outflow between the two analyzed periods, which may be associated with anthropogenic transformations of the catchment area. It may have resulted indirectly from changes in the structure of the catchment coverage, as found in the research by Szatten and Habel [66]. It was also confirmed by the increase in the value of C.V. in both analyzed periods. On the other hand, a slight decrease in mean annual flow in Smukała indicated the retention role of reservoirs in the lower part of the Brda, in a reduction of C.V value. Similar results were presented in research by Wang et al. [84], demonstrating significant flow alteration since the start of operation of the Three Gorges Dam on the Yangtze River. The changes may also indirectly result from increased evaporation caused by increased air temperature in recent decades. A study by Döll and Zhang [85] showed that in the second part of the 21st century, climate change effects on river regimes would outrank reservoirs. The calculated values of the flood-free season, especially for the Smukała station, indicated the role of reservoirs in equalizing the river outflow after changing the operating regime to run–of–river.

Low HA values for the Tuchola station for the first group of indicators demonstrated a natural course of the hydrological regime in the upper part of the Brda catchment area. Moderate discharge transformation for summer months showed the impact of flood episodes (Figure 3). In the case of the Smukała station, for the period up to 2000, a moderate discharge transformation was observed. This variation was the result of the hydropeaking operation of the cascade. The change of the operating regime to the run–of–river (since 2001) reduced the HA value for the second analyzed period (Figure 3). The average class of flow transformation below the water stages was recorded only for the summer months, which was related to the occurrence of rainfall and faster channel reaction time of the Brda system to changes in water flow.

The share of groundwater in the total runoff in the Brda catchment area is significant, as was shown in the studies by Jutrowska [63]. This fact is also indicated by the values of the Base Flow Index (BFI) (Table 2), illustrating the contribution of groundwater to river flow [86] and the lack of days with no water flow in the entire analyzed period. The decrease in the BFI value for Smukała in the studied periods from the level of 0.61 to 0.56 (Table 2) indicated a reduction in the impact of ground supply due to equalization of the water outflow below the cascade (run–of–river regime). The influence of the geological structure of the catchment area and the changes in the dominant land cover in the lower part of the catchment (from agricultural and forest areas to urban areas) were noticeable here [66,87], which affected the dynamics of the Brda River regime.

For the Tuchola station, an increase in the flow was observed between the analyzed periods, both for the minimum and maximum flows. It was related to the intensification of water outflow from the catchment area. The decrease in the minimum short-term flows (1-day–7-days) and maximum short-term flows (1-day–3-days) (Table 2) observed on the Smukała station indicated the equalization of water flow caused by the operation of hydropower plants. This fact was confirmed by the simultaneous increase in the minimum and maximum long-term (90-day) river flows (Table 2). Similar results were presented in studies by Magilligan and Nislow [45], Wang et al. [88], Pyron and Neumann [89], and Zhang et al. [90]. The overall reduction of peak minimum and maximum values were the most commonly defined effect of the dams functioning. This also resulted in a reduction of the nutrient exchange volume between rivers and floodplains. Besides, it is worth noting that the higher class of flow transformation for the Smukała site (indicators 7-days maximum–90-days maximum) (Figure 3) in the first analyzed period may determine the increased impact on the structuring of channel morphology below the cascade. It may result in the intensification of the erosion process below the water stages [53], described by Kondolf [24] as the phenomenon known as "hungry water".

Acceleration of the date of minimum discharge (Table 2) indicated a deepening water deficit problem (drought phenomenon), resulting in earlier occurrence of minimum flows. Acceleration was also observed in research by Lin et al. [48]. The acceleration of extreme water condition occurrence time also affects organisms' life cycles or spawning cues for migratory fish. It can influence the connectivity between the floodplain and the river channel, affecting especially fish that lay eggs in the river floodplain or oxbow lake [84]. The impact of the LBC reservoirs' presence was illustrated by the difference between dates of minimum and maximum discharge for two analyzed water gauges (Table 2). For the first analyzed period, for the minimum flows, it was nine days, indicating that the hydropower plant operated without changing the river regime even in the case of low water levels. For the maximum flows, the difference was 43 days, indicating the retention of the supply water flowing into the cascade in the case of high water flows. In the second analyzed period, these differences amounted to 16 and 19 days, respectively. This indicates longer water retention in the case of minimum flows and faster water outflow for the maximum flows, which can be associated with increased energy production in periods of high water levels on the Brda River.

The high value of HA for the fourth group's parameters indicates an extensive transformation of the hydrological regime due to the cascade of the lower Brda reservoirs (Figure 3). The parameters from this group had the highest RVA values of all analyzed indicators. The number and duration of low and high pulses indicated the Brda cascade's intervention regime in the first analyzed period (Table 2). This could have affected the artificial regulation of the exchange of nutrients and organic matter between the river and the floodplain, influencing the amount of sediment transport and the texture of the riverbed. Changes in sediment and nutrient transport after commission of the reservoir were reported by Ligon et al. [16] and Stanford and Ward [91]. According to Richter et al. [74], the fourth group of hydrologic alteration indices showed limited habitat availability for aquatic species, including water birds and other terrestrial organisms. The typical lowland nature of the deep and thermally stratified Koronowski Reservoir is a significant source of an indigenous supply of nutrients and suspended sediments from the areas near the river bed (reservoir). Hence, it is an essential indicator of the Brda River hydrological transformation regarding the balance between a river system's sediment supply and transport capacity [92,93]. On the other hand, in the lower, shallow Tryszczyn and Smukała reservoirs, this indicator's impact is only visible in the zone directly below the dam. The changes of the flow pulses can influence the downstream erosion of the channel and degradation of the bed below dams, as pointed out in the research of Leopold et al. [94]. A significant decrease in the value of the parameters mentioned above in the second period resulted from changing the operating system to a run–of–river regime (Table 2).

The change in the operating system was also reflected in the rise and fall rate of discharge and annual discharge number of reversals, which indicated a much more frequent fluctuation of flows below the analyzed water stages but with a smaller amplitude (Table 2). Similar results were reported by Chen et al. [47], Yang et al. [46], and Timpe and Kaplan [95].

The conducted research allows determining the conceptual model of river sediment and nutrient circulation for the cascade of reservoirs to select the most favorable operating regime of a hydropower plant for the hydrological continuum of the river.

In the first analyzed period, the SSL balance for LBC reservoirs showed its removal (Figure 6A), indirectly reflecting the influence of frequent and large water level/flow rate fluctuations (Figure 2) characteristic of the hydropeaking regime. With the start of the run–of–river regime for LBC reservoirs in 2001, SSL transport and accumulation conditions changed. The energetics of the suspended sediment transport environment decreased, and the occurrence of high (June) and low (August) flows shifted, resulting in increased sediment accumulation in LBC reservoirs. This is consistent with the observations carried out on artificial reservoirs [28–31,96], also according to LBC reservoir sedimentation [25,66]. On the other hand, the DSL balance in both observed regimes indicates the continued transit through the LBC reservoirs, where the decrease in removal (Figure 6B) was directly related to the reduction in the volume of flows; water mass being the only factor determining the transport of the dissolved load.

The observed figure of eight pattern loops for sediment transport for the studied sites (Figure 5A,B) resulted from both the changes taking place in the Brda catchment area and the influence of the change in the operating regime of the hydropower plant, as demonstrated by the IHA analysis. The recorded advance of the maximum water flow in the channel with the maximum value of the analyzed sediment transport indicators demonstrated the supply of matter from a remote upstream area. On the other hand, the anticipation of the maximum value of the studied sediment transport indicators concerning the maximum water flow in the river indicated that the near channel was the sediment source. In the case of the spring period, the transit of matter from the catchment area was dominant (Figure 5A,B), with a delay resulting from high retention capacity, i.e., the presence of the river-lake system in the catchment area above the LBC combined with the postglacial formations of the catchment area. Sources of sediment supply during the summer period, when the near channel was the sediment source, were anthropogenic surfaces. The hardened surfaces, taking into account climate change [97] and change in land use [66], directly correlated with SSL and DSL transport and floodplain area. The impact of the cascade of reservoirs is most clearly reflected in the course of the loop for the second analyzed period (regime II), where the loop was disturbed for maximum (spring) and minimum (summer) flows. For the Tuchola station, there were no such apparent disturbances in the loop as for the Smukała station (Figure 5A,B). This shows that the functioning of the reservoir cascade had a decisive influence on the dynamics and continuity of the Brda river sediment transport.

In the case of nutrients TN and TP, the situation is more complicated because, in the first analyzed period, the balance of TNL and TPL for LBC indicates their removal (Figure 6C,D). After changing the hydropower plant's operation to the run–of–river regime in 2001, a negative nutrient balance for LBC was still recorded. Additionally, Wang [36] indicated that the presence of reservoirs in the river's longitudinal profile, including thermal stratification of reservoir waters, has an essential control over the content and forms of nutrients in the water. In the case of phosphorus, its reduction in the waters outflowing from the LBC may be a common phenomenon, as phosphorus is usually absorbed by bottom sediments [41], and its more significant reduction was noticeable when the LBC dams operated in the run–of–river. The annual mean values of total nitrogen depend on both allochthonous and autochthonous supply. After changing the hydropower plant's operation to the run–of–river regime in 2001, TN increased slightly, enriching the outflowing waters with nitrogen (Figure 4C). This can be directly related to the source of nutrient supply and the flow nature of the cascade operation, which decreased the energy of the aquatic environment in reservoirs. Therefore, better conditions can be expected for the activity of the microflora and fauna of reservoirs and the mineralization of organic debris, which is also characteristic of other reservoirs [42]. Above the LBC, as a result of human activity, the nature and source of nutrients could have changed in the spring period; the supply being from the floodplain area to the upstream area (Figure 4C,D). Despite the notable change in both periods, the nutrient limiting the primary production was total nitrogen (TN:TP ratio < 16). This resulted in an extended delivery time of nutrients to the LBC reservoirs. Similar results were shown by Syvitski et al. [98] and Van Cappellen and Maavara [99]. The lacustrine character and thermal stratification of the waters of the largest LBC reservoir, combined with the change of the source of the Smukała station from the catchment area to the flood plain area for total nitrogen (Figure 4C,D), resulted in the fact that the volume of primary production inside the LBC became decisive for the nutrient balance. Additionally, the low TN:TP ratio stimulates the growth of cyanobacterial biomass and strengthens their advantage over other phytoplankton groups [100], which is unfavorable in the light of internal enrichment of water with nutrients. The total nitrogen balance showed its continuous removal (Figure 6C), while the balance of total phosphorus in the second analyzed period showed its accumulation (Figure 6D). This contributes to driving the internal primary production inside the LBC, directly influencing the ecological status of waters. The primary limiting production in LBC reservoirs in the first period was total nitrogen, while as a result of changes in the river regime resulting in the above, i.e., the described changes in nutrient supply sources, total phosphorus became the limiting nutrient (TN:TP ratio > 16). The ease of keeping phosphorus in the reservoir, confirmed by the research of Qin et al. [101], promotes the development of phytoplankton groups other than cyanobacteria [100], which improves the ecological status of waters.

5. Conclusions

Based on the results of conducted studies on the influence of hydropower plant regime change on suspended sediment transport dynamics on the example Lower Brda River Cascade (LBC), the following conclusions were drawn:

1. The commission of the reservoirs in the cascade system influenced the hydrological regime on the section of the reservoirs themselves and the river below. The hydropeaking operating regime resulted in high water flow dynamics, especially below the LBC, reflecting high HA values. The change of the operating regime to the run–of–river, more in tune with the habitat requirements of aquatic species and those associated with water, resulted in the equalization of water flow throughout the year, which was reflected in a decrease in the HA index.
2. The observed acceleration of the occurrence of the minimum and maximum flows in the Brda River below the LBC resulted directly from the change of the hydropower plant's operating regime to the run–of–river. It was also an indirect indicator of progressing climate change.
3. Functioning LBC disturbs the hydrological continuum of the Brda River. The change of hydropeaking to the run–of–river operating regime of the hydropower plant increased the accumulation of suspended sediment in reservoirs. In the case of dissolved load transport, a direct link to the operating regime was noted, but without a significant impact on the overall balance, indicating continuous load transit. In the case of nutrients, the change in the LBC operating regime resulted in changes in the course of biochemical processes taking place in the reservoirs, which translated into the TN and TP transport balance.
4. The observed figure-eight pattern loops show the influence of hydropower plant operation on the LBC's sediment and nutrient transport. This was especially noticeable for the summer period, when heavy rainfall caused an increased supply of sediment and nutrients due to the surface runoff from hardened areas. There were also changes in the matter supply sources, affecting nutrients and indicating a change of the nutrient primary limiting production in reservoirs from total nitrogen to total phosphorus.
5. The research pointed out the usefulness of analyses of river regimes and sediment and nutrient data for long-term observations of environmental pressures in relation to water management in fluvial schemes with dammed water systems.

Author Contributions: Conceptualization, D.S. and M.H.; methodology, D.S., M.H. and Z.B.; software, D.S.; validation, M.H.; formal analysis, Z.B.; investigation, D.S. and M.H.; resources, D.S. and M.H.; data curation, D.S.; writing—original draft preparation, D.S. and M.H.; writing—review and editing, M.H. and Z.B.; visualization, M.H. and D.S.; supervision, Z.B.; project administration, D.S.; funding acquisition, D.S., M.H. and Z.B. All authors have read and agreed to the published version of the manuscript.

Funding: This research and APC was funded by the Project Supporting Maintenance of Research Potential of the Institute of Geography at Kazimierz Wielki University [grant number BS/2016/N2].

Institutional Review Board Statement: Not applicable.

Informed Consent Statement: Not applicable.

Data Availability Statement: Publicly available data sets were analyzed in this study. These data can be found here: (i) hydrological data [https://danepubliczne.imgw.pl/ (accessed on 19 May 2020)]; (ii) water quality data is provided from the State Environmental Monitoring Program.

Acknowledgments: We would like only to thank the four anonymous reviewers for constructive comments that substantially improved our paper.

Conflicts of Interest: The authors declare no conflict of interest.

References

1. Zwoliński, Z. *Geomorficzne Dostosowywanie Się Koryta Parsęty do Aktualnego Reżimu Rzecznego*; Institute of Geography and Spatial Organization Polish Academy of Sciences: Wrocław/Warszawa/Kraków/Gdańsk/Łódź, Poland, 1989; p. 144.
2. Wilson, D.; Hannah, D.M.; McGregor, G.R. A large scale hydroclimatological perspective on western European river flow regimes. *Hydrol. Res.* **2012**, *44*, 809–833. [CrossRef]
3. Wrzesiński, D.; Sobkowiak, L. Detection of changes in flow regime of rivers in Poland. *J. Hydrol. Hydromech.* **2018**, *66*, 55–64. [CrossRef]
4. Vörösmarty, C.J.; Meybeck, M.; Fekete, B.; Sharma, K.; Green, P.; Syvitski, J.P.M. Anthropogenic sediment retention: Major global impact from registered river impoundments. *Glob. Planet. Chang.* **2003**, *39*, 169–190. [CrossRef]
5. Syvitski, J.P.M.; Milliman, J.D. Geology, geography, and humans battle for dominance over the delivery of fluvial sediment to the coastal ocean. *J. Geol.* **2007**, *115*, 1–19. [CrossRef]
6. Baldy, V.; Trémolières, M.; Andrieu, M.; Belliard, J. Changes in phosphorus content of two aquatic macrophytes according to water velocity, trophic status and time period in hardwater streams. *Hydrobiologia* **2006**, *575*, 343–351. [CrossRef]
7. Schindle, D.W. Eutrophication and recovery in experimental lakes—Implications for lake management. *Science* **1974**, *184*, 897–899. [CrossRef] [PubMed]
8. Allan, J.D. Landscapes and riverscapes: The influence of land use on stream ecosystems. *Annu. Rev. Ecol. Evol. Syst.* **2004**, *35*, 257–284. [CrossRef]
9. Michalak, A.M.; Anderson, E.J.; Beletsky, D.; Boland, S.; Bosch, N.S.; Bridgeman, T.B.; DePinto, J.V. Record-setting algal bloom in Lake Erie caused by agricultural and meteorological trends consistent with expected future conditions. *Proc. Natl. Acad. Sci. USA* **2013**, *110*, 6448–6452. [CrossRef] [PubMed]
10. Machesky, M.L.; Holm, T.R.; Slowikowski, J.A. Phosphorus speciation in stream bed sediments from an agricultural watershed: Solid-phase associations and sorption behavior. *Aquat. Geochem.* **2010**, *16*, 639–662. [CrossRef]
11. Revenga, C.; Brunner, J.; Henniger, N.; Kassem, K.; Payne, R. *Pilot Analysis of Global Ecosystems: Freshwater Systems*; World Resources Institute: Washington, DC, USA, 2000; p. 78.
12. Nilsson, C.; Reidy, C.A.; Dynesius, M.; Revenga, C. Fragmentation and Flow Regulation of the World's Large River Systems. *Science* **2005**, *308*, 405–408. [CrossRef]
13. Belletti, B.; Garcia de Leaniz, C.; Jones, J.; Bizzi, S.; Börger, L.; Segura, G.; Castelletti, A.; Bund, W.; Aarestrup, K.; Barry, J.; et al. More than one million barriers fragment Europe's rivers. *Nature* **2020**, *588*, 436–441. [CrossRef]
14. Walling, D.E. Human impact on land-ocean sediment transfer by the world's rivers. *Geomorphology* **2006**, *79*, 192–216. [CrossRef]
15. World Commission on Dams (WCD) Dams and Development. *A New Framework for Decision Making. The Report of the World Commission on Dams*; Taylor and Francis Group, Earthscan: London, UK, 2000; p. 446.
16. Ligon, F.K.; Dietrich, W.E.; Trush, W.J. Downstream ecological effects of dams. *BioScience* **1995**, *45*, 183–192. [CrossRef]
17. Directive 2001/77/EC of 27 September 2001 on the Promotion of Electricity Produced from Renewable Energy Sources in the Internal Electricity Market (O.J.EC L 283 of 27.10.2001). Available online: https://eur-lex.europa.eu/legal-content/en/ALL/?uri=CELEX%3A32001L0077 (accessed on 20 January 2021).
18. *Energy from Renewable Sources in 2018*; Statistics Poland: Warszawa, Poland, 2019; p. 90. Available online: https://stat.gov.pl/obszary-tematyczne/srodowisko-energia/energia/energia-ze-zrodel-odnawialnych-w-2018-roku,10,2.html (accessed on 9 July 2020).
19. Williams, G.P.; Wolman, M.G. *Downstream Effects of Dams on Alluvial Rivers*; United States Geological Survey, US Government Printing Office: Washington, DC, USA, 1984; p. 83.
20. Vörösmarty, C.J.; Sharma, K.P.; Fekete, B.M.; Copeland, A.H.; Holden, J.; Marble, J.; Lough, J.A. The storage and aging of continental run off in large reservoir systems of the world. *Ambio* **1997**, *26*, 210–219.
21. Obodovskyi, O.; Habel, M.; Szatten, D.; Rozlach, Z.; Babiński, Z.; Maerker, M. Assessment of the Dnieper Alluvial Riverbed Stability Affected by Intervention Discharge Downstream of Kaniv Dam. *Water* **2020**, *12*, 1104. [CrossRef]
22. Lisle, T.E.; Hilton, S. The volume of fine sediment in pools: An index of sediment supply in gravel-bed streams. *Water Resour. Bull.* **1992**, *28*, 371–383. [CrossRef]
23. Van Rijn, L. Sediment Transport. Part III: Bed forms and alluvial roughness. *J. Hydraul. Eng.* **1984**, *110*, 1733–1754. [CrossRef]
24. Kondolf, G. Hungry water: Effects of Dams and Gravel Mining on River Channels. *Environ. Manag.* **1997**, *21*, 533–551. [CrossRef] [PubMed]
25. Szatten, D.; Habel, M.; Pellegrini, L.; Maerker, M. Assessment of Siltation Processes of the Koronowski Reservoir in the Northern Polish Lowland Based on Bathymetry and Empirical Formulas. *Water* **2018**, *10*, 1681. [CrossRef]
26. Moilanen, A.; Hanski, I. On the use of connectivity measures in spatial ecology. *Oikos* **2001**, *95*, 147–151. [CrossRef]

27. Pringle, C.M.; Freeman, M.C.; Freeman, B.J. Regional effects of hydrologic alterations on riverine macrobiota in the New World: Tropical-temperate comparisons. *Bioscience* **2000**, *50*, 807–823. [CrossRef]
28. Amenuvor, M.; Gao, W.; Li, D.; Shao, D. Effects of Dam Regulation on the Hydrological Alteration and Morphological Evolution of the Volta River Delta. *Water* **2020**, *12*, 646. [CrossRef]
29. Urbaniak, M.; Kiedrzynska, E.; Zalewski, M. The role of a lowland reservoir in the transport of micropollutants, nutrients and the suspended particulate matter along the river continuum. *Hydrol. Res.* **2012**, *43*, 400–411. [CrossRef]
30. Babic-Mladenovic, M.; Kolarov, V.; Damjanovic, V. Sediment regime of the Danube River in Serbia. *Int. J. Sediment. Res.* **2013**, *28*, 470–485. [CrossRef]
31. Babiński, Z. Hydromorphological consequences of regulating the lower Vistula, Poland. *Regul. Rivers Res. Manag.* **1992**, *7*, 337–348. [CrossRef]
32. Friedl, G.; Wüest, A. Disrupting biogeochemical cycles—Consequences of damming. *Aquat. Sci.* **2002**, *64*, 55–65. [CrossRef]
33. Wang, X.; Yang, T.; Yong, B.; Krysanova, V.; Shi, P.; Li, Z.; Zhou, X. Impacts of climate change on flow regime and sequential threats to riverine ecosystem in the source region of the Yellow River. *Environ. Earth Sci.* **2018**, *77*, 465. [CrossRef]
34. Richter, B.D.; Davis, M.M.; Apse, C.; Konrad, C. A presumptive standard for environmental flow protection. *River Res. Appl.* **2012**, *28*, 1312–1321. [CrossRef]
35. Directive 2000/60/EC of the European Parliament and of the Council of 23 October 2000 Establishing a Framework for Community Action in the Field of Water Policy. Available online: https://eur-lex.europa.eu/eli/dir/2000/60/oj (accessed on 21 October 2019).
36. Wang, F. Impact of a large sub-tropical reservoir on the cycling of nutrients in a river. *Water Res.* **2020**, *186*, 116363. [CrossRef]
37. Merritt, D.M.; Scott, M.L.; Poff, N.L.; Auble, G.T.; Lytle, D.A. Theory, methods and tools for determining environmental flows for riparian vegetation: Riparian vegetation-flow response guilds. *Freshw. Biol.* **2009**, *55*, 206–255. [CrossRef]
38. Poff, N.L.; Allan, J.D.; Bain, M.B.; Karr, J.R.; Prestegaard, K.L.; Richter, B.D.; Sparks, R.E.; Stromberg, J.C. The natural flow regime. *BioScience* **1997**, *47*, 769–784. [CrossRef]
39. Finger, D.; Schmid, M.; Wüest, A. Comparing effects of oligotrophication and upstream hydropower dams on plankton and productivity in perialpine lakes. *Water Resour. Res.* **2007**, *43*, W12404. [CrossRef]
40. Harrison, J.A.; Seitzinger, S.P.; Bouwman, A.F.; Caraco, N.F.; Beusen, A.H.W.; Vörösmarty, C.J. Dissolved inorganic phosphorus export to the coastal zone: Results from a spatially explicit, global model. *Glob. Biogeochem. Cycles* **2005**, *19*, GB4S03. [CrossRef]
41. Bosch, N.S. The influence of impoundments on riverine nutrient transport: An evaluation using the Soil and Water Assessment Tool. *J. Hydrol.* **2008**, *355*, 131–147. [CrossRef]
42. Teodoru, C.; Wehrli, B. Retention of sediments and nutrients in the Iron Gate I Reservoir on the Danube River. *Biogeochemistry* **2005**, *76*, 539–565. [CrossRef]
43. Freeman, M.C.; Bowen, Z.H.; Bovee, K.D.; Irwin, E.R. Flow and habitat effects on juvenile fish abundance in natural and altered flow regimes. *Ecol. Appl.* **2001**, *11*, 179–190. [CrossRef]
44. Dettinger, M.D.; Diaz, H.F. Global characteristics of stream flow seasonality and variability. *J. Hydrometeorol.* **2000**, *1*, 289–310. [CrossRef]
45. Magilligan, F.J.; Nislow, K.H. Changes in hydrologic regime by dams. *Geomorphology* **2005**, *71*, 61–78. [CrossRef]
46. Yang, T.; Zhang, Q.; Chen, Y.D.; Tao, X.; Xu, C.; Chen, X. A spatial assessment of hydrologic alteration caused by dam construction in the middle and lower Yellow River, China. *Hydrol. Process.* **2008**, *22*, 3829–3843. [CrossRef]
47. Chen, Y.D.; Yang, T.; Xu, C.Y.; Zhang, Q.; Chen, X.; Hao, Z.C. Hydrologic alteration along the Middle and Upper East River (Dongjiang) basin, South China: A visually enhanced mining on the results of RVA method. *Stoch. Environ. Res. Risk Assess.* **2010**, *24*, 9–18. [CrossRef]
48. Lin, K.; Lian, Y.; Chen, X.; Lu, F. Changes in runoff and eco-flow in the Dongjiang River of the Pearl River Basin, China. *Front. Earth Sci.* **2014**, *8*, 547–557. [CrossRef]
49. Saraiva-Okello, A.M.L.; Masih, I.; Uhlenbrook, S.; Jewitt, G.P.W.; Van der Zaag, P.; Riddell, E. Drivers of spatial and temporal variability of streamflow in the Incomati River basin. *Hydrol. Earth Syst. Sci.* **2015**, *19*, 657–673. [CrossRef]
50. Mittal, N.; Bhave, A.G.; Mishra, A.; Singh, R. Impact of Human Intervention and Climate Change on Natural Flow Regime. *Water Resour. Manag.* **2016**, *30*, 685–699. [CrossRef]
51. Duan, W.; Guo, S.; Wang, J.; Liu, D. Impact of Cascaded Reservoirs Group on Flow Regime in the Middle and Lower Reaches of the Yangtze River. *Water* **2016**, *8*, 218. [CrossRef]
52. Gebremicael, T.G.; Mohamed, Y.A.; Zaag, P.; Hagos, E.Y. Temporal and spatial changes of rainfall and streamflow in the Upper Tekezē–Atbara river basin, Ethiopia. *Hydrol. Earth Syst. Sci.* **2017**, *21*, 2127–2142. [CrossRef]
53. Gierszewski, P.J.; Habel, M.; Szmańda, J.; Luc, M. Evaluating effects of dam operation on flow regimes and riverbed adaptation to those changes. *Sci. Total. Environ.* **2020**, *710*, 136202. [CrossRef] [PubMed]
54. Lu, X.X.; Li, S.; Kummu, M.; Padawangi, R.; Wang, J.J. Observed changes in the water flow at Chiang Saen in the lower Mekong: Impacts of Chinese dams? *Quat. Int.* **2014**, *336*, 145–157. [CrossRef]
55. Szmanda, J.; Gierszewski, P.; Habel, M.; Luc, M.; Witkowski, K.; Bortnyk, S.; Obodovskyi, O. Response of the Dnieper River fluvial system to the river erosion caused by the operation of the Kaniv hydro-electric power plant (Ukraine). *Catena* **2021**, 105265. [CrossRef]

56. Kim, B.S.; Kim, B.K.; Kwon, H.H. Assessment of the impact of climate change on the flow regime of the Han River basin using indicators of hydrologic alteration. *Hydrol. Process.* **2011**, *25*, 691–704. [CrossRef]
57. Wang, X.; Yang, T.; Wortmann, M.; Shi, P.; Hattermann, F.; Lobanova, A.; Aich, V. Analysis of multi-dimensional hydrological alterations under climate change for four major river basins in different climate zones. *Clim. Chang.* **2017**, *141*, 483–498. [CrossRef]
58. McDaniel, R.D.; O'Donnell, F.C. Assessment of Hydrologic Alteration Metrics for Detecting Urbanization Impacts. *Water* **2019**, *11*, 1017. [CrossRef]
59. Map of the Polish Hydrographic Division; Department of Hydrography and Morphology of River Channels Institute of Meteorology and Water Management. Available online: http://mapa.kzgw.gov.pl/ (accessed on 10 December 2016).
60. Galon, R. Morfologia doliny i sandru Brdy. *Stud. Soc. Scient. Tor.* **1953**, *C*, 1–6.
61. Galon, R. *Geomorfologia Polski*; Polish Scientific Publishers: Warszawa, Poland, 1972; p. 372.
62. Choiński, A. *Zróżnicowanie i Uwarunkowania Zmienności Przepływów Rzek Polskich*; Adam Mickiewicz University Press: Poznań, Poland, 1988; p. 99.
63. Jutrowska, E. *Antropogeniczne Zmiany Warunków Hydrologicznych w Dorzeczu Brdy*; Biblioteka Monitoringu Środowiska: Bydgoszcz, Poland, 2007; p. 128.
64. Lorenc, H. (Ed.) *Atlas klimatu Polski*; Institute of Meteorology and Water Management National Research Institute: Warszawa, Poland, 2005; p. 116.
65. *Atlas Hydrologiczny Polski*; Institute of Meteorology and Water Management National Research Institute: Warszawa, Poland, 1997; p. 95.
66. Szatten, D.; Habel, M. Effects of Land Cover Changes on Sediment and Nutrient Balance in the Catchment with Cascade-Dammed Waters. *Remote. Sens.* **2020**, *12*, 3414. [CrossRef]
67. *Studium Regionu Doliny Brdy*; Biuro Planów Regionalnych: Warszawa, Poland, 1953; p. 235.
68. Szatten, D. Wpływ zabudowy hydrotechnicznej na występowanie ekstremalnych stanów wody na przykładzie Brdy skanalizowanej. *Inżynieria Ekol.* **2016**, *46*, 55–60. [CrossRef]
69. Pietrucień, C. Stosunki hydrograficzne w rejonie Zalewu Koronowskiego. *Stud. Soc. Scient. Tor.* **1967**, *I*, 15–34.
70. Hoffman, A. (Ed.) *Załącznik do Założeń Projektu Elektrowni Wodnej Koronowo*; Zjednoczenie Energetyczne Okręgu Bydgosko-Toruńskiego: Gdańsk, Poland, 1950; p. 230.
71. Biuro Wojewody Kujawsko-Pomorskiego. Pozwolenie Wodnoprawne na Szczególne Korzystanie z Wód Rzeki Brdy Dla Potrzeb Elektrowni Wodnej Smukała. 2004. Available online: http://archiwum.kujawsko-pomorskie.pl/index.php?option=com_content&task=view&id=20512&Itemid=665 (accessed on 10 September 2020).
72. Water Law Act. Journal of Laws. 2017. Available online: http://isap.sejm.gov.pl/isap.nsf/DocDetails.xsp?id=WDU20170001566 (accessed on 20 January 2021).
73. Szatten, D. Zmiany stanów wody w Zbiorniku Koronowskim w latach 1996-2012. *Inżynieria Ekologiczna* **2015**, *44*, 204–209. [CrossRef]
74. Richter, B.D.; Baumgartner, J.V.; Powell, J.; Braun, D.P. A Method for Assessing Hydrologic Alternation within Ecosystems. *Conserv. Biol.* **1996**, *10*, 1163–1174. [CrossRef]
75. Richter, B.D.; Baumgartner, J.V.; Wigington, R.; Braun, D.P. How much water does a river need? *Freshw. Biol.* **1997**, *37*, 231–249. [CrossRef]
76. Richter, B.D.; Baumgartner, J.V.; Braun, D.P.; Powell, J. A spatial assessment of hydrologic alternation within a river network. *Regul. Rivers Res. Manag.* **1998**, *14*, 329–340. [CrossRef]
77. *Water Quality—Determination of Suspended Sediment Concentration (PN-EN 872)*; Polish Committee for Standardization: Warszawa, Poland, 2007; p. 8.
78. Wren, D.; Barkdoll, B.; Kuhnle, R.; Derrow, R. Field Techniques for Suspended-Sediment Measurement. *J. Hydraul. Eng.* **2000**, *126*, 97–104. [CrossRef]
79. Hermanowicz, W. *Fizyczno-Chemiczne Badanie Wody i Ścieków*; Arkady: Warszawa, Poland, 1999; p. 556.
80. *Water Quality—Determination of Nitrogen (PN-EN 12260:2004)*; Polish Committee for Standardization: Warszawa, Poland, 2014; p. 13.
81. *Water Quality—Determination of Phosphorus (PN-EN ISO 6878:2006)*; Polish Committee for Standardization: Warszawa, Poland, 2006; p. 23.
82. Redfield, A.C. The influence of organisms on the composition of sea-water. *Sea* **1963**, *2*, 26–77.
83. Łajczak, A. *Studium Nad Zamulaniem Wybranych Zbiorników Zaporowych w Dorzeczu Wisły*; Institute of Geography and Spatial Organization Polish Academy of Sciences: Warszawa, Poland, 1995; p. 105.
84. Wang, Y.; Wang, D.; Lewis, Q.W.; Wu, J.; Huang, F. A framework to assess the cumulative impacts of dams on hydrological regime: A case study of the Yangtze River. *Hydrol. Process.* **2017**, *31*, 3045–3055. [CrossRef]
85. Döll, P.; Zhang, J. Impact of climate change on freshwater ecosystems: A global-scale analysis of ecologically relevant river flow alterations. *Hydrol. Earth Syst. Sci.* **2010**, *7*, 1305–1342.
86. Bloomfield, J.P.; Allen, D.J.; Griffiths, K.J. Examining geological controls on Baseflow Index (BFI) using regression analysis: An illustration from the Thames Basin, UK. *J. Hydrol.* **2009**, *373*, 164–176. [CrossRef]
87. CORINE Land Cover 2006; Chief Inspectorate for Environmental Protection & Institute of Geodesy and Cartography. Available online: http://clc.gios.gov.pl/ (accessed on 6 April 2017).

88. Wang, H.; Yang, Z.; Saito, Y.; Liu, J.P.; Sun, X. Interannual and seasonal variation of the Huanghe (Yellow River) water discharge over the past 50 years: Connections to impacts from ENSO events and dams. *Glob. Planet. Chang.* **2006**, *50*, 212–225. [CrossRef]
89. Pyron, M.; Neumann, K. Hydrologic alterations in the Wabash River watershed, USA. *River Res. Appl.* **2008**, *24*, 1175–1184. [CrossRef]
90. Zhang, Q.; Xiao, M.; Liu, C.L.; Singh, V.P. Reservoir-induced hydrological alterations and environmental flow variation in the East River, the Pearl River basin, China. *Stoch. Environ. Res. Risk Assess.* **2014**, *28*, 2119–2131. [CrossRef]
91. Stanford, J.A.; Ward, J.V. An ecosystem perspective of alluvial rivers—Connectivity and the hyporheic corridor. *J. N. Am. Benthol. Soc.* **1993**, *12*, 48–60. [CrossRef]
92. Phillips, J.D. Sedimentation in bottomland hardwoods downstream of an east Texas dam. *Environ. Geol.* **2001**, *40*, 860–868.
93. Pitlick, J.; Wilcock, P. Relations between streamflow, sediment transport, and aquatic habitat in regulated rivers. In *Geomorphic Processes and Riverine Habitat*; Dorava, J.M., Montgomery, D.R., Palcsak, B.B., Fitzpatrick, F.A., Eds.; American Geophysical Union, Water Science and Application: Washington, DC, USA, 2001; Volume 4, pp. 185–198.
94. Leopold, L.B.; Wolman, M.G.; Miller, J.P. *Fluvial Processes Geomorphology*; Dover Publications: New York, NY, USA, 1992; p. 893.
95. Timpe, K.; Kaplan, D. The changing hydrology of a dammed Amazon. *Sci. Adv.* **2017**, *3*, 1700611. [CrossRef]
96. Babiński, Z.; Habel, M. Impact of a single dam on sediment transport continuity in large lowland rivers. In Proceedings of the 13th International Symposium on River Sedimentation, Stuttgart, Germany, 19–22 September 2016; Wieprecht, S., Haun, S., Weber, K., Noack, M., Terheiden, K., Eds.; Taylor & Francis CRP Press: Leiden, The Netherlands, 2016; pp. 975–982.
97. Vinnikov, K.Y. Global Warming Trend of Mean Tropospheric Temperature Observed by Satellites. *Science* **2003**, *302*, 269–272. [CrossRef]
98. Syvitski, J.P.M.; Vörösmarty, C.J.; Kettner, A.J.; Green, P. Impact of humans on the flux of terrestrial sediment to the global coastal ocean. *Science* **2005**, *308*, 376–380. [CrossRef] [PubMed]
99. Van Cappellen, P.; Maavara, T. Rivers in the Anthropocene: Global scale modifications of riverine nutrient fluxes by damming. *Ecohydrol. Hydrobiol.* **2016**, *16*, 106–111. [CrossRef]
100. Wilk, P.; Orlińska-Woźniak, P.; Gębala, J. Variability of nitrogen to phosphorus concentration ratio on the example of selected coastal river basin. *Sci. Rev. Eng. Environ. Sci.* **2017**, *26*, 55–65.
101. Qin, B.; Zhou, J.; Elser, J.J.; Gardner, W.S.; Deng, J.; Brookes, J.D. Water depth underpins the relative roles and fates of nitrogen and phosphorus in lakes. *Environ. Sci. Technol.* **2020**, *54*, 3191–3198. [CrossRef] [PubMed]

resources

MDPI

Article

Risk Assessment of Lack of Water Supply Using the Hydraulic Model of the Water Supply

Krzysztof Boryczko *, Izabela Piegdoń, Dawid Szpak and Jakub Żywiec

Department of Water Supply and Sewerage Systems, Faculty of Civil, Environmental Engineering and Architecture, Rzeszow University of Technology, Al. Powstancow Warszawy 6, 35-959 Rzeszow, Poland; piegi@prz.edu.pl (I.P.); dsz@prz.edu.pl (D.S.); j.zywiec@prz.edu.pl (J.Ż.)
* Correspondence: kb@prz.edu.pl; Tel.: +48-17-865-1427

Abstract: Modern management of water supply systems is based on a preventive strategy consisting of the prevention of failures and crisis situations. Water pipe failures resulting in a lack of water supply for a long period pose a threat to the water consumers safety. The aim of the work was to present the methodology and develop a risk map of lack of water supply to consumers. The article presents a failures simulation of the main pipes transporting treated water from the water treatment plant to the city carried out using the EPANET 2.0. software. The simulation results made it possible to determine the consequences of failures by determining the number of inhabitants (consumers) affected with lack of water supply as a result of failure of the main pipes near the water treatment plant WTP which, together with the failure rate, were used to prepare risk maps of lack of water supply. The developed method was presented on the water supply network located in Central and Eastern Europe. It was found that the highest risk of lack of water supply is related to the failure of the M3 main pipe, which transports water to the eastern and north-eastern parts of the city. It is recommended to modernize the M3 main pipe, which will reduce the number of failures resulting in a lack of water supply.

Keywords: water supply network; failures; risk maps; Epanet

Citation: Boryczko, K.; Piegdoń, I.; Szpak, D.; Żywiec, J. Risk Assessment of Lack of Water Supply Using the Hydraulic Model of the Water Supply. *Resources* **2021**, *10*, 43. https://doi.org/10.3390/resources10050043

Academic Editor: Diego Copetti

Received: 12 April 2021
Accepted: 30 April 2021
Published: 5 May 2021

1. Introduction

The operation of water supply systems (WSS) is aimed at providing consumers with water of appropriate quality, in the right amount and under the appropriate pressure [1–3]. Increasingly, the activity of water supply companies is based on a preventive strategy consisting of the prevention of failures and crisis situations. The basis of this strategy is a risk assessment based on the detailed identification of hazards and the validation of existing safety measures. This is consistent with the World Health Organization (WHO) guidelines on Water Safety Plans (WSP) [4,5]. Reducing the number of undesirable events increases the safety of water supply to consumers [1,3,6]. Various IT tools are used to support the work of water companies. IT tools for managing WSS can be divided into four groups: Geographic Information System (GIS), Supervisory Control And Data Acquisition (SCADA), Enterprise Resource Planning (ERP) and hydraulic models [7–23]. The integrated operation of these tools allows detailed information to be obtained that can be used for monitoring the work in real mode and and to archive the obtained data and their subsequent use. Managing the WSS, implemented by an integrated IT system is aimed, inter alia, at visualization of the WSS, determination of the current operating status, optimization and design of the WSS or the location of failures and hidden water leaks [7,9,10,15]. The risk assessment in the WSS can be performed based on data obtained from IT tools, especially from the hydraulic model, which will allow for supplementing and extending the risk matrix analysis method recommended for WSP [10,11,13,22,24–26].

One of the main difficulties in implementing WSP signaled by the industry is the inability to define risk and the uncertainty in assigning point weights to individual risk factors. The World Health Organization recommends, among others, using a two-parameter risk matrix that presents risk as the product of the probability of occurrence of undesirable events and losses resulting from their occurrence [4]. It seems that this is the correct approach. However, one should refrain from adopting the point weights of the input parameters subjectively without reference to information obtained from the monitoring system or simulations. Comprehensive risk assessment related to the functioning of the WSS should take into account both the events related to the lack of water supply and water pollution. The developed method focuses on the amount of water. The safety theory does not cover all technical failures, but only those that may pose a threat to human health and life. For this reason, the work focuses on failures of the main pipes near the water treatment plant (WTP), which may result in a long lack of water supply to a large part of consumers. Due to high pressure, large diameters, and hence, a large amount of water flowing out of the pipe during a failure, their effects are significant and difficult to remove. The developed method is based on data on the failures of the WSS and a hydraulic model (Epanet 2.0) that works on the basis of historical data. The Epanet (United States Environmental Protection Agency, Washington D.C., USA) software simulated the effects of failures on the main pipes which was expressed by the number of inhabitants who do not have water as a result of the pressure drop.

Risk maps are widely used in various fields and there is a lot of research on this topic. It is a simple and effective tool that can be used to quickly identify high-risk areas. Most often, the probability of undesirable events and its consequences are taken into account. In the broadly-understood water management, the most common are flood risk maps, which are used in many countries as one of the tools for environmental decisions. An extensive, multi-criteria approach to the assessment of flood risk and the creation of flood risk maps is presented in [27]. Risk maps in WSS are not widely used, so there is not much research on this topic. This is due to the fact that the WSS management based on IT tools is a relatively new approach, especially in developing countries. The maps can relates risk associated with quantity of supplied water and risk associated with quality of supplied water [28]. Until now, research on risk maps associated with quality of supplied water have been performed mainly in developing countries, where contaminated water causes many serious diseases. In work [24], mapping was performed and concentration of fluoride in drinking water was analyzed based on the GIS system. The exposure to trace metal contamination of drinking water sources in Pakistan was analyzed in the study [29]. Analysis associated with quantity of supplied water are often based on water pipe failures simulation in a hydraulic model and pressure drop analysis [15,22,25,26,28,30]. The aim of the work was to present the methodology and to develop a risk map of lack of water supply to consumers for a selected city in Central and Eastern Europe. The article presents a failures simulation of the main pipes transporting treated water from the WTP to the city using the Epanet 2.0 program. Based on the number of inhabitants (consumers) affected with lack of water supply as a result of failure of the main pipes near the WTP and the failure rate values for these pipes the risk of lack of water supply has been determined. On this basis a risk map was developed taking into account the three-level risk scale, i.e., low, medium and high.

2. Materials and Methods

2.1. Characteristics of the Research Object

The research object is the water supply network (WSN) located in south-eastern Poland. It supplies approximately 200,000 inhabitants with water and covers an area of 120 km^2. The area is administratively divided into 30 areas as shown in Figure 1.

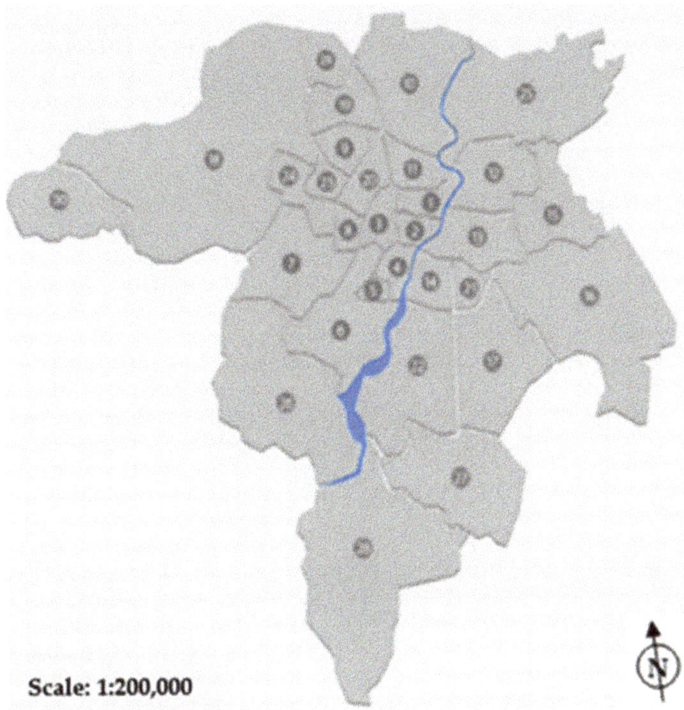

Figure 1. Administrative division of the city area covered by the analysis.

The total length of the WSN is 1025 km, of which 55 km is the main network, 605 km is the distribution network and 365 km are house connections. The system has been in use for 80 years, of which approximately 35% of pipes have been in use for less than 10 years, approximately 23% of pipes have been in use for 11–25 years, approximately 30% of pipes have been in use for 26–50 years, the oldest pipes have been in use for longer than 50 years make up about 12% of all causes. 43% of the network is made of PE, 32% are PVC pipes, 17% are cast iron pipes, 4% of pipes are made of steel and 4% of other materials. The network operates at 80% in a closed system. The network cooperates with two groups of water tanks: ZB1 and ZB2, which are located in the eastern and western parts of the city, with a total capacity of about 35,000 m^3. The scheme of the WSN has been shown in Figure 2. A skeleton of the WSN are 4 main pipes transporting treated water from the WTP to the city:

- M0 main pipe—supplies the north western part of the city and tanks ZB1,
- M1 main pipe—transports water to the central and northern parts of the city,
- M2 main pipe—transports water to the southern and central parts of the city,
- M3 main pipe—transports water from WTP to the eastern and north- -eastern parts of the city, supplying tanks ZB2.

The initial sections of these pipes, which are not connected to each other, play a key role. For this reason, a failure rate analysis was carried out for the initial sections of main pipes, which was the basis for the risk assessment of the lack of water supply to consumers in individual administrative areas. Table 1 presents the data on the initial main pipes sections. In Figure 2, the colors indicate the location of the initial main pipes sections: M0—green line, the M1—blue line, M2—red line, M3—pink line.

Figure 2. The scheme of the water supply network.

Table 1. Information on main pipe sections accepted for analysis.

Main Pipe	Diameter	Length	Material
	mm	km	
M0	1200	3.92	gray cast iron
M1	450	3.86	gray cast iron
M2	400	3.04	PE
M3	400	5.76	gray cast iron

2.2. The Risk Matrix for Lack of Water Supply

The failure simulation was made on the basis of the Authors' hydraulic model of the WSN [26]. The model has existed for several years and is constantly updated with new pipes and objects. The model was developed in the EPANET 2.0. software and has all the input parameters that allow for hydraulic modeling of the WSN:

- pipe lengths are scaled according to the real map,
- terrain ordinates and pipe diameters are consistent with the map,
- water partitioning in nodes was assumed according to the number of inhabitants,
- characteristics of pumps and tanks were adopted on the basis of information provided by the water company.

In environmental engineering (as in most engineering applications) the basic risk definition applies, which presents risk as a product of the probability of undesirable events occurrence and losses resulting from it. The paper proposes the use of the two-parameter risk matrix:

$$r_{ij} = P_i \cdot C_j \quad (1)$$

In the Equation (1) the "*Pi*" stands for point weight related to the probability of undesirable events occurrence ($i = 1, 2, \ldots, n$, where "n" is the number of the scale adopted for the probability parameter); the "*Cj*" stands for point weight related to the consequences (lack of water supply) related to the occurrence of undesirable event ($j = 1, 2, \ldots, m$ where "m" is the number of the scale adopted for the consequences parameter).

This approach is consistent with the WHO guidelines on WSP. According to the WHO guidelines [4], the risk associated with each hazard can be described by the probability of its occurrence (e.g., "low", "medium", "high") and the consequences of the hazard (e.g., "low", "medium", "high"). The potential impact on human health should be the most important element in the risk analysis, but other factors, such as continuity of water supply should also be taken into account. The water company should define in detail its understanding of the terms that are used. Risk factor point weights are often adopted on the basis of experience, knowledge and opinion of water company employees, good practices and technical literature. In order to avoid subjectivization of the risk assessment, one should strive to use the largest possible amount of numerical data that can be obtained from IT tools commonly used to manage WSS. The central point of the system is the GIS numerical map. Its use enables computer visualization of the WSN and precise location of failures on the map. The GIS system also enables the presentation of information from the monitoring system and the presentation of the results of hydraulic simulations.

It is proposed to adopt the values of P and C parameters on the basis of operational data obtained from the computer system for supervision over the water supply network, i.e., the results of the main pipes failure simulations (parameter C) and the failure rate determined on the basis of failure data (parameter P). The evaluation criteria for individual point weights are presented in Tables 2 and 3. The risk determined in accordance with (1) takes the values from the range: 1 to 25. The risk matrix is presented in Table 4. A standard three scale of risk levels is proposed:

- tolerated risk – 1 ÷ 3,
- controlled risk – 4 ÷ 12,
- unacceptable risk – 15 ÷ 25.

Table 2. Criteria and point weights for parameter P.

Probability	Failure Rate λ [Failure · km^{-1} · year $^{-1}$]	P
very low	<0.3	1
low	0.3–0.5	2
medium	0.5–0.75	3
high	0.75–1.0	4
very high	>1.0	5

Table 3. Criteria and point weights for parameter C.

Consequences	Population without Water	C
very low	<100	1
low	100–500	2
medium	500–2000	3
high	2000–5000	4
very high	>5000	5

Table 4. Risk matrix.

P	C				
	1	**2**	**3**	**4**	**5**
1	1	2	3	4	5
2	2	4	6	8	10
3	3	6	9	12	15
4	4	8	12	16	20
5	5	10	15	20	25

In order to determine the value of the consequences parameter (parameter C), the failure rate was determined, which takes into account the number of failures and the length of the analyzed main pipes [13,30]:

$$\lambda = \frac{n(\Delta t)}{L \cdot \Delta t} \left[failure \cdot km^{-1} \cdot year^{-1}\right] \tag{2}$$

In the Equation (2) the "n (Δt)" stands for the number of failures in the time interval Δt; the "L" stands for the length [km] of examined pipes in the time interval Δt and the "Δt" is the considered period of time in years.

Due to the greatest consequences for water consumers, only the failures at the initial sections of main pipes that are most severe for water consumers are considered. The range of the analyzed main pipes is shown in Figure 2.

3. Results

3.1. Water Main Pipes Failure Simulation

For the analysis, the existing hydraulic model water distribution system WDS [26] was used, which was extended with water supply pipes for newly built housing estates, and was verified and updated with model input data, e.g., water consumption, length and diameter of pipes, roughness of pipes, volume of tanks, operating parameters of the pumping station. The developed model was calibrated by comparing the pressure measurement results from several measurement points on the main pipes of the water supply network with the results of the analysis using the Epanet software. The maximum difference in the pressure values did not exceed 1.5 m. As standard, pressure measurements on the analyzed water supply network at measuring points are carried out for external hydrants for the needs of the National Fire Department.

The work shows the simulation of the failure of the initial sections of the M0, M1, M2 and M3 pipes (complete closure of the flow for a period of 24 h) carried out using the EPANET 2.0. software. These pipes were selected because of the biggest consequences of their failure. Figure 3 shows the pressure distribution in the water supply network during normal failure-free operation. In the areas of low pressure value (i.e., areas <10 m in Figure 3) in the northern and eastern part of the city are located tall buildings powered by a hydrophore. Therefore, it does not affect the continuity of water supply. The water company is in charge of 39 local hydrophore units. From these facilities, 31 are currently in use, while 8 are out of service. These objects were not included in the model, hence, the red areas in Figure 3. The developed model is the first in the history hydrualic model for the analyzed city water supply network and includes the main pumping station and the most important pipes. It is sufficient to carry out a reliable simulation of the effects of main pipes failure and to present the developed methodology. Currently the detailed hydraulic model integrated with other IT tools used in the enterprise (GIS and SCADA systems), which can be used in real conditions and run a simulation based on historical data is being developed. The real-time model should be understood as a mathematical model powered by a SCADA system, which uses the recorded values as boundary conditions for calculations. The model will binclude all main and distribution pipes and for household connections greater than or equal to DN80. It will be one of the main elements of the supervision computer system of the water supply network.

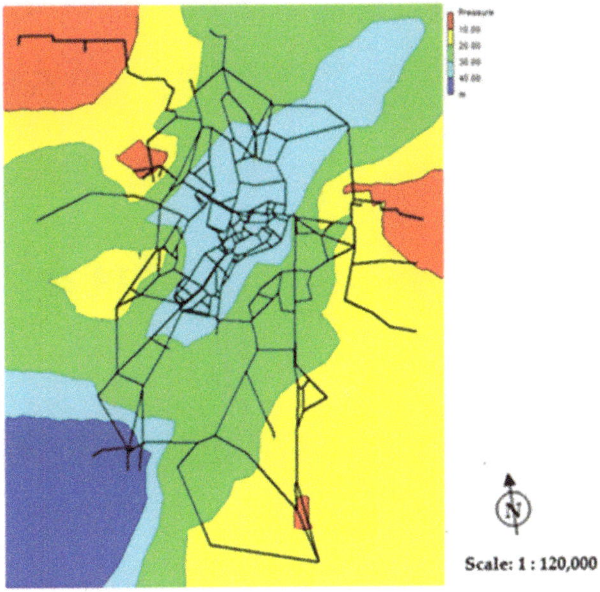

Figure 3. Pressure distribution during normal conditions.

The simulation results of the failure of four main pipes are shown in Figure 4. A decrease in pressure value below the required level causes limitation or suspension of water supply (no possibility of water consumption). The performed simulation did not show a pressure value decrease below the required pressure level in the central and northern parts of the city. This is due to the fact that the analyzed network was built as the looped network, and the city is supplied by four main pipes. Switching off 1 of 4 pipes does not cause a significant pressure value decrease and is not noticeable for most of the consumers. The inhabitants of areas located in the southern and eastern part of the city are the most exposed to risk of lack of water supply.

3.2. Failure Analysis of Water Main Pipes

Table 5 presents the values of failure rate, calculated with formula (1) for the main pipes M0, M1, M2 and M3 in the years 2004 ÷ 2018 and their basic statistics. The results of the analysis are presented graphically in Figure 5.

Table 5. Statistics of the main pipes M0–M3 failure rates.

Statistics	**Failure Rate λ [1·km^{-1}·year^{-1}]**			
	M0	**M1**	**M2**	**M3**
average	0.02	0.93	1.38	1.32
median	0	0.78	1.32	1.22
minimum	0	0.26	0.33	0.87
maximum	0.26	1.81	3.62	2.43
standard deviation	0.07	0.42	0.98	0.47
lower quartile (25%)	0	0.52	0.98	0.87
upper quartile (75%)	0	1.17	1.97	1.56

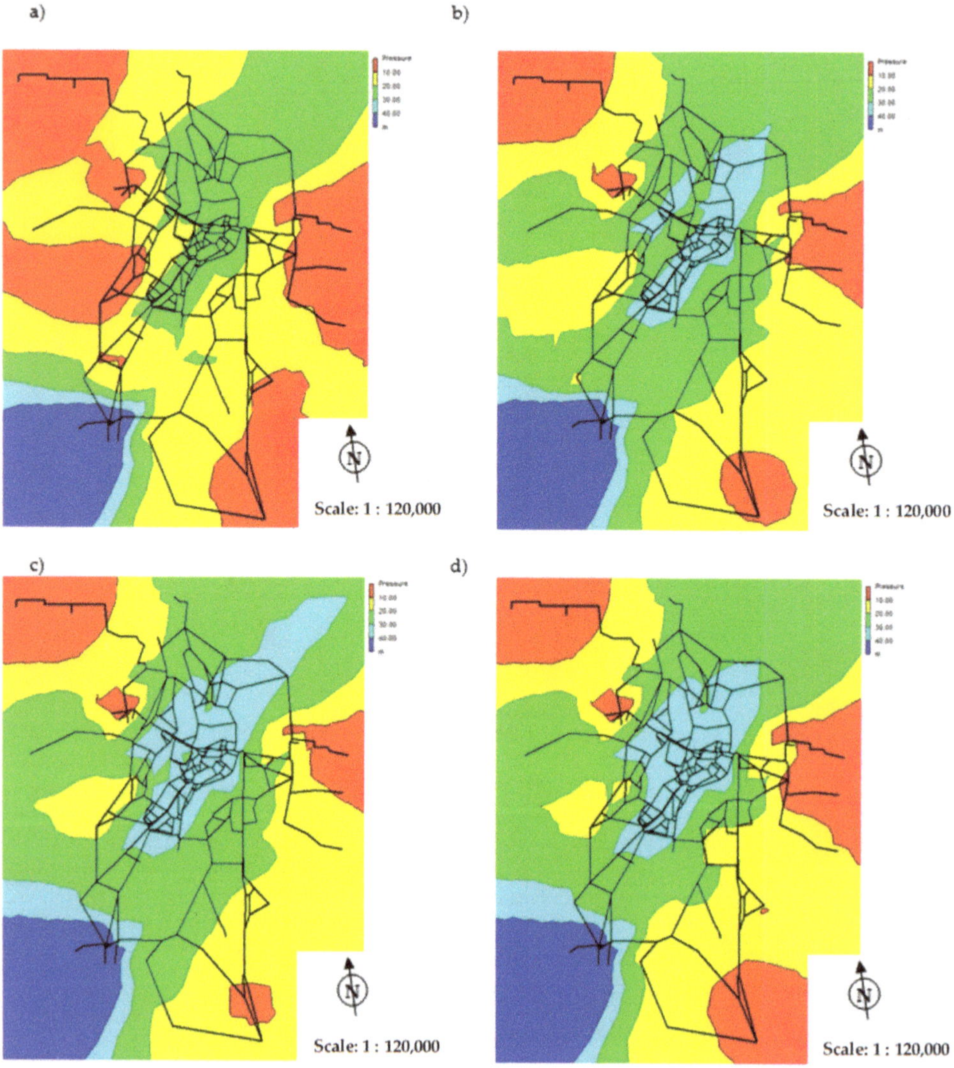

Figure 4. Pressure distribution during failure of the main pipes: (**a**) M0 pipe; (**b**) M1 pipe; (**c**) M2 pipe; (**d**) M3 pipe.

In the analyzed period, 232 main pipes failures occurred, which is: about 0.5% on the M0 pipe, about 23.5% on the M1 pipe, about 27% on the M2 pipe, and about 49% on the M3 pipe of all examined failures. The most common cause of water network failure is unsealing in gray cast iron pipes. There was only one failure on the M0 pipe in the tested period of 15 years, which proves it is in very good technical condition. The highest unit failure rate was observed for the M2 pipe, i.e., λ_{avg} = 1.38 fail./km·year and for the M3 pipe i.e., λ_{avg} = 1.32 fail./km·year. The average value of the failure rate determined for the M1 pipe is λ_{avg} = 0.93 fail./km·year. The determined λ values for M1, M2 and M3 pipes differ from the generally accepted limit value, which for the main network λ = 0.3 fail./km·year. Failure of the initial sections of main pipes, due to the high values of the failure rates in the analyzed city is a significant source of risk of the lack of water supply, thus, the adopted assumption to simulate failures of these pipes is correct.

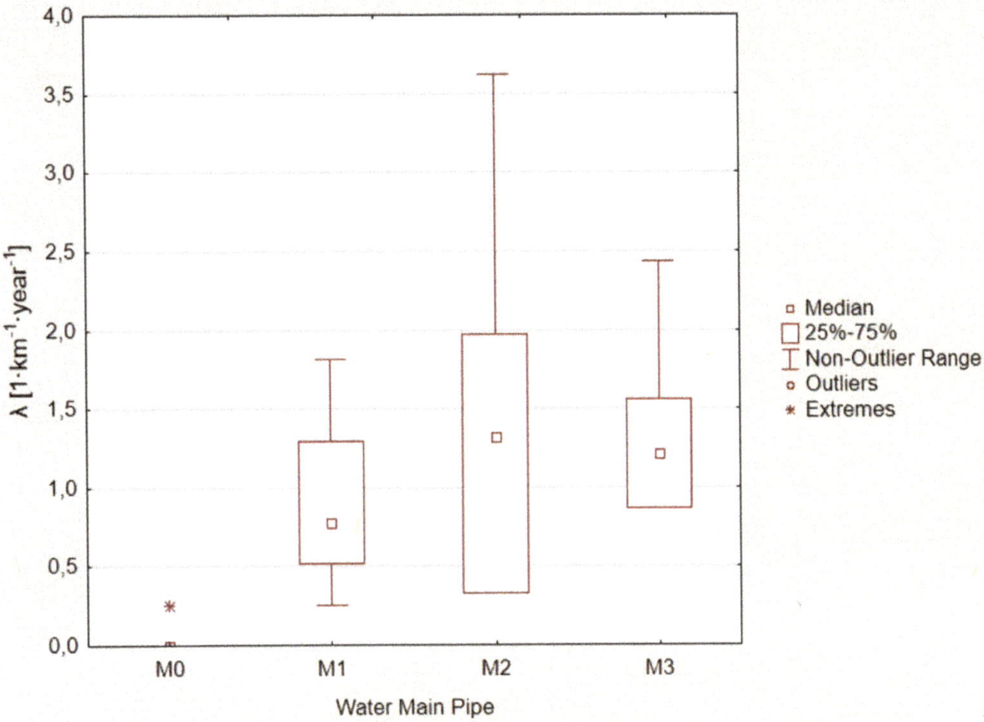

Figure 5. Failure rates of the M0, M1, M2 and M3 pipes.

3.3. The Risk of Lack of Water Supply to the Consumers

In accordance with the proposed methodology for determining the risk of the lack of water supply in a crisis situation, presented in point 2 of this study, the value of the risk of the lack of water supply was estimated for each of the city's residential areas. The failure of the M0, M1, M2 and M3 main pipes near the WTP was selected for the quantitative analysis of the risk related to the lack of water supply. The scale for the P parameters was adopted in accordance with Table 1. Probability of failure of the:

- M0 pipe: λ_{avg} = 0.02 fail./km·year → P = 1 (very small),
- M1 pipe: λ_{avg} = 0.93 fail./km·year → P = 4 (high),
- M2 pipe: λ_{avg} = 1.38 fail./km·year → P = 5 (very high),
- M3 pipe: λ_{avg} = 1.32 fail./km·year → P = 5 (very high).

The pressure distribution in water supply network obtained as a result of the simulation of the failure of the M0, M1, M2 and M3 main pipes (shutdown of the flow for a 24 h) in the area near to the WTP carried out in the EPANET 2.0 program are presented in Section 3.1. On the basis of the pressure in the network and the height of the buildings, the number of inhabitants affected with the lack of water supply was determined, as shown in Table 6. For parameter C (number of consumers affected with lack of water supply), point weights were adopted in accordance with Table 2. The risk value was calculated based on the formula (1). The water supply network supplying the area was divided into areas according to the city's administrative division.

Table 6. Number of inhabitants (consumers) affected with lack of water supply as a result of failure of the M0, M1, M2 and M3 main pipes near the WTP.

Residential Area	The Number of Inhabitants (Consumers)	Number of Inhabitants Affected with Lack of Water Supply —Failure of M0 Pipe	Parameter C Point Weight—Failure of M0 Pipe	Number of Inhabitants Affected with Lack of Water Supply—Failure of M1 Pipe	Parameter C Point Weight—Failure of M1 Pipe	Number of Inhabitants Affected with Lack of Water Supply—Failure of M2 Pipe	Parameter C Point Weight—Failure of M2 Pipe	Number of Inhabitants Affected with Lack of Water Supply—Failure of M3 Pipe	Parameter C Point Weight—Failure of M3 Pipe
1	4377	0	1	0	1	0	1	0	1
2	3358	0	1	0	1	0	1	0	1
3	4897	0	1	0	1	0	1	0	1
4	7530	0	1	0	1	0	1	0	1
5	4524	0	1	0	1	0	1	0	1
6	8604	0	1	0	1	0	1	0	1
7	3262	320	2	300	2	0	1	0	1
8	6876	0	1	0	1	0	1	0	1
9	11,322	0	1	0	1	0	1	0	1
10	4850	0	1	0	1	0	1	0	1
11	8791	0	1	0	1	0	1	0	1
12	3986	0	1	0	1	0	1	0	1
13	10,344	0	1	0	1	0	1	0	1
14	13,910	0	1	0	1	0	1	0	1
15	6106	0	1	0	1	0	1	820	3
16	7059	0	1	0	1	0	1	560	3
17	8350	0	1	0	1	0	1	840	3
18	9685	0	1	0	1	0	1	0	1
19	10,622	0	1	0	1	0	1	0	1
20	5318	0	1	0	1	0	1	0	1
21	4167	0	1	0	1	0	1	400	2
22	12,706	0	1	0	1	0	1	0	1
23	7716	0	1	0	1	0	1	0	1
24	5718	0	1	0	1	0	1	0	1
25	2323	0	1	0	1	0	1	0	1
26	3529	1060	3	700	3	0	1	0	1
27	2549	760	3	260	2	130	2	1000	4
28	959	480	2	100	2	60	1	450	2
29	6008	0	1	0	1	0	1	0	1
30	626	220	2	0	1	0	1	0	1

The map of risk of the lack of water supply for the four considered cases (failure of the M0, M1, M2 and M3 pipes) is presented in Figure 4.

Based on Figure 6, it was found that the area 27 is the most exposed to the risk of a lack of water supply. This area consist of single-family houses, small service buildings and schools. The highest risk of lack of water supply is related to the failure of the M3 main pipe, which transports water to the eastern and north-eastern parts of the city and supplies the ZB2 reservoirs. This is due to the fact that the M3 main pipe has a high failure rate (it is made of gray cast iron) and it is the only pipe supplying the eastern part of the city for which a high risk of lack of water supply was obtained. The water supply company should implement appropriate measures to reduce the risk. Due to the inconvenience for water consumers, modernization of the M3 main pipe should be considered. The analyzed fragment of the M3 pipe is 5.76 km long, which is about 0.5% of the total length of the network. Therefore, it seems that such action should be feasible for the water supply company. However, this decision should be preceded by an assessment of the company's finances and the nuisance of the renovation work for people and the environment. It is recommended that trenchless technology is used. The risk of lack of water supply in the case of a failure of the M0 main pipe is small. The recommended action is to monitor the existing pipe condition, flow rate and pressure measurement and monitoring, which will allow for quick detection of emergency situations. The average risk of a lack of water supply was obtained in the case of simulating the failure of M1 and M2 pipes. Despite the high failure rate of these pipes, which proves that they are not in the best technical condition, their location does not pose a high risk of lack of water supply. However, it should be expected that their technical condition will deteriorate and their modernization should be considered in the longer term.

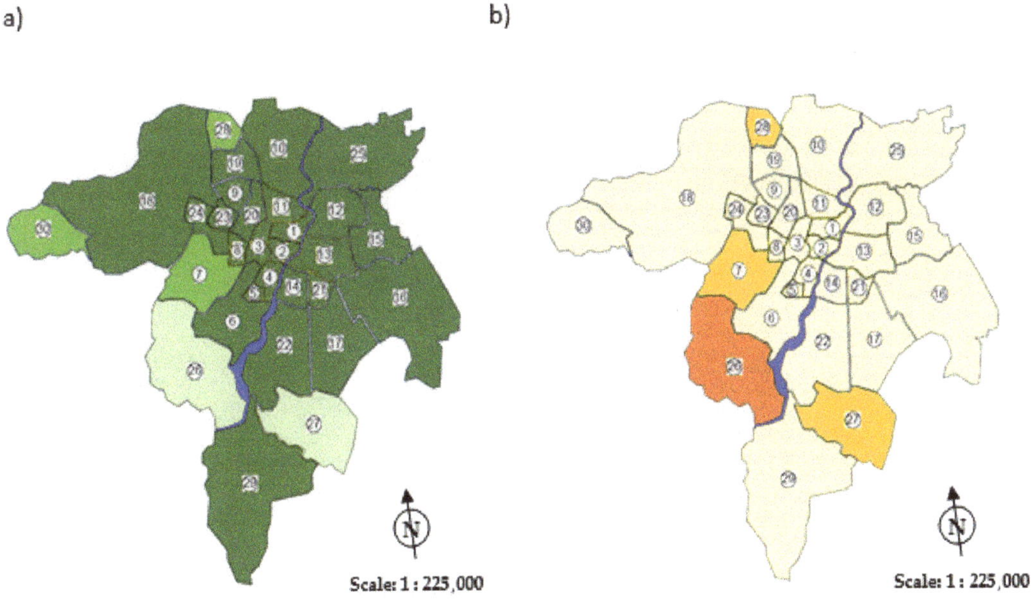

Figure 6. *Cont.*

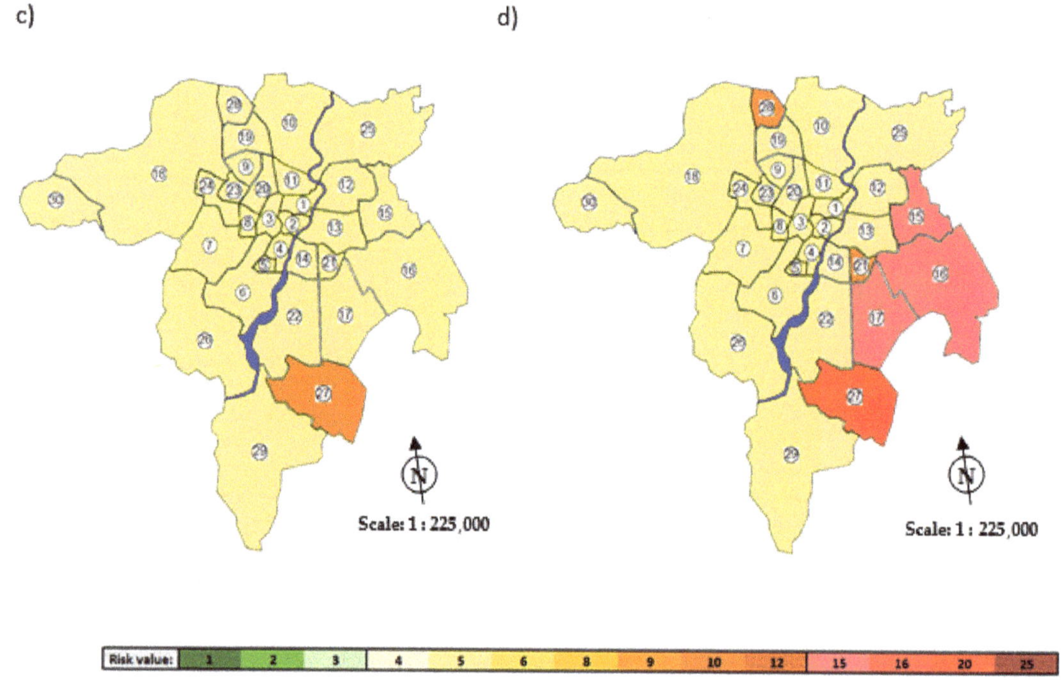

Figure 6. Map of risk of water supply lack in case of failure of the: (**a**) M0 main pipe; (**b**) M1 main pipe; (**c**) M2 main pipe; (**d**) M3 main pipe.

4. Conclusions and Perspectives

It is possible to simulate the closure (i.e., failure) of individual pipes in the water supply network, and then a comparative analysis of the pressure distribution in this network in failure-free conditions and during failure. In this way, it is possible to determine the consequence of failures on individual pipes in relation to the area where the pressure value dropped below the required level, the duration of this pressure reduction and the potential number of consumers affected by the limitation of water supply.

The presented method is based on failure-data of the water supply network and a simulating hydraulic model. The usefulness of the method depends on the reliability of the input data. The proposed approach is in line with the WHO guidelines on risk analysis in water supply systems.

This study focuses on the quantitative aspect of drinking water supply, the comprehensive analysis should also take into account the qualitative aspect, which can also be done in EPANET software (e.g., EPANET-MSX) [16,23,24,29]. Based on the results obtained, strategies can be developed to prevent main pipes failures that cause long-term water supply outages. The simulation results indicated that the highest risk was posed by the failure of the M3 main pipe, which led to the conclusion that it requires renovation. The analyzed network was made as a looped network, and the city is supplied from four main pipes, thus, for the most of housing estates, there is no high risk of lack of the water supply, even during a long-term failure of the city's main water pipes.

The method is fully reproducible and can be used to analyze the operation of other water supply networks. It can be especially useful for water supply networks with one intake or a small number of pipes supplying the city, i.e., small water supply systems. The method can be the basis of an operational strategy aimed at maintaining water supplies, which are now becoming reliability-oriented, i.e., monitoring, surveys and preventive maintenance.

Author Contributions: All authors equally contributed to the development of this manuscript. All authors have read and agreed to the published version of the manuscript.

Funding: This research received no external funding.

Acknowledgments: We thank the reviewers for their feedback, which helped to improve the quality of the manuscript.

Conflicts of Interest: The authors declare no conflict of interest.

References

1. Pietrucha-Urbanik, K.; Rak, J. Consumers' Perceptions of the Supply of Tap Water in Crisis Situations. *Energies* **2020**, *13*, 3617. [CrossRef]
2. Roeger, A.; Tavares, A.F. Do Governance Arrangements Affect the Voluntary Adoption of Water Safety Plans? An Empirical Study of Water Utilities in Portugal. *Water Resour. Manag.* **2020**, *34*, 1757–1772. [CrossRef]
3. Szpak, D.; Tchórzewska-Cieślak, B. The Use of Grey Systems Theory to Analyze the Water Supply Systems Safety. *Water Resour. Manag.* **2019**, *33*, 4141–4155. [CrossRef]
4. *Guidelines for Drinking-Water Quality*, 4th ed.; WHO: Geneva, Switzerland, 2011.
5. *Water Safety Plan Manual: Step-By-Step Risk Management for Drinking-Water Suppliers*; WHO: Geneva, Switzerland, 2009.
6. Kordana, S.; Słyś, D. Decision Criteria for the Development of Stormwater Management Systems in Poland. *Resources* **2020**, *9*, 20. [CrossRef]
7. Saravanan, K.; Anusuya, E.; Kumar, R. Real-time water quality monitoring using Internet of Things in SCADA. *Environ. Monit. Assess.* **2018**, *190*, 556. [CrossRef]
8. Fontenot, E.; Ingeduld, P.; Wood, D. Real Time Analysis of Water Supply and Water Distribution Systems. In Proceedings of the World Water & Environmental Resources Congress, Philadelphia, PA, USA, 23–26 June 2003.
9. Weber, C.; Bunn, S. Water distribution optimization: Taking scada one step forward. In Proceedings of the World Environmental and Water Resources Congress 2009: Great Rivers, Kansas City, MO, USA, 17–21 May 2009.
10. Stańczyk, J.; Burszta-Adamiak, E. The Analysis of Water Supply Operating Conditions Systems by Means of Empirical Exponents. *Water* **2019**, *11*, 2452. [CrossRef]
11. Aral, M.M.; Maslia, M.L. Evaluation of human exposure to contaminated water supplies using GIS and modelling. In Proceedings of the Application of geographic information systems in hydrology and water resources management (Proc. HydroGIS'96 conference), Vienna, Austria, 16–19 June 1996; pp. 243–252.
12. Yan, B.; Su, X.R.; Chen, Y.Y. Functional Structure and Data Management of Urban Water Supply Network Based on GIS. *Water Resour. Manag.* **2009**, *23*, 2633–2653. [CrossRef]
13. Kubat, B.; Kwietniewski, M. Modelling and visualization of failure rate of a water supply network using the regression method and GIS. *Desalination Water Treat.* **2020**, *186*, 19–28. [CrossRef]
14. Tchórzewska-Cieślak, B.; Rak, J.R.; Szpak, D. Method of Analysis and Assessment of ICT System Safety in a Water Company. *J. Konbin* **2019**, *49*, 301–318. [CrossRef]
15. Lu, R.Q.; Zhang, H.W.; Niu, Z.G.; Peng, W.J. Study on an arrangement method for pressure monitoring points in urban water supply network by nonlinear mapping theory. *J. Hydraul. Eng.* **2010**, *41*, 25–29.
16. Seyoum, A.G.; Tanyimboh, T.T.; Siew, C. Assessment of water quality modelling capabilities of EPANET multiple species and pressure-dependent extension models. *Water Supply* **2013**, *13*, 1161–1166. [CrossRef]
17. Agunwamba, J.C.; Ekwule, O.R.; Nnaji, C.C. Performance evaluation of a municipal water distribution system using WaterCAD and Epanet. *J. Water Sanit. Hyg. Dev.* **2018**, *8*, 459–467. [CrossRef]
18. Menapace, A.; Pisaturo, G.R.; De Luca, A.; Gerola, D.; Righetti, M. EPANET in QGIS framework: The QEPANET plugin. *J. Water Supply Res. Technol.* **2019**, *69*, 1–5. [CrossRef]
19. Sun, F.; Chen, J.-N.; Zeng, S.-Y. Development and application of a multi-species water quality model for water distribution systems with EPANET-MSX. *Huanjing Kexue* **2008**, *29*, 3360–3367. [PubMed]
20. Ramana, G.V.; Chekka, V.S.S.S. Validation and Examination of Existing Water Distribution Network for Continuous Supply of Water Using EPANET. *Water Resour. Manag.* **2018**, *32*, 1993–2011. [CrossRef]
21. Mohapatra, S.; Sargaonkar, A.; Labhasetwar, P.K. Distribution Network Assessment using EPANET for Intermittent and Continuous Water Supply. *Water Resour. Manag.* **2014**, *28*, 3745–3759. [CrossRef]
22. Duan, X.; Zong, Y.; Hao, K.; Huang, D.; Li, Y. Research of hydraulic reliability of water supply network based on the simulation of EPANET. *IOP Conf. Series Earth Environ. Sci.* **2019**, *349*, 012042. [CrossRef]
23. Zhang, F.-E.; Chen, C.; Li, Y.; Dong, L.; Tu, B.; Zheng, M.; Zhou, R. The investigation of relationship between residual chlorine simulation attenuation and nodes water age in water distribution system by EPANET. In Proceedings of the 2011 International Symposium on Water Resource and Environmental Protection, Xi'an, Shaanxi Province, China, 20–22 May 2011; Volume 1.
24. Aslani, H.; Zarei, M.; Taghipour, H.; Khashabi, E.; Ghanbari, H.; Ejlali, A. Monitoring, mapping and health risk assessment of fluoride in drinking water supplies in rural areas of Maku and Poldasht, Iran. *Environ. Geochem. Heal.* **2019**, *41*, 2281–2294. [CrossRef]

25. Boryczko, K.; Tchorzewska-Cieslak, B. Analysis and assessment of the risk of lack of water supply using the EPANET program. Environmental Engineering IV. In Proceedings of the Conference on Environmental Engineering IV, Lublin, Poland, 3–5 September 2012.
26. Boryczko, K.; Tchórzewska-Cieślak, B. Analysis of risk of failure in water main pipe network and of delivering poor quality water. *Environ. Prot. Eng.* **2014**, *40*. [CrossRef]
27. Meyer, V.; Scheuer, S.; Haase, D. A multicriteria approach for flood risk mapping exemplified at the Mulde river, Germany. *Nat. Hazards* **2009**, *48*, 17–39. [CrossRef]
28. Piegdoń, I.; Tchórzewska-Cieślak, B. Methods of visualizing the risk of lack of water supply. In Proceedings of the Safety and Reliability: Methodology and Application (Proc. Eur. Saf. Reliab. Conf., ESREL), Wroclaw, Poland, 14–18 September 2014.
29. Bhowmik, A.K.; Alamdar, A.; Katsoyiannis, I.; Shen, H.; Ali, N.; Ali, S.M.; Bokhari, H.; Schäfer, R.B.; Eqani, S.A.M.A.S. Mapping human health risks from exposure to trace metal contamination of drinking water sources in Pakistan. *Sci. Total. Environ.* **2015**, *538*, 306–316. [CrossRef] [PubMed]
30. Pietrucha-Urbanik, K.; Studziński, A. Qualitative analyscis of the failure risk of water pipes in terms of water supply safety. *Eng. Fail. Anal.* **2019**, *95*, 371–378. [CrossRef]

resources

MDPI

Article

An Approach to Analysing Water Consumers' Acceptance of Risk-Reduction Costs

Barbara Tchórzewska-Cieślak [1], Katarzyna Pietrucha-Urbanik [1,*] and Emilia Kuliczkowska [2]

[1] Department of Water Supply and Sewerage Systems, Faculty of Civil, Environmental Engineering and Architecture, Rzeszow University of Technology, Al. Powstancow Warszawy 6, 35-959 Rzeszow, Poland; cbarbara@prz.edu.pl

[2] Department of Geotechnics, Geomatics and Waste Management, Faculty of Environmental, Geomatic and Energy Engineering, Kielce University of Technology, Al. 1000-lecia Panstwa Polskiego 7, 25-314 Kielce, Poland; emkulicz@tu.kielce.pl

* Correspondence: kpiet@prz.edu.pl; Tel.: +48-17-865-1703

Received: 18 August 2020; Accepted: 4 November 2020; Published: 6 November 2020

Abstract: The proper operation of a water supply system (WSS) requires constant investment. The priority is to provide residents with high quality potable water, in the required quantity and pressure, in accordance with the applicable regulations. The paper presents an assessment of the potential inherent operational risk of a WSS in support of the risk-based investment management process. It is of high importance to invest in the operational safety as it concerns both producers and consumers. The investment engenders additional costs that should partially be supported by the consumers. Thus, the paper presents a methodology to analyse consumers' readiness to accept water supply services' additional costs. The proposed methods may underpin a comprehensive program for risk-based investment management and operational decision-making. The case study and the approach in this article concern one particular regional WSS, based on information collected from water consumers. The assessment suggests a willingness to tolerate additional costs in view of enhancing the performance of the water supply services.

Keywords: water supply system; risk analysis; risk management; safety of water supply

1. Introduction

A global water supply safety system represents a collection of organisations, institutions, technical systems, educational, and control measures, whose aim is to ensure the safety of those who consume water. A safety management system tends to be introduced at the local level [1,2], and in today's world typically, or even necessarily, embraces risk management [3]. In the context of a water supply company, this is a multi-stage procedure that aims at improving system security, especially regarding the supply of drinking water in both its quantitative and qualitative aspects [4]. Key aspects here are risk analysis, risk assessment, and decision-making regarding acceptability and temporary control [5,6].

At the outset, it is necessary to recall how risk derives from a lack of knowledge of events that may occur in the human environment as a process operator [7,8]. On that basis, today's active risk management entails the identification and analysis of the causes of risk, limiting losses, and building a strategy of success. The effects to be achieved are risk control and risk reduction to tolerable levels [9,10].

In the case of water distribution systems, the most important factors are production, logistics, and research and development [11]. Where these are integrated, risk management can be coordinated [12]. The risk associated with the production and distribution of drinking water relates in particular to the likelihood of adverse events and the extent of possible damage [13]. During the phase in which water is produced, risk can be assigned to either strategic or operational categories [14].

Strategic production risk is long-term, relating to the investment decisions water supply companies take, not least regarding the quality of the water source as conceived broadly (activities in compliance with provisions on zones for the protection of intakes) [15]. Operational production risk, in turn, refers to current disruptions in the production and distribution of water, and is of a short-term nature [16,17].

Logistic risk is determined in the planning phase of water supply operations, and relates to its management methods [18,19]. The purpose of logistic protection in crisis situations (e.g., flood or drought) is to ensure appropriate organisation and operational effectiveness within the framework of crisis response, in a manner adequate to the level of the threat. In the Polish case specifically, this should be in accordance with the country's Crisis Management Act. The most important logistical safeguards limiting risk include, among other things [20–22]:

- Developing emergency plans in the event of failures of various kinds;
- Securing the work of means of communication and an alarm notification network;
- Arranging the supply of water from alternative sources (e.g., water delivery to housing estates in water tankers);
- Providing basic medical and sanitary measures (e.g., for water supply);
- Providing resources as part of external support (e.g., in the supply of bottled water to the population);
- Organising preventive inspections aimed at assessing the state of emergency protection and identifying potential new causes of failures;
- Developing a crew-training system in the event of a crisis, including periodic civil-defence exercises.

Risk management should be treated as a process inseparable from the management of the entire water supply company, as related to the development of risk response methods, i.e., the preparation of organisational infrastructure in support of risk management [23]. Risk identification is based on the selection of representative emergency events that may occur during system operation, including initiating events capable of inducing a so-called domino effect [24]. Risk assessment is then a process of qualitative and quantitative analysis using methods adequate to a given type of risk and determining the criterion value for the adopted risk scale. Risk scale may be a three-level scale distinguishing tolerated, controlled, and unacceptable risk, or a five-level one in which the risk area is additionally distinguished as neglected and absolutely unacceptable [25]. Risk control and reduction involves the implementation of procedures as well as the registration and evaluation of the results of changes [26]. This helps the company to define its policy and implies a process for the implementation of measures in order to obtain an acceptable risk level at an acceptable cost [27,28]. In the literature, the following indicators can be used to analyse risk-reduction costs [29,30]:

- Assumed cost of preventing undesirable events, the so-called ICAF—Implied Cost of Averting a Fatality:

$$\mathrm{ICAF} = \Delta\mathrm{Cost}/\Delta \mathrm{r}, \tag{1}$$

where ΔCost is the cost of protection, or the prevention of undesirable events; Δr is the degree of risk reduction.

$$\Delta \mathrm{r} = \mathrm{r_p} - \mathrm{r_k}, \tag{2}$$

where $\mathrm{r_p}$ is the initial value of risk and $\mathrm{r_k}$ is the risk value following the introduction of additional protective and preventative actions.

- An indicator called the Cost of Unit Risk Reduction—CURR:

$$\mathrm{CURR} = \frac{\Delta\mathrm{Cost} - \Delta\mathrm{EB}}{\Delta \mathrm{r}}, \tag{3}$$

where ΔEB are Economic Benefits—profit related to risk reduction.

A water supply company's priority is to ensure the continuity of drinking water supply of adequate quality. Ageing water distribution systems and growing quality requirements demand large financial outlays. Consequently, this impacts on the cost of water supply services. The process of communicating with water consumers, and of marketing and information activities, represents an integral aspect of the management of a water supply company, including its risk management. The approach presented here gives a view to a water supply company in the context of management and informs consumers about the functioning of water utilities. Water-supply companies are obliged to analyse risk, to develop water safety plans, and thus, to pursue modernisation and risk reduction. Given recent epidemiological threats related to COVID-19, additional procedures are in place, and should also gain the acceptance of consumers. All these activities have an impact on the price of water in the context of risk reduction and increasing levels of safety. Therefore, consumers have to be informed of company pricing politics. In line with the approach presented here, there is a clear indication as to whether the means of informing, communicating with, and explaining to consumers are appropriate, and why these costs are incurred. Company managers should assess the acceptability of their modernisation actions to reduce risk and the corresponding cost by the consumers.

2. Criteria for Risk Acceptance as an Element of Risk Management

Risk assessment in the context of a WSS consists of risk analysis and risk evaluation [31,32]. The former should also assess the functional limitations of individual WSS subsystems [33]. More generally, the results of the risk analysis represent input into risk evaluation, whose purpose is to decide whether risks are within tolerable limits, or whether there should be reduction via [34]:

- Systemic solutions (e.g., modernisation of the system, alternative water sources kept in constant readiness, emergency capacity in water tanks, correction of water treatment technology, redesign of the water supply network, alternative energy sources, reserving strategic network facilities, and the introduction of remote supervision and system control using GIS);
- Technical and complementary protective measures (the introduction of a multi-barrier system, i.e., an early, delayed, or late warning system);
- Information for water consumers.

Figure 1 presents detailed risk management procedures for a water supply company.

The definition of risk-acceptability criteria should primarily take account of aspects related to the safety of water consumers, and technical or technical/economic analysis. Such criteria are used for decisions that are made on running the system (e.g., regarding renovation, modernisation, and authorisation for use). According to [35], a system can be considered safe if the level of risk generated during its operation does not exceed certain limits.

In situations where criteria regarding acceptable risk have not yet been established, the risk value at the given moment can be considered a measure of safety [36]. This is the basis to determine the criteria that offer a value of tolerated risk. The risk acceptability level and risk-assessment methods are often subject to legal regulations regarding specific technical systems [37–39], e.g., transport-related or industrial. Given that, the more exigent the risk-acceptance criteria are, the more extensive the security and protection measures are [40–42]. Risk-acceptability criteria are then an important element to a water supply company's financial policy [43].

Determination of criterion values in risk assessment denotes the application of the so-called As Low As Reasonably Practicable (ALARP) principle, the assumption being that the level of risk should be seen in this way, with the "reasonably" aspect implying justification of an economic and technical nature. The ALARP principle was first introduced in the United Kingdom. Accordingly, it was considered as an unacceptable value for risk of death: for individual employees ($r = 0.001$) and for the population ($r = 0.0001$). The ALARP principle assumes that the entire risk range is divided into three areas:

- The area below the lower limit, i.e., of widely tolerated risk;
- The area above the upper limit, i.e., of unacceptable risk;
- The ALARP area between the two given boundaries.

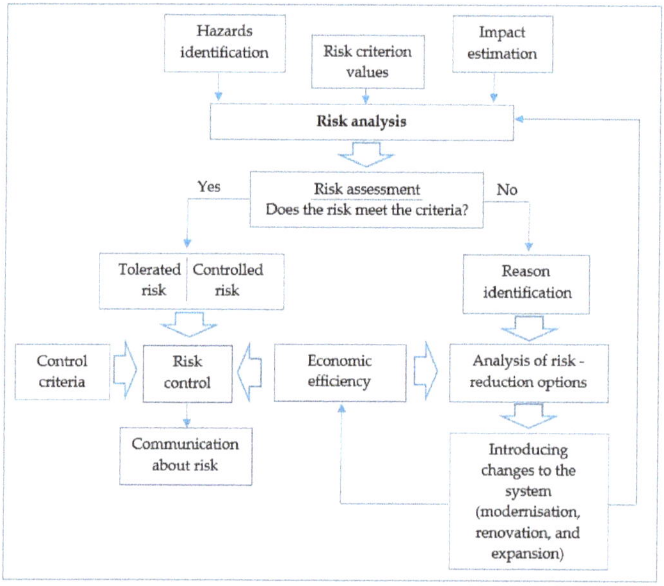

Figure 1. Risk management procedures in a water supply company.

Figure 2 proposes criteria values for the individual risks faced by water consumers, with reference to the ALARP concept.

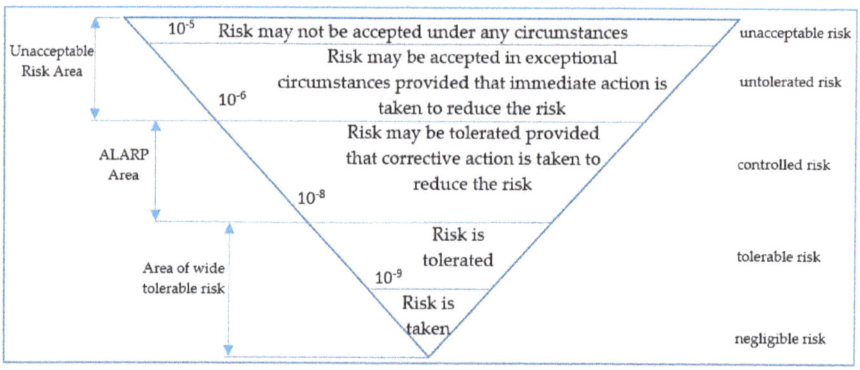

Figure 2. Proposed criteria for the accepted risk to water consumers.

The risk-reduction process should include a cost–benefit analysis. The level of risk to be determined is that at which the costs of further reduction are disproportionately high in relation to potential benefits. The Health and Safety Executive (HSE) guidelines introduce the concept of the statistical cost of avoiding fatalities, which, in accordance with the given guidelines, is estimated at around GBP (British Pound Sterling) 1 million.

3. Materials and Methods

3.1. Materials

The distinguished water supply system is located in Poland's Podkarpacie Province. The supplied city is on the right bank of the Vistula in the southeastern part of the country, covers a total area of 130 km^2, and lies at 50°02′01″ N 22°00′17″ E. In 2019, the city had an urbanisation rate of 1546 inhabitants/km^2 and is at between 197 and 384 m a.s.l.

The city is supplied with water from the river using a bank chamber water intake with a capacity of 84,000 m^3/d. Water is treated in two modernised Water Treatment Plants and meets the quality requirements for water intended for human consumption. In total, about 190,000 residents of the city and nearby towns are using the distinguished water supply system currently. In 2019, the average daily production of treated water was approx. 34,600 m^3/day, with customers' demand for water met fully in this capacity.

The research was carried out in relation to the assessment of water supply services by water consumers (Table 1), which were used to determine the quality indicators. The conducted tests form an element of regional studies and are preliminary partial tests used to estimate the assessments included in the assumed indicator. The selection of a representative random sample was assumed according to guidelines presented in [35]. The research was conducted on a sample of 200 people (of legal age over the age of 18), prepared in line with recommendations contained in the literature. Statistical analysis was performed with the Statsoft software, version 12 [44]. The highest reliability of the questionnaire occurred with ten questions. The further elimination of questions does not improve the result, but worsens it, so it is the optimal set of questions for the formulated set of 10 questions. None of the people who completed the questionnaire were excluded. Reliability statistics showed that the Cronbach alpha value has a high reliability (consistency) for the questionnaire. Differences concerning mean values for variables were tested with the Student *t*-test, and for proportions, the Z test was applied. In order to evaluate the equality of distributions, a Kolmogorov–Smirnov test was performed. A *p*-value of less than 0.05 was considered statistically significant. No statistically significant differences between gender and age group were observed.

Table 1. Point weights for individual Acceptable Risk Index (ARI) parameters.

Parameter	Description of the Parameter	Point Weight
Degree of drinking water quality assessment Q_i	big	3
	average	2
	small	1
Degree of water prices acceptance by consumers P_j	good	3
	average	2
	bad	1
Degree of water consumers' trust T_k	high	3
	average	2
	low	1

3.2. Assumptions Underpinning the Method Analysing Consumers' Acceptance of Risk-Reduction Costs

As the risk-reduction process requires financial investment, an impact on the price of drinking water is exerted and should be accepted by water consumers. The level of acceptance for a water company's expenditure related to risk management depends on various factors, such as: a quality-of-life indicator, the awareness of water consumers regarding the risks arising from lack of risk management procedures, and the degree of confidence in the water supply company. Experts in the sector of water supply system management tested, adjusted, and validated a survey through a pilot study. The question was selected in such a way as to gather information about quality of services related to the use of public water supply systems, in particular, reliable and safe access to drinking water (in accordance with applicable regulations). An important element in this respect is also the subjective assessment of water consumers, their sense of safety, and their trust in the water supply company as to the quality of services provided, which translates directly into the acceptance of actions (including acceptance of the water price) undertaken by the water utility (e.g., modernization, renovation, etc.)

The use of so-called Acceptance Risk Index (ARI) achieves an evaluation of the level of acceptance by water consumers of the costs incurred by a supply company, as it implements risk management methods [35].

The Acceptance Risk Index can be described as a function of subjective assessments related to the assessment by consumers (users) of the functioning of a water supply company:

$$ARI_{ijk} = Q_i \times P_j \times T_k, \tag{4}$$

The adopted methodology proposed the following assessments included in the ARI_{jik}:

- Q_i is a consumers' assessment of drinking water quality, for i = 1, 2, 3;
- P_j is an assessment related to the acceptance of water prices by consumers, for j = 1, 2, 3;
- T_k is an assessment related to the degree of trust and knowledge of water consumers as regards activities carried out by a water supply company, including the trust that tap water is safe for health, for k = 1, 2, 3.

The next stage of the procedure is to classify ARI values into individual sets that characterize the obtained values of Q_i, P_j, and T_k on the adopted point weights presented in the Table 1.

The obtained values of Q_i, P_j, and T_k take values from a three-point scale from 1 to 3 on the basis of the performed assessment related to the consumers' assessment [35].

The set of possible ARI values, ARI = $\{ARI_{ijk}\}$, can be represented in the form of the matrix.

In this way, the data matrix M_{ARI} for the ARI indicator is as follows:

$$M_{ARI} = \left|ARI_{ijk}\right| = \begin{vmatrix} ARI_{111} & ARI_{211} & ARI_{311} \\ ARI_{112} & ARI_{212} & ARI_{312} \\ ARI_{113} & ARI_{213} & ARI_{313} \\ ARI_{121} & ARI_{221} & ARI_{321} \\ ARI_{122} & ARI_{222} & ARI_{322} \\ ARI_{123} & ARI_{223} & ARI_{323} \\ ARI_{131} & ARI_{231} & ARI_{331} \\ ARI_{132} & ARI_{232} & ARI_{322} \\ ARI_{133} & ARI_{233} & ARI_{333} \end{vmatrix} \tag{5}$$

The ARI index assumes values from 1 to 27.

The classification proposed for acceptance levels has:

- High-scale (when ARI takes values equals or higher than 12), $ARI_h = \{ARI_{322}, ARI_{223}, ARI_{323}, ARI_{232}, ARI_{322}, ARI_{233}, ARI_{333}\}$.
- Average-scale (when ARI takes values from 6 to 12), $ARI_a = \{ARI_{312}, ARI_{213}, ARI_{313}, ARI_{321}, ARI_{222}, ARI_{123}, ARI_{231}, ARI_{331}, ARI_{132}, ARI_{133}\}$.
- Low-scale (when ARI takes values less than 6), $ARI_l = \{ARI_{111}, ARI_{211}, ARI_{311}, ARI_{112}, ARI_{212}, ARI_{113}, ARI_{121}, ARI_{221}, ARI_{122}, ARI_{131}\}$.

The assessment of the analysis is in turn as follows:

- If the ARI level is classified as high on the scale provided, this denotes acceptance by consumers of the costs incurred by the water supply company.
- If, on a given scale, the level of ARI is classified as medium, this means that the costs incurred by the water supply company are tolerated by consumers. However, the company should pursue an information campaign to convince consumers of the necessity of the new measures being applied.

- If the ARI level is classified as low on the scale provided, this means that the costs incurred by the water supply company are not accepted by water consumers. The company should verify its action plan and make adjustments in the design phase.

3.3. Results of Research

The results of the research related to consumers' assessments of the quality of water supply services were used to determine the ARI value. The structuring of answers is as presented in Tables 2–5 [35], while the results of the research (quality assessment of water supply services) are included in Figures 3–12.

Table 2. Scheme of possible answers from consumers of water as regards the assessed quality of supply services regarding the parameters of ARI—degree of drinking-water quality assessment.

Degree of Drinking-Water Quality Assessment—Q_i	
No. → Question	1. How do you assess the quality of tap water at your place of residence?
Assessment	○ very bad ○ bad ○ sufficient ○ good ○ very good ○ I have no opinion

Table 3. Scheme of possible answers from consumers of water as regards the assessed quality of supply services regarding the parameters of ARI—degree of the acceptance of water prices by consumers.

Degree of the Acceptance of Water Prices by Consumers—P_j	
No. → Question	2. Do you accept the price of water at your place of residence?
Assessment	○ definitely not ○ not ○ rather not ○ rather yes ○ yes ○ I have no opinion

Table 4. Scheme of possible answers from consumers of water as regards the assessed quality of supply services regarding the parameters of ARI—degree of trust of water consumers.

Degree of Trust of Water Consumers—T_k	
No. → Question Assessment	3. How do you assess the work of the company supplying water at your place of residence? ○ very bad ○ bad ○ sufficient ○ good ○ very good ○ I have no opinion
No. → Question Assessment	4. In the event of water in your water supply network being absent or contaminated, would you know where the nearest alternative water source is? ○ definitely not ○ not ○ rather not ○ rather yes ○ yes ○ I have no opinion
No. → Question Assessment	5. Do you trust the company, that the tap water is safe for your health? ○ definitely not ○ not ○ maybe not ○ maybe yes ○ yes ○ I have no opinion
No. → Question Assessment	6. Do you accept the activities of the water supply company carried out in order to modernise, expand and renovate the water supply system at your place of residence? ○ definitely not ○ not ○ rather not ○ rather yes ○ yes ○ I have no opinion

Table 5. Scheme of possible answers from consumers of water as regards the assessed quality of supply services regarding the parameters of ARI—consumer knowledge regarding the city's water supply.

	Consumer Knowledge Regarding the City's Water Supply
No. → Question Assessment	7. Have you experienced any nuisance related to the lack of tap water? ○ definitely not ○ not ○ rather not ○ rather yes ○ yes ○ I have no opinion
No. → Question Assessment	8. Have you experienced any problems related to the consumption of tap water of inadequate quality? ○ definitely not ○ not ○ rather not ○ rather yes ○ yes ○ I have no opinion
No. → Question Assessment	9. Do you save water? ○ definitely not ○ not ○ rather not ○ rather yes ○ yes ○ I have no opinion
No. → Question Assessment	10. How do you assess the quality of water resources in the region? ○ definitely not ○ not ○ rather not ○ rather yes ○ yes ○ I have no opinion

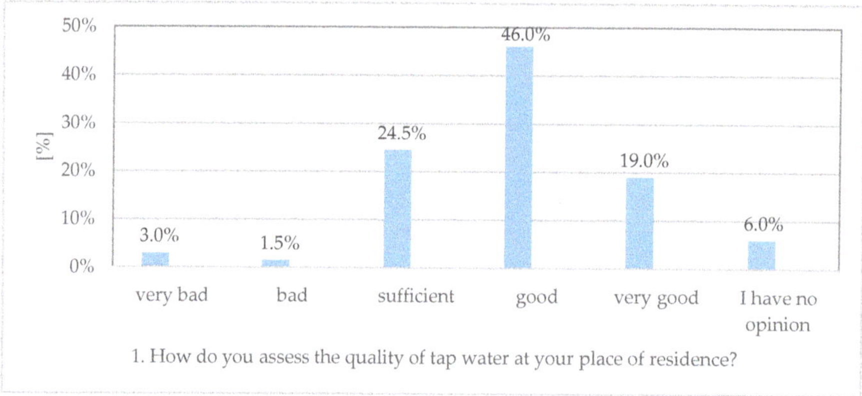

Figure 3. Results for answers to Question 1.

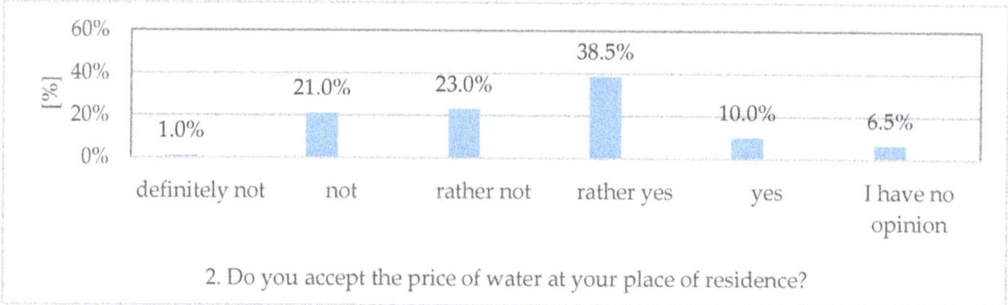

Figure 4. Results for answers to Question 2.

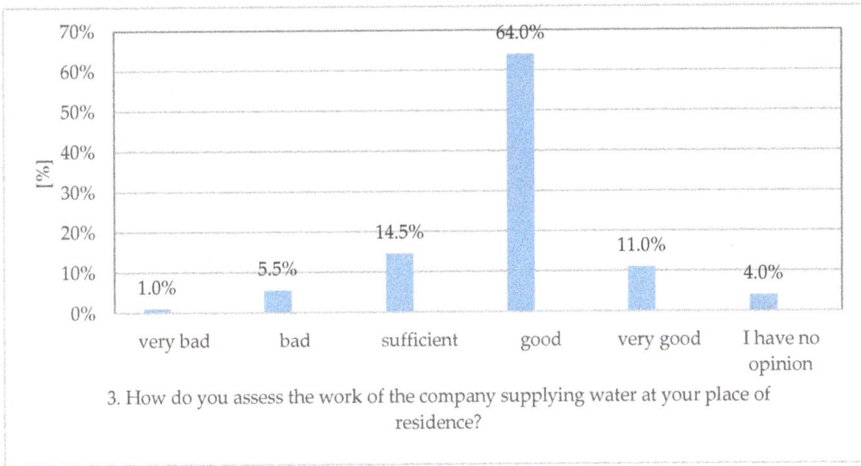

Figure 5. Results for answers to Question 3.

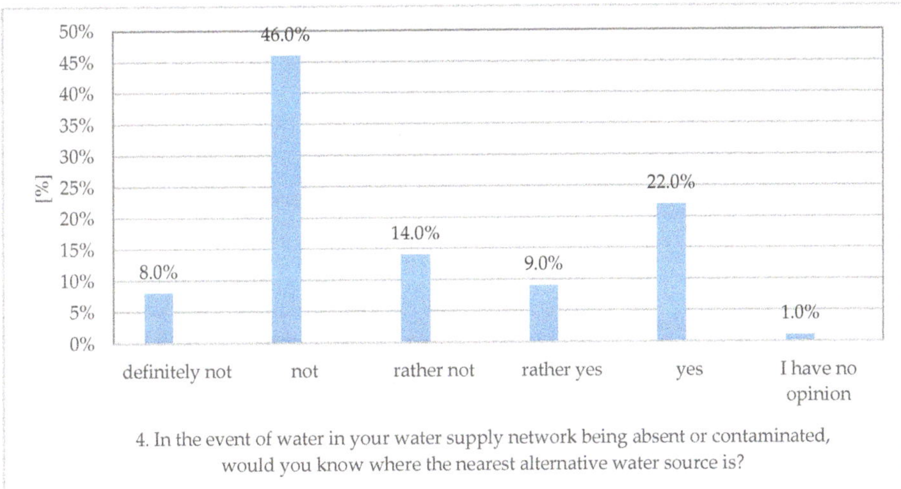

Figure 6. Results for answers to Question 4.

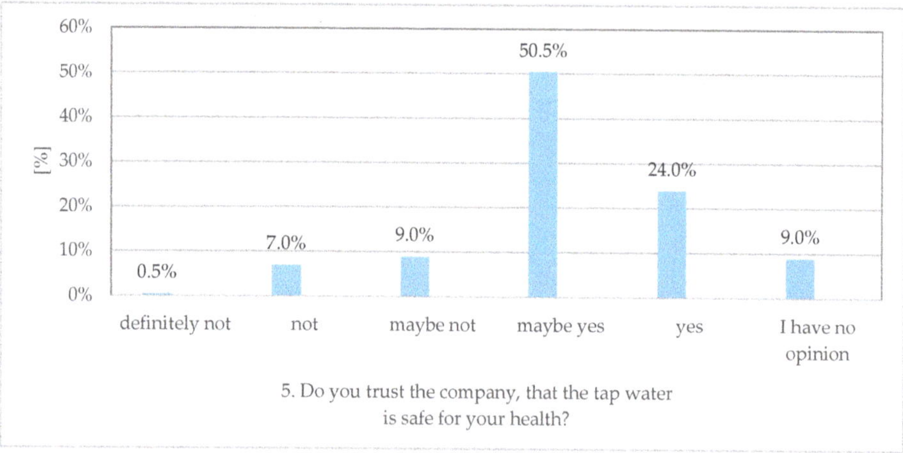

Figure 7. Results for answers to Question 5.

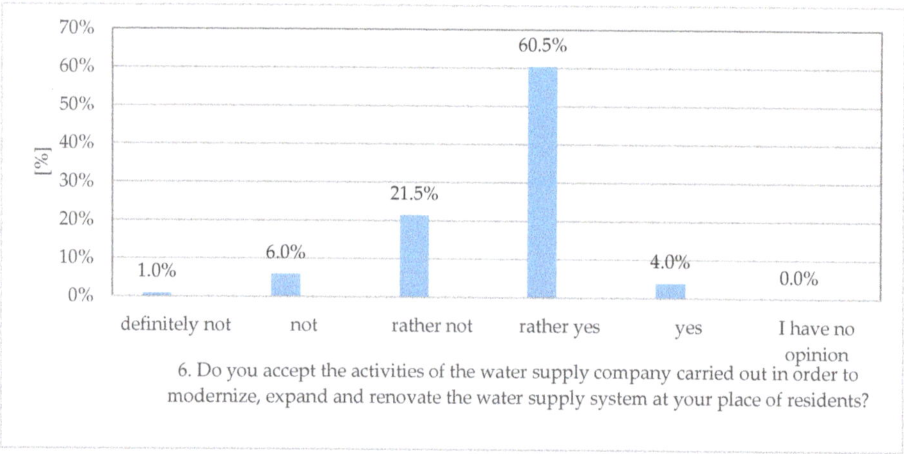

Figure 8. Results for answers to Question 6.

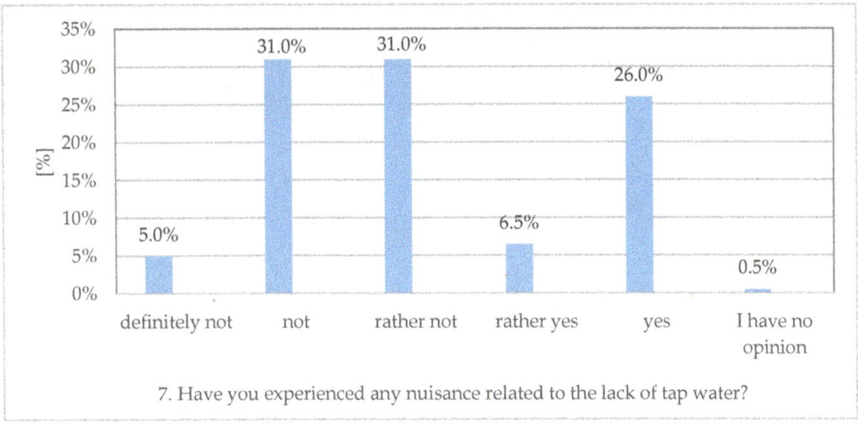

Figure 9. Results for answers to Question 7.

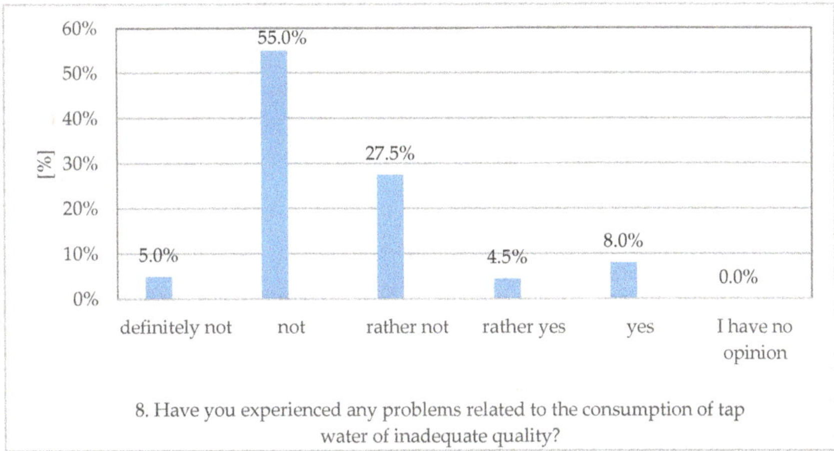

Figure 10. Results for answers to Question 8.

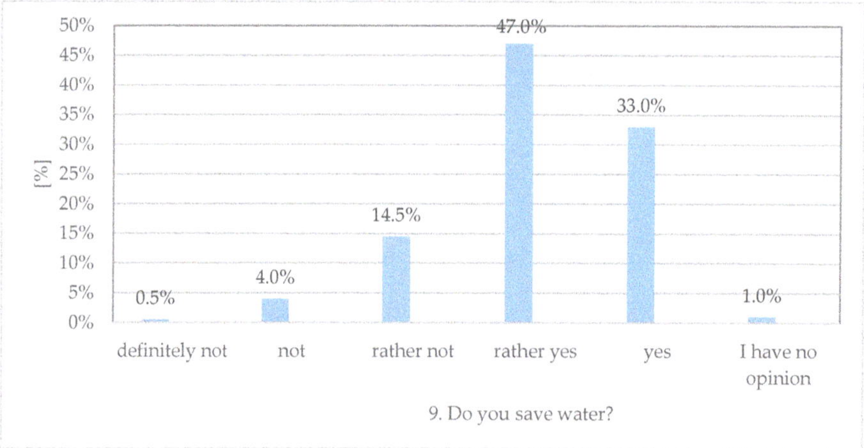

Figure 11. Results for answers to Question 9.

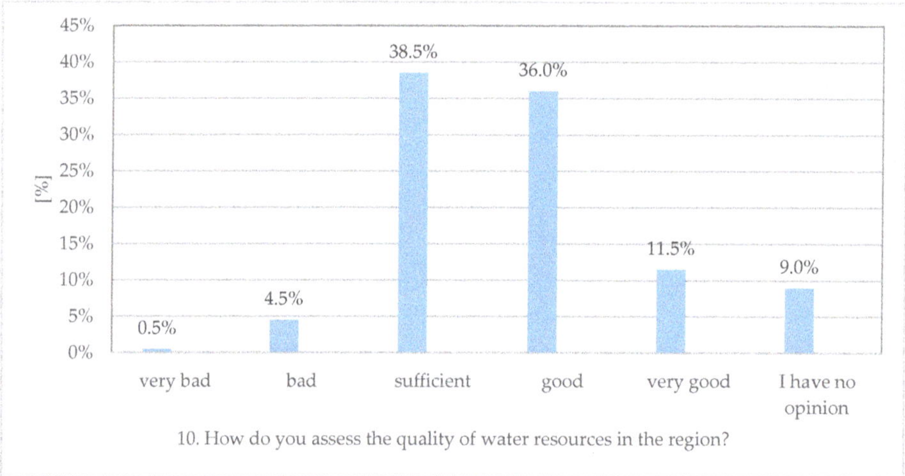

Figure 12. Results for answers to Question 10.

Answers to Questions 1–6 were used directly to determine the ARI indicator. Questions from 7 to 10 were asked to obtain approximate information on consumer knowledge regarding the city's water supply.

The results of the research allowed the following conclusions to be drawn:

- Around 65% of respondents assess the quality of tap water, the work of the water supply company (75%), and the quality of water resources (47%) as good and very good;
- Around 67% of respondents have not experienced a related nuisance with a lack—or poor quality—of tap water (87%);
- Over 80% of respondents declare that they save water;
- Around 68% of respondents do not know where alternative water sources are located.

To generate the ARI water supply quality assessment index, individual assessments were estimated based on the results of a research:

- According to the results of the research (Figure 3), around 65% of respondents assess water quality in the city as good and very good. Based on this result, in accordance with previous guidelines presented in the Table 1, the assessment of level was defined as good ($Q_1 = 1$).
- Answers to Question 2 (Figure 4) indicate that about 48.5% of respondents approve of the price of water in the city, and about 44% of them do not accept it, or "rather" do not accept it. On this basis, in line with the data in Table 1, it was assumed that assessment is at the average level ($P_2 = 2$),
- Consumer confidence and communication of the water supply company with water consumers: To estimate this assessment, the answers to Questions 3–6 included in Table 4 were analysed (Figures 5–8). Based on these research results, it can be concluded that about 74.5% of respondents accept and evaluate the activities of the water supply company favourably or very favourably. The results also showed that about 60% of those answering did not know about alternative water sources in the city. Based on the results of the analyses, it was assumed that this assessment assumes the average value ($T_3 = 2$).

The ARI was calculated in line with Formula (4) and is equal to 4.

According to the presented guidelines, this indicator is in the low range, from which it follows that water consumers in the serviced area tolerate the costs incurred by the company with respect to modernisation, protection, and repairs of the water distribution system in view of reducing the risk of failure. The analysis showed that the water supply company should pay more attention to the need to inform consumers about the existing risk and behaviour at the time of its occurrence, protection options, crisis prevention, as well as arrangements for water prices and the need to reduce water losses and save water. The results of the research contained in [45] indicate that the reliability of water supply expressed in terms of its quality and quantity is an important factor in its assessment by water consumers. Additionally, the test results contained in [46,47] indicate that the continuity of the water supply is an important factor in the assessment of water by consumers, who are willing to pay more for a reduction in the frequency and duration of interruptions in water supply. The aim of the water supply system's safe operation is to counteract against lack of water or its bad quality threatening the health of municipal water pipe users and to supervise this action using processes and information resources in the given operating conditions, in compliance with the valid law, and with economic justification [48]. However, legal regulations do not control the operation of water supply companies and conditions of receiving water with certain deficiencies [49]. Therefore, in countries where there are WSS, the proposed approach allows one to orientate on consumer opinions and to check consumer satisfaction about WSS operations. In developing waterworks, a pro-consumer attitude also requires improvement in consumer service, but the situation differs from systems, in which continuous water supply is taken for granted [50].

4. Conclusions

Society expects high standards in the sphere of social and economic life. While quality of life is a subjective and hard-to-measure concept, one standard should undoubtedly be reliable and safe access to clean water. This often denotes high costs to water supply companies as they seek to minimise the risk associated with the possibility of various adverse events in the water supply system arising.

The presented research-based method of analysing the acceptance by water consumers of the costs incurred by enterprises in risk reduction should be part of an appropriate policy that an enterprise pursues in the context of consultation with the local community. It can also constitute an important step towards ensuring the safety of water consumers and should therefore be a fundamental element in the strategy pursued by water utilities. Detailed procedures should be consulted with a wide range of experts from various fields. The costs of changes and improvements should be taken account of, but priority should always be given to providing consumers with water that is safe for their health.

An important element of accident risk management in a water supply company should be the analysis of consumer acceptance of the actions taken to reduce risk, as these influence the price of water. On this basis, water utilities can implement information management procedures.

The survey will always be subjective, as the scope of interpretation is wide. This is a proposal, and questions should be modified for each country, city, municipality, etc., in line with population size. Local governments preparing such a survey can modify and adapt it to local conditions in other ways, given that each system and community is different. Such methods can also in fact be used as other local government surveys are conducted. As various types of threats arise today, water utilities will have to introduce a range of modernisation procedures, at the same time ensuring their actions remain acceptable to the public. It is not possible to create a survey for every case. This process can be pursued based on appropriate research. The proposed method allows assessment of the level of public acceptance.

Author Contributions: All authors equally contributed to the development of this manuscript. All authors have read and agreed to the published version of the manuscript.

Funding: This research received no external funding.

Acknowledgments: We thank the reviewers for their feedback, which helped to improve the quality of the manuscript.

Conflicts of Interest: The authors declare no conflict of interest.

References and Notes

1. Kaplan, S.; Garrick, B.J. On the quantitative definition of risk. *Risk Anal.* **1981**, *1*, 11–27. [CrossRef]
2. Merkel, W.; Castell-Exner, C. Managing risk under normal operation and in crisis situations. *Water Util. Manag. Int.* **2010**, *9*, 19–22.
3. Pollard, S.J.T.; Strutt, J.E.; Macgillivray, B.H.; Hamilton, P.H.; Hrudey, S.E. Risk analysis and management in the water utility sector-a review of drivers, tools and techniques. *Process Saf. Environ. Prot.* **2008**, *82*, 1–10.
4. Drinking Water Directive (Council Directive 98/83/EC of 3 November 1998 on the Quality of Water Intended for Human Consumption), with Its Latest Amendments Including Commission Directive (EU) 2015/1787 of 6 October 2015. Available online: https://ec.europa.eu/environment/water/water-drink/legislation_en.html (accessed on 6 November 2020).
5. Council Directive 2008/114/EC of 8 December 2008 on the Identification and Designation of European Critical Infrastructures and the Assessment of the Need to Improve Their Protection, OJ L 345, 23.12.2008. Available online: https://eur-lex.europa.eu/LexUriServ/LexUriServ.do?uri=OJ:L:2008:345:0075:0082:EN:PDF (accessed on 6 November 2020).
6. Molinos-Senante, M.; Maziotis, A.; Sala-Garrido, R. Estimating the cost of improving service quality in water supply: A shadow price approach for England and Wales. *Sci. Total Environ.* **2016**, *539*, 470–477. [CrossRef] [PubMed]
7. Chen, B.P.-T.; Chen, C.-S. Feasibility Assessment of a Water Supply Reliability Index for Water Resources Project Planning and Evaluation. *Water* **2019**, *11*, 1977. [CrossRef]
8. Rogers, J.W.; Garrick, E.; Louis, G.E. Risk and opportunity in upgrading the US drinking water infrastructure system. *J. Environ. Manag.* **2008**, *87*, 26–36. [CrossRef]
9. World Health Organization. *Guidelines for Drinking-Water Quality*, 4th ed.; World Health Organization: Geneva, Switzerland, 2011.
10. Tchórzewska-Cieślak, B.; Pietrucha-Urbanik, K. Failure risk analysis in the collective water supply systems in crisis situations. *J. Pol. Saf. Reliab. Assoc.* **2013**, *1*, 129–136.
11. Mays, L.W. *The Role of Risk Analysis in Water Resources Engineering*; Arizona State University: Tempe, AZ, USA, 2005.
12. Tanyimboh, T.T.; Burd, R.; Burrows, R.; Tabesh, M. Modelling and reliability analysis of water distribution systems. *Water Sci. Technol.* **1999**, *39*, 249–255. [CrossRef]
13. Hrudey, S.E. Drinking water quality–a risk management approach. *Water* **2001**, *26*, 29–32.
14. Schramm, V.B.; Schramm, F. An Approach for Supporting Problem Structuring in Water Resources Management and Planning. *Water Resour. Manag.* **2018**, *9*, 2955–2968. [CrossRef]
15. Su, H.-T.; Tung, Y.-K. Comparisons of Risk-based Decision Rules for the Application of Water Resources Planning and Management. *Water Resour. Manag.* **2014**, *12*, 3921–3935. [CrossRef]

16. Lee, S.; Oak, S.; Jung, D.; Jun, H. Development of Failure Cause–Impact–Duration (CID) Plots for Water Supply and Distribution System Management. *Water* **2019**, *11*, 1719. [CrossRef]
17. Tchórzewska-Cieślak, B. Estimation of acceptance of the costs of risk associated with the operation of the water supply system. *Ochr. Srodow.* **2007**, *7*, 69–72.
18. Shinstine, D.S.; Ahmed, I.; Lansey, K. Reliability/availability analysis of municipal water distribution networks: Case Studies. *J. Water Resour. Plan. Manag.* **2002**, *128*, 140–151. [CrossRef]
19. Kordana, S.; Słyś, D. Decision Criteria for the Development of Stormwater Management Systems in Poland. *Resources* **2020**, *9*, 20. [CrossRef]
20. Quimpo, R.; Wu, S. Condition assessment of water supply infrastructure. *J. Inf. Syst.* **1997**, *3*, 15–20. [CrossRef]
21. Grbčić, L.; Lučin, I.; Kranjčević, L.; Družeta, S. A Machine Learning-based Algorithm for Water Network Contamination Source Localization. *Sensors* **2020**, *20*, 2613. [CrossRef] [PubMed]
22. Martin-Candilejo, A.; Santillán, D.; Iglesias, A.; Garrote, L. Optimization of the Design of Water Distribution Systems for Variable Pumping Flow Rates. *Water* **2020**, *12*, 359. [CrossRef]
23. Cao, Y.; Zhang, W.; Ren, J. Efficiency Analysis of the Input for Water-Saving Agriculture in China. *Water* **2020**, *12*, 207. [CrossRef]
24. Rak, J. Selected problems of water supply safety. *Environ. Prot. Eng.* **2009**, *2*, 23–28.
25. Rak, J.; Tchórzewska-Cieślak, B. Review of matrix methods for risk assessment in water supply system. *J. Konbin* **2006**, *1*, 67–76.
26. Dettori, M.; Pittaluga, P.; Busonera, G.; Gugliotta, C.; Azara, A.; Piana, A.; Arghittu, A.; Castiglia, P. Environmental Risks Perception Among Citizens Living Near Industrial Plants: A Cross-Sectional Study. *Int. J. Environ. Res. Public Health* **2020**, *17*, 4870. [CrossRef] [PubMed]
27. Di Bona, G.; Forcina, A.; Falcone, D.; Silvestri, L. Critical Risks Method (CRM): A New Safety Allocation Approach for a Critical Infrastructure. *Sustainability* **2020**, *12*, 4949. [CrossRef]
28. Tiganoaia, B.; Niculescu, A.; Negoita, O.; Popescu, M. A New Sustainable Model for Risk Management—RiMM. *Sustainability* **2019**, *11*, 1178. [CrossRef]
29. Maziotis, A.; Villegas, A.; Molinos-Senante, M. The cost of reducing unplanned water supply interruptions: A parametric shadow price approach. *Sci. Total Environ.* **2020**, *719*, 137487. [CrossRef]
30. Molinos-Senante, M.; Sala-Garrido, R. How much should customers be compensated for interruptions in the drinking water supply? *Sci. Total Environ.* **2017**, *586*, 642–649. [CrossRef]
31. EN 15975-1. Security of drinking water supply. Guidelines for risk and crisis management. Part 1 Crisis management.
32. EN 15975-2. Security of drinking water supply. Guidelines for risk and crisis management. Part 2. Risk management.
33. Haimes, Y.Y.; Li, Y. *Risk Modeling, Assessment and Management*; Wiley: New York, NY, USA, 1998.
34. Hastak, H.; Baim, E. Risk factors affecting management and maintenance cost of urban infrastructure. *J. Inf. Syst.* **2001**, *7*, 67–75. [CrossRef]
35. Tchórzewska-Cieślak, B. *Methods for Analyzing and Assessing the Risk of Failure of the Water Distribution Subsystem*; Rzeszów University of Technology Publishing House: Rzeszow, Poland, 2011.
36. Grey-Gardner, G. Implementing risk management for a water supplies, a catalyst and incentive for change. *Rangel. J.* **2008**, *30*, 149–156. [CrossRef]
37. Geng, Z.Q.; Chen, N.; Han, Y.M.; Ma, B. An improved intelligent early warning method based on MWSPCA and its application in complex chemical processes. *Can. J. Chem. Eng.* **2020**, *98*, 1307–1318. [CrossRef]
38. Duan, W.; He, B.; Nover, D.; Yang, G.; Chen, W.; Meng, H.; Zou, S.; Liu, C. Water Quality Assessment and Pollution Source Identification of the Eastern Poyang Lake Basin Using Multivariate Statistical Methods. *Sustainability* **2016**, *8*, 133. [CrossRef]
39. Geng, Z.; Hu, X.; Han, Y.; Yanhua, Z. A Novel Leakage-Detection Method Based on Sensitivity Matrix of Pipe Flow: Case Study of Water Distribution Systems. *J. Water Resour. Plan. Manag. ASCE* **2019**, *145*, 04018094. [CrossRef]
40. Geng, Z.Q.; Wang, Z.; Hu, H.X.; Han, Y.M.; Lin, X.Y.; Zhong, Y.H. A fault detection method based on horizontal visibility graph-integrated complex networks: Application to complex chemical processes. *Can. J. Chem. Eng.* **2019**, *97*, 1129–1138. [CrossRef]

41. Bondoc, I. European Regulation in the Veterinary Sanitary and Food Safety Area, a Component of the European Policies on the Safety of Food Products and the Protection of Consumer Interests: A 2007 Retrospective. Part One: The Role of European Institutions in Laying Down and Passing Laws Specific to the Veterinary Sanitary and Food Safety Area. *Universul Juridic Supliment* **2016**, *5*, 12–15.
42. Bondoc, I. European Regulation in the Veterinary Sanitary and Food Safety Area, a Component of the European Policies on the Safety of Food Products and the Protection of Consumer Interests: A 2007 Retrospective. Part Two: Regulations. *Universul Juridic Supliment* **2016**, *5*, 16–19.
43. Ezell, B.; Farr, J.; Wiese, I. Infrastructure risk analysis of municipal water distribution system. *J. Inf. Syst.* **2000**, *6*, 118–122. [CrossRef]
44. StatSoft, Inc. STATISTICA (Data Analysis Software System), Version 12. Available online: www.statsoft.com (accessed on 15 August 2020).
45. Shuang, Q.; Liu, H.J.; Porse, E. Review of the Quantitative Resilience Methods in Water Distribution Networks. *Water* **2019**, *11*, 1189. [CrossRef]
46. Pietrucha-Urbanik, K.; Rak, J.R. Consumers' Perceptions of the Supply of Tap Water in Crisis Situations. *Energies* **2020**, *13*, 3617. [CrossRef]
47. Fang, L.; Wu, F. Can Water Rights Trading Scheme Promote Regional Water Conservation in China? Evidence from a Time-Varying DID Analysis. *Int. J. Environ. Res. Public Health* **2020**, *17*, 6679. [CrossRef] [PubMed]
48. Quach, S.; Thaichon, P.; Hewege, C. Triadic relationship between customers, service providers and government in a highly regulated industry. *J. Retail. Consum. Serv.* **2020**, *55*, 102148. [CrossRef]
49. Franceys, R.W.A.; Gerlach, E. Consumer involvement in water services regulation. *Util. Policy* **2011**, *19*, 61–70. [CrossRef]
50. Shin, S.; Lee, S.; Judi, D.R.; Parvania, M.; Goharian, E.; McPherson, T.; Burian, S.J. A Systematic Review of Quantitative Resilience Measures for Water Infrastructure Systems. *Water* **2018**, *10*, 164. [CrossRef]

resources

MDPI

Review

Inventory of Good Practices of Sustainable and Circular Phosphorus Management in the Visegrad Group (V4)

Marzena Smol [1,*], Paulina Marcinek [1,*], Zuzana Šimková [2], Tomáš Bakalár [2], Milan Hemzal [3], Jiří Jaromír Klemeš [3], Yee Van Fan [3], Kinga Lorencz [4], Eugeniusz Koda [5] and Anna Podlasek [5]

1 Mineral and Energy Economy Research Institute, Polish Academy of Sciences, J. Wybickiego 7A St., 31-261 Kraków, Poland
2 Faculty of Mining, Ecology, Process Control and Geotechnologies, Technical University of Košice, Park Komenského 19, 04 001 Košice, Slovakia
3 Sustainable Process Integration Laboratory–SPIL, NETME Centre, Faculty of Mechanical Engineering, Brno University of Technology–VUT Brno, Technická 2896/2, 616 69 Brno, Czech Republic
4 Bay Zoltán Nonprofit Ltd. for Applied Research, Kondorfa utca 1 St., H-1116 Budapest, Hungary
5 Institute of Civil Engineering, Warsaw University of Life Sciences (WULS-SGGW), Nowoursynowska 159 St., 02-776 Warsaw, Poland
* Correspondence: smol@meeri.pl (M.S.); marcinek@meeri.pl (P.M.)

Abstract: The most important raw material needed for food production is phosphorus (P), which cannot be replaced by other elements. P is listed as a Critical Raw Material (CRM) for the European Union (EU). It is an element essential for human nutrition and is used for fertiliser production. The key importance of P for human life is evidenced by the fact that if there were not enough P in fertilisers, we would only be able to feed 1/3 of the world's population. Unfortunately, in Visegrad Group (V4) countries, Poland, Slovakia, Czech Republic, and Hungary, there is a lack of mineral deposits of phosphate rock. Therefore, there is a strong need to cover the demand for the P by importing from countries of varying stability, both economic and political, such as Russia, China, or Morocco. It is risky; if the borders for deliveries of goods are closed, it may be impossible to meet the needs of P. On the other hand, V4 countries have large secondary P resources in P-rich waste, which are lost due to P is not recovered on an industrial scale. The paper presents the importance of P raw materials in V4, the revision of primary and secondary P sources that can be used in agricultural systems, as well as the structure of import and export of P raw materials in these countries. In addition, examples of good phosphorus recovery practices in the V4 countries are presented. They include a list of initiatives dedicated to the sustainable management of P resources, and examples of P recovery projects. Implementation of P recovery for internal P-rich waste in V4 could ensure the safety of food production in this region. Such and similar initiatives may contribute to faster independence of the V4 countries from the import of P raw materials.

Keywords: phosphorus; resources; critical raw materials; Visegrad Group; V4; sustainable management

Citation: Smol, M.; Marcinek, P.; Šimková, Z.; Bakalár, T.; Hemzal, M.; Klemeš, J.J.; Fan, Y.V.; Lorencz, K.; Koda, E.; Podlasek, A. Inventory of Good Practices of Sustainable and Circular Phosphorus Management in the Visegrad Group (V4). *Resources* **2023**, *12*, 2. https://doi.org/10.3390/resources12010002

Academic Editor: Diego Copetti

Received: 15 November 2022
Revised: 12 December 2022
Accepted: 21 December 2022
Published: 30 December 2022

1. Introduction

Phosphorus (P) is one of the most important nutrients needed to sustain life [1,2], with properties that cannot be replaced by any other element [3]. Moreover, P is non-renewable [4,5]. In 2014, the European Commission (EC) indicated phosphorus rock as one of the most important critical raw materials (CRMs) for the European economy [2], and P was placed on the CRMs list [6]. Then, in 2017 and in 2020, phosphate rock and P were also included in the updated lists of CRMs [7,8].

P has no metallic properties. P is classified as flu of non-metals [9]. P is the basic nutrient responsible for the growth of all living organisms with properties that cannot be substituted [10]. Moreover, P is the third major component (after potash and nitrogen) used in industrial fertilisers. P represents a crucial element of the food security system [11].

To reduce its dependence on external markets, the European Union (EU) has, in recent years, emphasised the need to look for alternative sources of P. One option is to recover this valuable raw material from selected waste streams. This approach is in line with the assumptions of the circular economy (CE), the EU economic model, which underlines that the transformation towards CE could bring significant economic and environmental benefits for the Member States, including V4. Activities in the field of recovery of raw materials, including P, are also part of the new EU strategy: the European Green Deal [12]. The initiatives for more sustainable management of P raw materials are part of the proposed Farm to Fork Strategy based on the principle of creating a fair, healthy, and environmentally friendly food system [13]. The transition to more sustainable food systems has already begun [14], but with current food production methods (based only on primary raw materials), feeding a population (in which there may be a threat of disruption in the delivery of these materials, e.g., as a result of closing borders, introducing restrictions on imports from selected countries) remains a challenge. Food production continues to pollute air, water, and soil, contribute to biodiversity loss and climate change, and consume vast amounts of natural resources, while a large proportion of the food produced is wasted. Poor quality food contributes to obesity and diseases such as cancer. The farm to fork strategy will address the use of fertilisers in agriculture [13], with a strong emphasis on the recovery of nutrients from waste [15].

The specific actions for the sustainable management of biogenic raw material resources (such as phosphorus) have been undertaken for many years in various regions of the EU and in individual Member States [3]. So far, however, no analyses have been conducted on the central region of the EU, which comprises the Visegrad Group consisting of four countries: Poland, the Czech Republic, Hungary, and Slovakia [16]. This region does not have phosphorus deposits, therefore, the demand for P raw materials (necessary for the production of fertilisers and food) is met only by imports. In the face of the threats of the 21st century, such as a pandemic, it seems reasonable to take action to ensure the safety of P raw materials in this region, as well as to intensify activities to subsidise P from available waste streams. Currently, the COVID pandemic is the greatest threat to modern economies. The V4 countries were the first in the EU to introduce restrictions to prevent the spread of the virus, which proves their great responsibility to residents [17]. Citizens stayed home, and the only thing they needed to survive was food. It showed that the greatest challenge is to ensure people's safety and access to food requires the provision of raw materials for food production.

The paper presents the importance of sustainable management of P raw materials in V4 countries. Primary and secondary P sources that can be used in agricultural systems are listed; import and export are presented. In addition, examples of good P recovery practices in the V4 countries are presented. The structure of the paper is as follows:

- clarify the importance of the recovery of P raw materials from waste in the context of V4 countries;
- overview of possible sources of P raw materials in V4 countries;
- overview of good examples of sustainable management of P raw materials in V4 countries;
- conclusions.

2. Materials and Methods

This section provides a description of the materials and methods that have been used in the study. The research framework is shown in Figure 1. There are four individual phases in this research. The first phase included a description of the case study region. The second phase included an overview of possible sources of P raw materials in V4 countries coming from primary and secondary sources. The third phase contained an overview of good examples of sustainable management of P raw materials in all V4 countries. The last phase covered conclusions from the study and recommendations for further research.

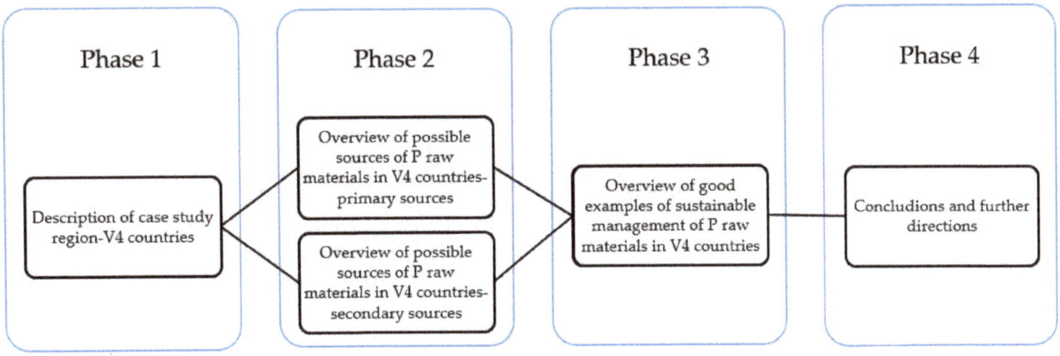

Figure 1. Scheme of the research framework.

In all phases of the work, a comprehensive analysis (desk research) of selected documents was used as a research method. The review covered numerous peer-reviewed scientific articles directly related to the subject of the flow of P raw materials in the V4 countries. The selection of the analysed literature was based on the following keywords: "phosphorus", "resources", "critical raw materials", "CRMs", "Visegrad Group", "V4", "sustainable management", "Czech Republic", "Hungary", "Poland", "Slovakia", "sewage sludge", "sewage sludge ash". The reviewed publications were searched on scientific platforms such as Elsevier Scopus and ScienceDirect, Multidisciplinary Digital Publishing Institute (MDPI), and Google Scholar. An important source of data was also a statistic published by Eurostat (the official statistic of the EU).

The initial results of this review were presented in the document "Portfolio of Phosphorus Friends in Europe", which was developed as part of the project "How to stay alive in V4? Phosphorus Friends Club builds V4's resilience (PhosV4)", financed by the Visegrad Fund (project no. 22110364). In this document, project partners contained information about the importance of P raw materials in securing the supply of food in V4 countries, the use of P raw materials in the food sector, and P raw materials flow in V4 countries. It is worth noticing that the identification of P recovery potential and good practices of P recovery in V4 countries is a research gap for which detailed data are not available at the moment. Therefore, it is an interesting research area that should be developed and studied. All results were discussed by project partners during the consortium meetings, and further directions for research were jointly designed.

3. Results

3.1. Case Study Region

The case study region in this paper is the V4 group, which contains the following four countries: the Czech Republic, Hungary, Poland, and Slovakia. The number of residents in this region it is above 63 M in 2022, with the higher number in Poland (37,654,247 people), followed by the Czech Republic (10,516,707 people), Hungary (9,689,010 people), and Slovakia (5,434,712 people) (Figure 2) [18].

Over the 11 years, the EU population has grown by around 6,887,000 citizens. Population growth occurred in Slovakia (by 43,000 in 2022 compared to 2011) and in the Czech Republic (population increase of 30,000 people). On the other hand, a decrease in the number of respondents took place in Poland (409,000 people less in 2022 compared to 2011) and in Hungary (a decrease in the population of 297,000 people in the analysed period).

The area of the EU countries currently covers 4,215,000 km^2, 5.2% of which is occupied by the V4 countries. Poland is the largest country belonging to the group of V4 countries, with an area of 312,700 km^2. Next is Hungary, with an area of 93,000 km^2, the Czech Republic at 78,900 km^2, and Slovakia, at 49,000 km^2, is the country in the region with the smallest area [19].

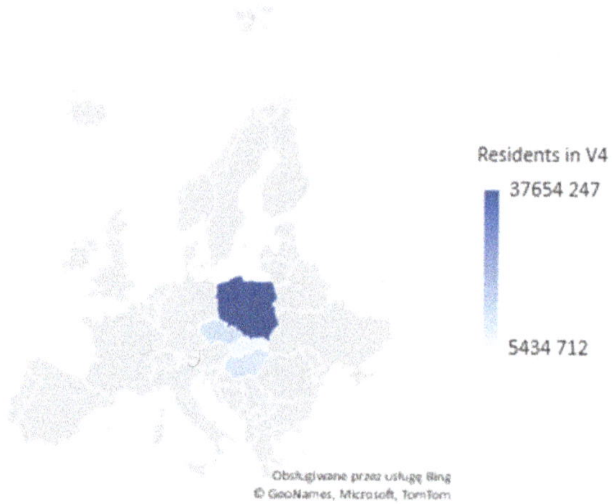

Figure 2. Residents in V4 [18].

The changes in population over 11 years (2011–2022) in the V4 countries is presented in Table 1.

Table 1. Population in V4 (in thousand) (data from [18]).

Year	2011	2012	2013	2014	2015	2016	2017	2018	2019	2020	2021	2022
EU-27 countries	439,942	440,553	441,258	442,884	443,667	444,803	445,534	446,209	446,559	447,485	447,001	446,829
Hungary	9986	9932	9909	9877	9856	9830	9798	9778	9773	9770	9731	9689
Poland	38,063	38,064	38,063	38,018	38,006	37,967	37,973	37,977	37,973	37,958	37,840	37,654
Slovakia	5392	5404	5411	5416	5421	5426	5435	5443	5450	5458	5460	5435
Czech Republic	10,487	10,505	10,516	10,512	10,538	10,554	10,579	10,610	10,650	10,694	10,495	10,517

3.2. *Primary and Secondary P Raw Materials in V4 Countries*

In the EU, P resources are limited, which means that most of the P in the EU is imported. The EU imports around 6 million mg of natural phosphate annually and around 1.2 million mg of P fertilisers from Russia, Morocco, and Tunisia [20]. EC presents 100% as the reliance percentage on P imports and 84% as the reliance percentage on phosphate rock [8]. The structure of global producers and main EU sourcing countries of phosphate rock is presented in Figure 3.

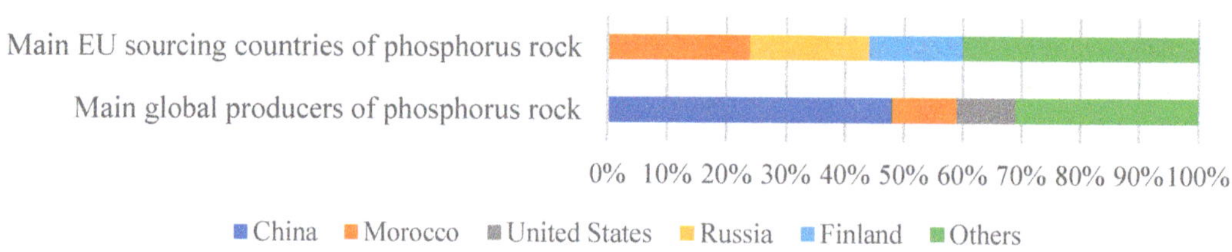

Figure 3. Structure of global producers and main EU sourcing countries of phosphate rock, data from [21].

Based on the available data, there are no P deposits mined in V4 countries at present. In Poland, there are phosphorite deposits that were mined in the past. P occurs at the north-eastern margin of the Holy Cross Mts. (vicinities of Radom-Iłża-Annopol-Gościeradów-Modliborzyce) in the form of calcium phosphate-rich nodules in the various types of sediment [22]. The exploitation of phosphate phosphorites began in the country between the First and the Second World Wars. Currently, due to economic aspects, no deposits are exploited. The last exploited phosphorite deposit, located in Chałupki, was closed in 1961, and 10 years later, the same was also done in Annopol [23]. The limit values of the parameters that describe the phosphorites deposit in Poland define that [22]:

- the maximum depth of deposits documentation is 400 m below the surface,
- the minimum P_2O_5 content in calcium phosphate-rich nodules is 15%,
- the minimum affluence of calcium phosphate-rich nodules is 1800 kg/m^2.

Qualitative parameters of the main phosphorites occurrences in Poland are presented in Table 2.

Table 2. Quality parameters of documented phosphate deposits (data from [23]).

Deposit	Calcium Phosphate-Rich Nodules (mm)	P_2O_5 Content in Calcium Phosphate-Rich Nodules (%)	Affluence of Calcium Phosphate-Rich Nodules (kg/m^2)	Affluence Versus Actual Limiting Parameters (%)
Annopol	>10	13.5	568	32
Burzenin	>2	18.1	385	21
Chałupki	>10	14.9	354	21
Gościeradów	>2	15.2	496	28
Iłża-Krzyżanowice	>2	18.6	791	44
Iłża—Chwałowice	>2	22.3	891	50
Iłża—Łęczany	>2	18.6	654	36
Iłża—Walentynów	>2	19.9	470	26
Radom—Dąbrówka Warszawska	>2	16.5	upper series: 317 lower series: 460	upper series: 18 lower series: 26
Radom—Krogulcza	>2	19.1	upper series: 218 lower series: 504	upper series: 12 lower series: 28
Radom—Wolanów	>2	15.4	upper series: 170 lower series: 447	upper series: 9 lower series: 25

The index describing the abundance largely deviates from the boundary values of the parameters that define the deposit. Deposits are flooded, which results in their potential exploitation. In addition, railway lines and high-voltage lines, roads or buildings were built in their areas through significant parts of the deposits. In extreme cases, it may cause the resources available for exploitations reduction as much as 50–80%.

All deposits from which phosphate rock was obtained in Poland were removed from the national resource balance in 2006. Currently, the domestic demand for phosphate rock raw materials is fully covered by imports, e.g., from Morocco, Algeria, and Egypt, where the availability of the described raw materials is much greater and more economical [22]. The phosphate rock import quantity to Poland during the last 18 years is presented in Figure 4. The largest amount of the P was imported in 2004, and the lowest amount in 2009, which is directly related to the global economic crisis that occurred in 2008.

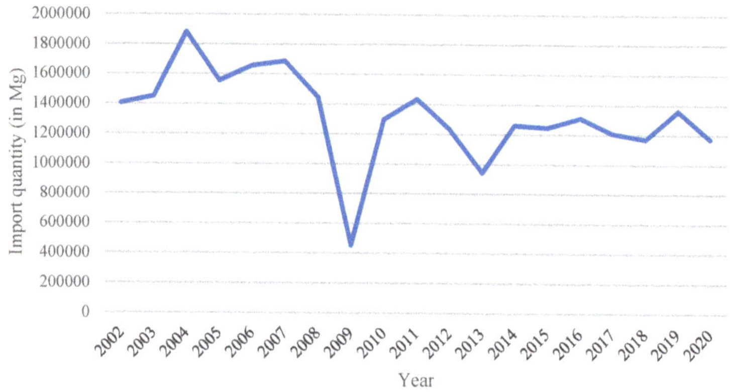

Figure 4. Phosphate rock import quantity to Poland, data from [24].

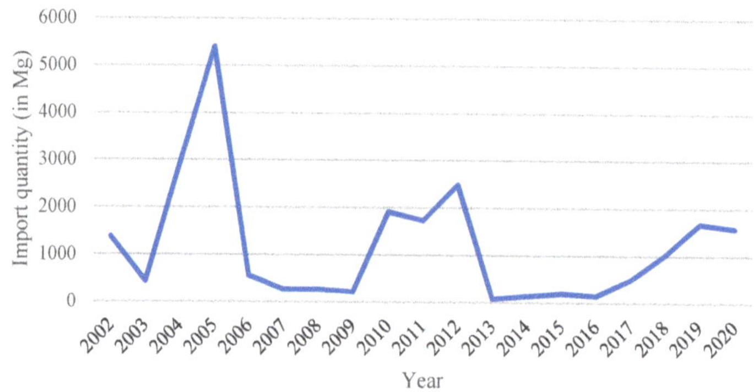

Figure 5. Phosphate rock import quantity to Slovakia, data from [24].

In the Czech Republic, there are no P deposits. The domestic demand for phosphate rock raw materials is fully covered by imports. The phosphate rock import quantity to the Czech Republic during the last 18 years is presented in Figure 6. The largest amount of the P was imported in 2008, and the lowest amount was in 2012.

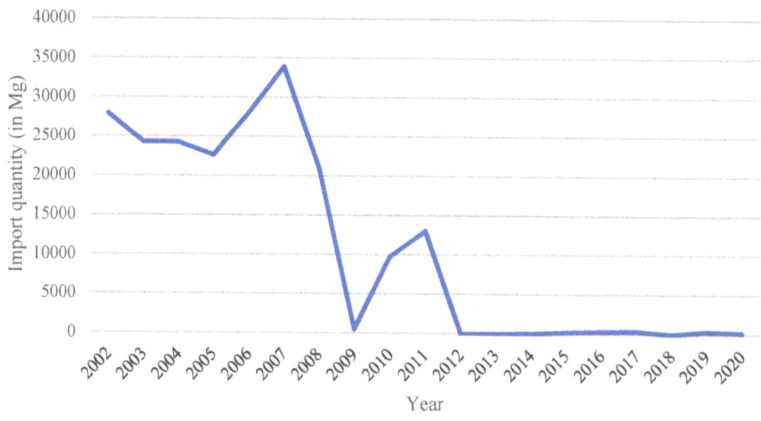

Figure 6. Quantity of imported phosphate rocks in the Czech Republic, data from [24].

In Slovakia, there is one deposit of phosphate; however, it also is not mined at the moment. The demand for phosphate raw materials is fully covered by imports. The phosphate rock import quantity to Slovakia during the last 18 years is presented in Figure 5. The highest amount of P raw materials was imported to Slovakia from Italy (68%) and the Czech Republic (31%). There is also limited import from Germany (0.2%), the United Kingdom, Belgium, Japan, and other countries (<0.1%) [25,26].

In Hungary, there are five sedimentary phosphate deposits, but there is no information available on what phosphorus contents there are and whether they will ever be used. The phosphate rock import quantity to Hungary during the available 10 years is presented in Figure 7. The largest amount of the P was imported in 2020, and the lowest amount was in 2014.

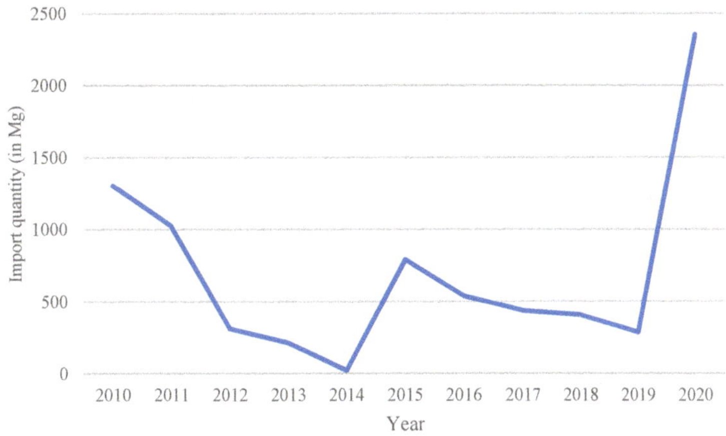

Figure 7. Quantity of imported phosphate rocks in Hungary, data from [24].

The vital importance of P and its growing deficiency influenced the dynamic development of science in the area of P recovery from different waste materials [27]. In V4 countries, there is high potential for the recovery of P from secondary sources such as:

- wastewater and sewage sludge (municipal and industrial) [28],
- sewage sludge ash [29],
- pig slurry [30],
- meat and bone meal [31],
- industrial waste [32],
- biomass [33].

The P contents in selected waste streams are shown in Table 3, which includes values of P concentration in sewage sludge, ashes from sewage sludge, animal manure, and compost from plant waste.

Table 3. P concentration in presented types of waste [mg/kg].

Types of Waste	P Concentration [mg/kg]				References
Sewage sludge	13,200	123,000	26,100	65,400	[34–37]
Sewage sludge ash	46,200	60,697	127,351	112,425	[37–40]
Animal manure	21,400	30,600	32,700	29,500	[41–44]
Compost from plant waste	40,900	89,000	83,000	78,000	[35,45–47]

Currently, household waste containing large amounts of P (mainly sewage sludge) could cover around 20–30% of the demand for phosphate fertilisers in the EU when recycled. However, this investment potential is still largely untapped in European countries [48], despite the fact that such an approach is in line with the assumptions of CE, in which waste generated should be treated as secondary raw materials. In V4 countries, municipal wastewater treatment plants (WWTPs) have the greatest potential for P recovery because P can theoretically be recovered at every stage of the treatment process, i.e., from sewage and leachates in the liquid phase, from dehydrated sewage sludge, from the solid phase of ashes after thermal transformation of municipal sewage sludge. In the successive stages of wastewater treatment and sewage sludge treatment, a smaller volume of the substrate used for P recovery is observed, while at the same time P concentration per unit volume is increasing [49]. The efficiency of P recovery from different substrates at WWTPs is equal to [50]:

- 45–55% for wastewater-outflow from the treatment plant,
- 45–50% for sedimentary liquid – leachate,
- 50–60% for dehydrated sewage sludge,
- >90% for sewage sludge ash.

The major part of P in substrates in WWTPs is transferred to sludges (up to 90%) [40]. Therefore, sludge and sludge ash are the most promising P-rich residues. Sewage sludge production and disposal from urban wastewater in V4 countries are presented in Table 4. There is an increasing amount of sewage sludge generated in V4 countries. A higher amount of SS is observed in Poland, which corresponds to the highest population in this country, followed by Hungary, the Czech Republic, and Slovakia. In total, 107,845 thousand mg of SS was produced in V4 countries in 2019 [51].

Table 4. Sewage sludge production and disposal from urban wastewater in V4 countries in 2009–2019, in thousand mg (data from [51]).

Country	2009	2010	2011	2012	2013	2014	2015	2016	2017	2018	2019
	Thousand mg										
Czech Republic	20,720	19,630	21,790	26,330	26,010	23,859	21,024	20,671	22,327	22,822	22,109
Hungary	14,930	17,034	16,833	16,060	17,047	16,312	17,770	21,796	26,684	23,366	22,789
Poland	56,330	52,670	51,920	53,330	54,030	55,600	56,800	56,833	58,445	58,307	57,464
Slovakia	5858	5476	5872	5871	5743	5688	5624	5305	5452	5593	5483
Total V4	97,838	94,810	96,415	101,591	102,830	101,459	101,218	104,605	112,908	110,088	107,845

There are several technologies for P recovery from SS, including P extraction by wet chemical methods under acid and alkali conditions [50]. However, to date, there is no reported industrial plant that is recovering P raw materials from SS in V4 countries. Therefore, further initiatives (as economic, environmental, law, or social) that support P recovery technologies implementation from SS should be developed.

The highest efficiency of P recovery was reported for the ashes generated in the process of thermal treatment of sewage sludge (>90%). In the V4 group, only Poland is equipped with municipal sewage sludge incineration plants (so-called mono-incineration plants). The detailed inventory of SSA generated in Poland was reported in [52]. The current capacity of 11 mono-incineration plants is equal to 160,300 mg d.w. of SS per year. The highest amount of ashes is generated in Warsaw and Cracow. The most important player is a mono-incineration plant in Warsaw (the capital of Poland) that produced >10,000 mg d.w. in 2018 (38% of total SSA generated in Poland). There are also significant amounts of ashes in Cracow (18% of the total in Poland), Łódź (14%), Gdańsk (14%), and Gdynia (12%). The rest of the installations produced less than 10% (6%—Gdynia, 5%—Szczecin, 3%—Kielce, 2%—Bydgoszcz, 1%—Olsztyn). In total, in 2018, 24,510 of fly ash and 24,510 mg of bottom slag and ash were produced. They potentially can be used in phosphate fertiliser production; however, for economic reasons, there is no industrial processing and production of P fertilisers from this waste stream. To protect the utility value of ashes, they have to be stored selectively and then directed to P recovery. This supports the possibility of turning waste into a resource if certain conditions are met. Despite the high content of P raw materials in the ashes, it is usually present in chemically bound forms, which makes its availability to plants difficult. In addition, the ashes may contain significant amounts of impurities, including heavy metals, which limits the possibility of their use in the production of fertilisers without prior treatment. In order to increase the bioavailability of phosphorus and reduce the content of heavy metals, this waste should be subjected to chemical and thermochemical processing. The most promising methods of phosphorus recovery from ashes are chemical methods (using phosphorus extraction-wet methods) and thermochemical methods (separation of P fraction at high temperatures 1000–2000 °C and conversion of phosphorus into forms available for plants) [52].

3.3. *Good Practices of P Recovery in V4 Countries*

This section includes an inventory of selected examples of good practices of sustainable P management in V4 countries. They include a list of innovative solutions enabling the recovery of P raw materials from different waste streams.

3.3.1. Good Practices of P Recovery in Poland

In Poland, many activities are undertaken that are dedicated to sustainable and circular management of P. There are several projects in the country, the aim of which is, inter alia, P recovery. Moreover, many companies take measures to support the acquisition of P from secondary sources. Table 5 shows examples of good P management practices in Poland.

Table 5. Examples of good practices P management in Poland.

Company/Project Name	Description of Good Practices	References
Jarocin Waterworks Company	The Jarocin Waterworks Company has signed a contract for carrying out an investment under the project 'Modernisation and Extension of WWTP Jarocin'. The project includes the implementation of five tasks (with a total value of 60 M EUR), supported by co-financing from the EU. The largest investment in the project is the construction of a station for the recovery of raw materials, such as nitrogen, P, and biogas, at the sewage treatment plant in Cielcza. This would allow it to recover between 100 and 200 kg of fertiliser per year. The water and wastewater management project implemented in Jarocin was recognised with a prestigious award at the international Wex Global 2018 conference, which took place in Lisbon. In the years that follow, the introduction on the market of technologies for the recovery of P will be planned, in particular in the wastewater sector.	[53]
Azoty Group "Fosfory"	Azoty Group "Fosfory" Sp. z o. o. are one of the leaders in the fertiliser and chemical industry in Europe. The highest quality of products and complete customer satisfaction are their priority. By producing agricultural fertilisers, they strive to maximise the benefits of buyers and maintain all environmental protection requirements. Group is a producer of mineral fertilisers that are widely used in agriculture, vegetable cultivation, and horticulture. Their offers also included chemical products. The Azoty Group "Fosfory" taking advantage of the location within the Gdansk port and access to the Chemikow Wharf in use, it imports some raw materials for the production of fertilisers by sea. With its experience in maritime trading, the Azoty Group "Fosfory" also conducts a wide range of services and reloading as well as sea freight of loose and liquid bulk goods in export and import.	[54]

Table 5. *Cont.*

Company/Project Name	Description of Good Practices	References
Sewage Treatment Plant—Tarnowskie Wodociagi	Tarnowskie Wodociągi Sp. z o.o. provides services in the field of collection and treatment of municipal wastewater. In 2007, on the premises of Sewage Treatment Plant-Tarnowskie Wodociągi Sp. z o.o. the construction of a sewage sludge drying plant was started. This investment was completed in 2008. The construction of the dryer is another step towards even more complete use of sewage sludge and reducing its mass four times. Ultimately, it is planned to utilise sewage sludge along with recovery of P compounds from the ashes.	[55]
Project "Sustainable management of phosphorus in the Baltic region (InPhos)"	Project "Sustainable management of phosphorus in the Baltic region" (InPhos) received funding from the European Institute of Innovation and Technology (EIT)—a body of the EU, under the Horizon 2020 program. The main objective of the InPhos project was to develop a strategy for sustainable P management (including identification of the P recovery potential) in the Baltic Sea Region by a working group of experts from the Baltic countries—Poland, Germany, Sweden, Finland, Latvia, Lithuania, Estonia, and Italy.	[56]
Project "Market ready technologies for P-recovery from municipal wastewater (PhosForce)."	The main objective of the "Market ready technologies for P-recovery from municipal wastewater" (PhosForce) project is to develop innovative technology for the recovery of P from wastewater. The Struvia® technology has been used to recover P in the form of struvite crystals from wastewater generated in municipal waste disposal facilities.	[57]
Project "Towards Circular Economy in wastewater sector: Knowledge transfer and identification of the recovery potential for Phosphorus in Poland (CEPhosPOL)."	The main goal of the "Towards Circular Economy in wastewater sector: Knowledge transfer and identification of the recovery potential for Phosphorus in Poland" (CEPhosPOL) project was to conduct research works focused on the identification of the recovery potential for P in Poland and the development of the sustainable model of the P management, based on the circular economy assumptions. The project was implemented under the Mieczysław Bekker programme for young researchers, financed by the National Academic Exchange Agency (NAWA).	[58]
Project "Polish Fertilisers form Ash (PolFerAsh)"	The main goal of the Polish project "Polish Fertilisers form Ash" (PolFerAsh) was to develop an environmentally-friendly technology for sewage sludge ash utilisation as a source of fertilisers and construction materials. The project has been conducted in the Cracow University of Technology and Mineral and Energy Economy Research Institute of the Polish Academy of Sciences in Poland and has received the founding from the National Centre for Research and Development.	[59]

3.3.2. Good Practices of P Recovery in Slovakia

In Slovakia, the company that produces fertilisers is Duslo, a.s. [60], which has become a fertiliser producer on a European and on a global scale. In addition, the country has undertaken actions aimed at sustainable P management, for example, through the project "Drinking water supply, sewerage and wastewater treatment" [61] or the Slovak Grant Agency for Science (Grant No. 1/0563/15) [62]. These activities are presented in Table 6.

3.3.3. Good Practices of P Recovery in the Czech Republic

In the Czech Republic, there are projects, institutions and organisations that support the sustainable development of P management. It is worth noticing that there is a national platform dedicated to P management–Czech Phosphorus Platform [63], which is an organisation that allows its members to act in the field of, inter alia, reducing dependence on imports and recycling of P from waste, from crop and livestock production in agriculture, from industry and municipal sewage. The activities of the Czech community in the field of sustainable P management are presented in Table 7.

3.3.4. Good Practices of P Recovery in HUNGARY

The leading Hungarian fertiliser partner network is called Genezis. This partner network includes five large companies, the activities and best practices of which are presented in Table 8.

Table 6. Examples of good practices P management in Slovakia.

Company/Project Name	Description of Good Practices	References
Duslo, a.s-company dealing with fertilisers	The biggest company dealing with fertilisers is Duslo, a.s., a member of the AGROFERT Group. It is one of the most significant companies in the chemical industry in Slovakia. It has developed into a manufacturer of fertilisers of European significance and a global supplier of rubber chemicals. It is a producer of polyvinyl acetate and polyacrylic glues and dispersions that it supplies to the global market. The company s product portfolio includes: industrial fertilisers, rubber chemicals, dispersions and glues, products of magnesium chemistry, and special products.	[60]
Project "Drinking water supply, sewerage and wastewater treatment."	The "Drinking water supply, sewerage and wastewater treatment" project contributed to reducing pollution and improving wastewater collection. The project also brought drinking water to people struggling to find regular or reliable supplies. As part of the project, the existing facilities were modernised and a new central pumping station was constructed. Improvements to existing facilities included making it easier to remove nitrogen and P from the water. These actions resulted in a radical increase in the capacity and efficiency of the existing wastewater plant.	[61]
Slovak Grant Agency for Science (Grant No. 1/0563/15)	The research project was carried out as planned research projects of the Department of Applied Ecology, Sumy State University, connected with subjects "Reduction of technogenic loading on the environment of enterprises of chemical, machine-building industry and heat and power engineering" according to the scientific and technical program of the Ministry of Education and Science of Ukraine (state registration No 0116U006606). The project focused, inter alia, on the biochemical treatment of sewage sludge and phosphogypsum under conditions reducing sulphates with the release of P. A schematic model of the dephosphatation process under the conditions of anaerobic stabilisation of sewage sludge and phosphogypsum was developed.	[62]

Table 7. Examples of good practices P management in the Czech Republic.

Company/Project Name	Description of Good Practices	References
Fosfa, a.s.	Fosfa, a.s. is an innovative Life Science company, the largest processor of yellow P in Europe and a successful exporter. After the successful resumption of phosphoric acid production, the company decided to invest in the production of special applications based on P and detergents. At present, Fosfa products are for food and alcohol industrial applications. The production scope of the company consists of product groups: sodium phosphates, potassium phosphates, ammonium phosphates, and thermal phosphoric acid. During production, the company keeps principles of sustainable development and footprint reduction strongly.	[64]
Lovochemie, a.s.	Lovochemie, a.s., is the largest producer of fertilisers in the Czech Republic. Its production program has significantly contributed to the development of Czech agriculture. The company decided to invest in the production of special applications based on P and detergents. Currently, the company produces NPK fertilisers. Lovochemie is trying to find long-term sustainable sources of P to replace current raw materials in future.	[65]
Czech Phosphorus Platform	Czech Phosphorus Platform is an organisation that brings together private companies, government agencies, academic institutions, and individuals. The organisation creates conditions for various activities of members in the area of recycling, circular economy, waste management, sustainable agriculture, and water management to reduce dependence on imports and to recycle P from waste, from crop and animal production in agriculture, from industrial and municipal wastewater.	[63]
Cleaning of the Brno lake from excess P	The Brno lake is the largest reservoir in Brno, measures 10 km in length, and the flooded area is 259 ha. The main problem of Brno lake for a long time was green cyanobacteria, which polluted the entire water area and made recreation impossible. Water purification and treatment in the lake began in 2007. The project on how to stop and improve the gradual deterioration of water quality at Brno lake, especially from flushing water, is called "Implementation of Measures at the Brno Valley Reservoir", which aims to reduce the effects of excessive eutrophication on water. The aeration system, in combination with ferric sulphate dosing, ensures the precipitation of P, which sinks to the bottom and becomes (so far) its harmless part. Results show an improvement in water quality in the lake. Applied systems are used to precipitate P, which is the main food for bacteria and most often enters the water with rainwater from fields where farmers use it as fertiliser. These measures have proved very successful over the years and therefore continue during the next stage of the project, "Implementation of measures at the Brno Valley Reservoir, IVth stage 2023–2027 ". The project is managed and implemented by the Moravia River Basin District. Its staff monitors the state of the water, monitors the health of aquatic animals, and generally finds out how the dam is doing thanks to continuous care. The next significant necessary steps are the removal of precipitated P from the bottom of the Brno Lake and the recovery of P by recycling.	[66]

Table 8. Examples of good practices P management in Hungary.

Company/Project Name	Description of Good Practices	References
Bige Holding Ltd.	Bige Holding Ltd. privatised the Tiszamenti Vegyiművek Rt. in 1997. Following the investment in 2004, Bige Holding Ltd. produces compacted NPK products from the Genesis fertiliser range, as well as sulphuric acid, phosphoric acid, and cryolite. The plant, which operates with compacting technology, produces reliably high-quality NPK and PK fertilisers without chemical reaction and drying process, with both meso- and micro-nutrient content in the quality required by the customer. The particle size and strength of the fertilisers are produced to meet today's modern European quality standards. The environmental impact of the new technology is minimal.	[67]
Nitrogénművek Zrt.	Nitrogénművek Zrt. in Pétfürdő is the one Hungarian nitrogen fertiliser company with ammonia and fertiliser production capacities. The range includes nitrogen fertilisers, complex NPK fertilisers, foliar and nutrient fertilisers. Chemical products and industrial gases generated during the fertiliser production process are also sold. The main task of the company is to meet the long-term demand for fertilisers in Hungarian agriculture. The current market share of Nitrogénművek Zrt. in the domestic fertiliser market is about 60%.	[68]
Péti Nitrokomplex Ltd.	Péti Nitrokomplex Ltd. is owned by Nitrogénművek Zrt, which was founded in 1991 by the self-establishment of the research and development part of the plant. The main goal of the company is to meet the needs of its customers and to adapt to the principles of environmentally friendly, integrated crop production, i.e., the rational supply of nutrients according to the area and the needs of the plan.	[69]
Nádudvar Agrochemical Ltd.	Nádudvar Agrochemical Ltd. operates a world-class state-of-the-art liquid fertiliser service system. The primary objective of the agricultural plant is economic production, which has necessitated the application of state-of-the-art methods in the crop production sector. To achieve this goal, created one of the most advanced liquid fertiliser plants of the time. The company's services include consultancy, transport, the setting up of transit depots, the provision of a group of application machines and the development of using technology.	[70]
Nzrt-Trade Ltd.	Nzrt-Trade Ltd. is a fertiliser supplier in the eastern part of Hungarian agriculture; as a member of the Bige Holding Group, it has a significant R&D activity in the production of fertilisers. The company has links with several research institutes and universities, which carry out the crop certification of its products and the basic research work necessary for their development.	[71]

4. Conclusions

Sustainable management of mineral resources is an important element of functioning of European countries, including the V4 countries. P is one of the most important elements that belongs to the group of CRMs and is an essential element of human nutrition. Moreover, P cannot be replaced by another element. What is more, there is a problem with limiting P resources and planetary boundary for phosphorus is clearly exceeded. The V4 countries do mine P raw materials, and they satisfy the demand with imports. It is possible to replace the current imports of the V4 countries with raw materials from secondary sources, such as:

- industrial wastewater,
- biomass,
- industrial waste,
- others.

Currently, the identification of P recovery potential and good practices of P recovery in V4 countries is a research gap for which detailed data are not available. Moreover, despite the current economic crisis, fertilizers from primary sources are still cheaper than fertilizers from secondary sources. For this reason, in the V4 countries, the topic of good practices in the context of obtaining alternative fertilizers from secondary sources is not popular. Nevertheless, V4 countries have taken steps to broaden the knowledge of P raw materials in society. Initiatives that disseminate information on P raw materials include organisations promoting innovative solutions for the extraction and sustainable management of P or projects related to P raw materials in which countries participate. Such projects include "How to stay alive in V4? Phosphorus Friends Club builds V4's resilience", whose main goal is to increase the knowledge and awareness of the importance of P raw materials for food production in the V4 countries. The project also aims to develop a strategy for the sustainable management of P, which will contribute to ensuring a sufficient amount of P for food production. It also includes various awareness-raising events such as a workshop and a follow-up conference. Project products as a P management roadmap in V4 countries will accelerate the implementation of P recovery and ensure the safety of food production during and after the COVID pandemic. Such and similar initiatives may contribute to faster independence of the V4 countries from the import of P raw materials.

Author Contributions: Conceptualisation, M.S., P.M. and J.J.K..; methodology, M.S. and P.M.; investigation, M.S.; P.M.; Z.Š.; T.B.; M.H.; J.J.K.; Y.V.F.; K.L.; E.K. and A.P.; resources, M.S.; P.M.; Z.Š.; T.B.; M.H.; J.J.K.; Y.V.F.; K.L.; E.K. and A.P.; writing—original draft preparation, M.S. and P.M.; writing—review and editing, M.S. and P.M.; visualisation, M.S. and P.M.; supervision, M.S. and J.J.K. All authors have read and agreed to the published version of the manuscript.

Funding: This research received no external funding.

Data Availability Statement: The datasets generated during and/or analysed during the current study are available from the corresponding author on reasonable request.

Acknowledgments: The study was developed under the project: PhosV4—"How to stay alive in V4? Phosphorus Friends Club builds V4's resilience", no. 22110364 (2021–2023), which is financed by Visegrad Fund.

Conflicts of Interest: The authors declare no conflict of interest.

References

1. Wieczorek, D.; Żyszka-Haberecht, B.; Kafka, A.; Lipok, J. Phosphonates as Unique Components of Plant Seeds—A Promising Approach to Use Phosphorus Profiles in Plant Chemotaxonomy. *Int. J. Mol. Sci.* **2021**, *22*, 11501. [CrossRef] [PubMed]
2. Smol, M.; Adam, C.; Kugler, S.A. Thermochemical Treatment of Sewage Sludge Ash (SSA)—Potential and Perspective in Poland. *Energies* **2020**, *13*, 5461. [CrossRef]
3. Smol, M.; Preisner, M.; Bianchini, A.; Rossi, J.; Hermann, L.; Schaaf, T.; Kruopienė, J.; Pamakštys, K.; Klavins, M.; Ozola-Davidane, R.; et al. Strategies for Sustainable and Circular Management of Phosphorus in the Baltic Sea Region: The Holistic Approach of the InPhos Project. *Sustainability* **2020**, *12*, 2567. [CrossRef]
4. Yu, H.; Lu, X.; Miki, T.; Matsubae, K.; Sasaki, Y.; Nagasaka, T. Sustainable Phosphorus Supply by Phosphorus Recovery from Steelmaking Slag: A Critical Review. *Resour. Conserv. Recycl.* **2022**, *180*, 106203. [CrossRef]
5. Sun, H.; Mohammed, A.N.; Liu, Y. Phosphorus Recovery from Source-Diverted Blackwater through Struvite Precipitation. *Sci. Total Environ.* **2020**, *743*, 140747. [CrossRef] [PubMed]
6. European Commission. *Communication from the Commission: On the Review of the List of Critical Raw Materials for the EU and the Implementation of the Raw Materials Initiative*; COM No. 297; European Commission: Brussels, Belgium, 2014.
7. European Commission. *Communication from the Commission: On the 2017 List of Critical Raw Materials for the EU*; COM No. 490; European Commission: Brussels, Belgium, 2017.
8. European Commission. *Communication from the Commission: Critical Raw Materials Resilience: Charting Path towards Greated Secrity and Sustainability*; COM No. 474; European Commission: Brussels, Belgium, 2020.

9. Akhtar, M.S.; Oki, Y.; Adachi, T. Intraspecific Variations of Phosphorus Absorption and Remobilization, P Forms, and Their Internal Buffering in Brassica Cultivars Exposed to a P-Stressed Environment. *J. Integr. Plant Biol.* **2008**, *50*, 703–716. [CrossRef]
10. Fotyma, M.; Igras, J.; Kopiński, J. Production and Environment Conditions of Fertilization Management in Poland (Produkcyjne i Środowiskowe Uwarunkowania Gospodarki Nawozowej w Polsce). *Stud. Rap. IUNG-PIB* **2009**, *14*, 187–206. [CrossRef]
11. Czuba, R. *Mineral Fertilization of Crops (Nawożenie Mineralne Upraw)*; Zakłady Chem. "POLICE" S.A.: Puławy, Poland, 1996; Volume 413.
12. Phosphorus. Available online: www.britannica.com/science/phosphorus-chemical-element (accessed on 21 October 2022).
13. European Commission. *Communication from the Commission: A Farm to Fork Strategy for a Fair, Healthy and Environmentally-Friendly Food System*; COM No. 381; European Commission: Brussels, Belgium, 2020.
14. Sgroi, F.; Marino, G. Environmental and Digital Innovation in Food: The Role of Digital Food Hubs in the Creation of Sustainable Local Agri-Food Systems. *Sci. Total Environ.* **2022**, *810*, 152257. [CrossRef]
15. Peydayesh, M.; Bagnani, M.; Soon, W.L.; Mezzenga, R. Turning Food Protein Waste into Sustainable Technologies. *Chem. Rev.* **2022**. [CrossRef]
16. Kochanek, E. The Role of Hydrogen in the Visegrad Group Approach to Energy Transition. *Energies* **2022**, *15*, 7235. [CrossRef]
17. Han, E.; Tan, M.M.J.; Turk, E.; Sridhar, D.; Leung, G.M.; Shibuya, K.; Asgari, N.; Oh, J.; García-Basteiro, A.L.; Hanefeld, J.; et al. Lessons Learnt from Easing COVID-19 Restrictions: An Analysis of Countries and Regions in Asia Pacific and Europe. *Lancet* **2020**, *396*, 1525–1534. [CrossRef] [PubMed]
18. Population on 1 January 2022. Available online: https://ec.europa.eu/eurostat/web/products-datasets/-/TPS00001 (accessed on 13 November 2022).
19. Main Indicators of the Visegrad Group Countries. Available online: https://stat.gov.pl/statystyka-miedzynarodowa/porownania-miedzynarodowe/publikacje-zawierajace-dane-z-zakresu-porownan-miedzynarodowych/publikacje-biezace/main-indicators-of-the-visegrad-group-countries,3,1.html (accessed on 14 November 2022).
20. Rahimpour Golroudbary, S.; El Wali, M.; Kraslawski, A. Rationality of Using Phosphorus Primary and Secondary Sources in Circular Economy: Game-Theory-Based Analysis. *Environ. Sci. Policy* **2020**, *106*, 166–176. [CrossRef]
21. Smol, M. Transition to Circular Economy in the Fertilizer Sector—Analysis of Recommended Directions and End-Users' Perception of Waste-Based Products in Poland. *Energies* **2021**, *14*, 4312. [CrossRef]
22. Burkowicz, A.; Galos, K.; Guzik, K.; Kamyk, J.; Kot-Niewiadomska, A.; Lewicka, E.; Smakowski, T.; Szlugaj, J. *Minerals Yearbook of Poland 2013*; Mineral and Energy Economy Research Institute of the Polish Academy of Sciences, Department of Mineral Policy, Polish Geological Institute—National Research Institute: Warsaw, Poland, 2014; ISBN 978-83-7863-378-5.
23. Polish Geological Institute—National Research Institute. The Balance of Mineral Resources Deposits in Poland as of 31.12.2021. Available online: http://surowce.pgi.gov.pl (accessed on 10 November 2022).
24. Food and Agriculture Organization of the United Nations. Available online: https://www.fao.org/faostat/en/#data/RFB (accessed on 10 November 2022).
25. Bakalár, T.; Pavolová, H.; Šimková, Z.; Bednárová, L. Phosphorus Management in Slovakia—A Case Study. *Sustainability* **2022**, *14*, 10374. [CrossRef]
26. Statistical Office of the Slovak Republic: Phosphorus Export/Import, Livestock Inventory 2010–2019. Available online: http://datacube.statistics.sk (accessed on 11 November 2022).
27. Smol, M. The Importance of Sustainable Phosphorus Management in the Circular Economy (CE) Model: The Polish Case Study. *J. Mater. Cycles Waste Manag.* **2019**, *21*, 227–238. [CrossRef]
28. Egle, L.; Rechberger, H.; Krampe, J.; Zessner, M. Phosphorus Recovery from Municipal Wastewater: An Integrated Comparative Technological, Environmental and Economic Assessment of P Recovery Technologies. *Sci. Total Environ.* **2016**, *571*, 522–542. [CrossRef]
29. Guedes, P.; Couto, N.; Ottosen, L.M.; Ribeiro, A.B. Phosphorus Recovery from Sewage Sludge Ash through an Electrodialytic Process. *Waste Manag.* **2014**, *34*, 886–892. [CrossRef]
30. Klem-Marciniak, E.; Hoffmann, K.; Huculak-Wczka, M.; Hoffamann, J.; Makara, A.; Kowalski, Z. Fertilizer Properties of Filter Sludge from Pig Slurry. *Przem. Chem.* **2015**, *94*, 2158–2161.
31. Staron, P.; Banach, M.; Kowalski, Z. Assessment of an Application of Ashes Produced by Incineration of Poultry Industry Waste as a Rich Phosphorus Source. *Przem. Chem.* **2013**, *92*, 1142–1144.
32. Kulczycka, J.; Kowalski, Z.; Smol, M.; Wirth, H. Evaluation of the Recovery of Rare Earth Elements (REE) from Phosphogypsum Waste—Case Study of the WIZÓW Chemical Plant (Poland). *J. Clean. Prod.* **2016**, *113*, 345–354. [CrossRef]
33. Johansson, S.; Ruscalleda, M.; Colprim, J. Phosphorus Recovery through Biologically Induced Precipitation by Partial Nitritation-Anammox Granular Biomass. *Chem. Eng. J.* **2017**, *327*, 881–888. [CrossRef]
34. Krzywy, E.; Iewska, A.; Włoszczyk, C. Assessment of the Chemical Composition and Fertilizing Value of Sewage Sludge and Composts Produced from Municipal Sewage Sludge (Ocena Składu Chemicznego i Wartosci Nawozowej Osadu Sciekowego Oraz Kompostow Wyprodukowanych z Komunalnego Osadu Sciekow. *Zesz. Probl. Postępow Nauk Rol.* **2004**, *499*, 165–171.
35. Gondek, K.; Filipek-Mazur, B. The Content and Uptake of Micronutrients by Oats under the Conditions of Fertilization with Composts of Various Origins in Terms of Feed Value and Environmental Impact (Zawartosc i Pobranie Mikroelementow Przez Owies w Warunkach Nawozenia Kompostami R. *Woda-Srodowisko-Obsz. Wiej.* **2005**, *1*, 81–93.

36. Poluszunska, J.; Slezak, E. Possibilities of Phosphorus Recovery from Sewage Sludge (Mozliwosci Odzysku Fosforu z Osadow Sciekowych). *Pr. Inst. Ceram. Mater. Bud.* **2015**, *21*, 7–21.
37. Tabatabai, M.A.; Frankenberger, W.T. Variability of Chemical Properties of Sewage Sludges in Iowa. *Res. Bull.* **1979**, *36*, 13.
38. Mattenberger, H.; Fraissler, G.; Brunner, T.; Herk, P.; Hermann, L.; Obernberger, I. Sewage Sludge Ash to Phosphorus Fertiliser: Variables Influencing Heavy Metal Removal during Thermochemical Treatment. *Waste Manag.* **2008**, *28*, 2709–2722. [CrossRef] [PubMed]
39. Lynn, C.J.; Dhir, R.K.; Ghataora, G.S.; West, R.P. Sewage Sludge Ash Characteristics and Potential for Use in Concrete. *Constr. Build. Mater.* **2015**, *98*, 767–779. [CrossRef]
40. Herzel, H.; Krüger, O.; Hermann, L.; Adam, C. Sewage Sludge Ash—A Promising Secondary Phosphorus Source for Fertilizer Production. *Sci. Total Environ.* **2016**, *542*, 1136–1143. [CrossRef]
41. Gondek, K.; Filipek-Mazur, B. Accumulation of Micronutrients in Oat Biomass and Their Availability in Soil Fertilized with Plant Waste Compost (Akumulacja Mikroelementow w Biomasie Owsa Oraz Ich Dostepnosc w Glebie Nawozonej Kompostem z Odpadow Roslinnych). *Acta Agrophysica* **2006**, *8*, 579–590.
42. Marcato, C.; Pinelli, E.; Pouech, P.; Winterton, P.; Guiresse, M. Particle Size and Metal Distributions in Anaerobically Digested Pig Slurry. *Bioresour. Technol.* **2008**, *99*, 2340–2348. [CrossRef]
43. Gollehon, N.; Caswell, M.; Ribaudo, M.; Kellogg, R.; Lander, C.; Letson, D. *Confined Animal Production and Manure Nutrients*; Agriculture Information Bulletin Number 771; United States Department of Agriculture (USDA), Economic Research Service: Washington, DC, USA, 2001.
44. Risberg, K.; Cederlund, H.; Pell, M.; Arthurson, V.; Schnürer, A. Comparative Characterization of Digestate versus Pig Slurry and Cow Manure—Chemical Composition and Effects on Soil Microbial Activity. *Waste Manag.* **2017**, *61*, 529–538. [CrossRef] [PubMed]
45. Halik, J.; Chowaniak, M.; Poczesna, E.; Polak, E. Agrochemical Assessment of the Fertilizing Value of Composts (Agrochemiczna Ocena Wartosci Nawozowej Kompostow). *Mater. II Ogolnopolskiej Młodziezowej Konf. Nauk.* **2004**, 96–100.
46. Gondek, K. Accumulation of Heavy Metals in Oat Fertilized with Compost (Akumulacja Metali Ciezkich w Owsie Nawozonym Kompostami). *Acta Agrophysica* **2007**, *10*, 89–102.
47. Chiew, Y.L.; Spångberg, J.; Baky, A.; Hansson, P.-A.; Jönsson, H. Environmental Impact of Recycling Digested Food Waste as a Fertilizer in Agriculture—A Case Study. *Resour. Conserv. Recycl.* **2015**, *95*, 1–14. [CrossRef]
48. European Commission. *Communication from the Commission: Consultative Communication on the Sustainable Use of Phosphorus*; COM No. 517; European Commission: Brussels, Belgium, 2013.
49. Smol, M.; Adam, C.; Anton Kugler, S. Inventory of Polish Municipal Sewage Sludge Ash (SSA)—Mass Flows, Chemical Composition, and Phosphorus Recovery Potential. *Waste Manag.* **2020**, *116*, 31–39. [CrossRef]
50. Smol, M. Phosphorus Extraction and Sludge Dissolution. In *Industrial and Municipal Sludge*; Elsevier: Amsterdam, The Netherlands, 2019; pp. 657–677.
51. Eurostat Sewage Sludge Production and Disposal from Urban Wastewater (in Dry Substance (d.s)). Available online: http://ec.europa.eu/eurostat/web/products-datasets/-/ten00030 (accessed on 14 November 2022).
52. Smol, M. Inventory of Wastes Generated in Polish Sewage Sludge Incineration Plants and Their Possible Circular Management Directions. *Resources* **2020**, *9*, 91. [CrossRef]
53. The Plant in Jarocin Will Produce Agricultural Fertiliser from Sediments. Available online: www.portalkomunalny.pl/oczyszczalnia-w-jarocinie-wyprodukuje-nawoz-rolniczy-z-osadow-374791 (accessed on 6 November 2022).
54. Azoty Group Fosfory Sp. z o.O. Available online: www.fosfory.pl (accessed on 6 November 2022).
55. Sewage Treatment Plant-Tarnowskie Wodociągi Sp. z o.O. Available online: www.tw.tarnow.pl/new/pliki/10_1.html (accessed on 6 November 2022).
56. Project "Sustainable Management of Phosphorus in the Baltic Region (InPhos)". Available online: www.min-pan.krakow.pl/projekty/en/2019/07/17/english-zrownowazone-zarzadzanie-fosforem-w-regionie-baltyckim/ (accessed on 6 November 2022).
57. Project "Market Ready Technologies for P-Recovery from Municipal Wastewater (PhosForce)". Available online: www.min-pan.krakow.pl/projekty/en/2019/07/17/technologia-odzysku-fosforu-ze-sciekow-komunalnych/ (accessed on 6 November 2022).
58. Project "Towards Circular Economy in Wastewater Sector: Knowledge Transfer and Identification of the Recovery Potential for Phosphorus in Poland (CEPhosPOL)". Available online: www.min-pan.krakow.pl/projekty/en/2019/07/16/program-stypendialny-im-mieczyslawa-bekkera/ (accessed on 6 November 2022).
59. Smol, M.; Kulczycka, J.; Lelek, Ł.; Gorazda, K.; Wzorek, Z. Life Cycle Assessment (LCA) of the Integrated Technology for the Phosphorus Recovery from Sewage Sludge Ash (SSA) and Fertilizers Production. *Arch. Environ. Prot.* **2020**, *46*, 42–52. [CrossRef]
60. Duslo, A.S. Available online: https://www.duslo.sk/en/about-us (accessed on 6 November 2022).
61. Project "Drinking Water Supply, Sewerage and Wastewater Treatment". Available online: www.ec.europa.eu/regional_policy/et/projects/major/slovakia/major-water-management-project-makes-waves (accessed on 6 November 2022).
62. Chernysh, Y.; Balintova, M.; Plyatsuk, L.; Holub, M.; Demcak, S. The Influence of Phosphogypsum Addition on Phosphorus Release in Biochemical Treatment of Sewage Sludge. *Int. J. Environ. Res. Public Health* **2018**, *15*, 1269. [CrossRef]
63. Czech Phosphorus Platform. Available online: www.fosforovaplatforma.cz/english/ (accessed on 6 November 2022).
64. Fosfa, A.S. Available online: www.web.fosfa.cz/en/ (accessed on 6 November 2022).
65. Lovochemie, A.S. Available online: www.lovochemie.cz/en (accessed on 6 November 2022).

66. Portfolio of Phosphorus Friend in Europe. Available online: https://www.phosv4.eu/project-products/portfolio-of-phosphorus-friends-in-europe-mapping-service (accessed on 12 November 2022).
67. Bige Holding Ltd. Available online: www.bigeholding.hu/english/index.html (accessed on 6 November 2022).
68. Nitrogénművek Zrt. Available online: www.nitrogen.hu/en (accessed on 6 November 2022).
69. Péti Nitrokomplex Ltd. Available online: www.nitrokomplex.hu/24-2/ (accessed on 6 November 2022).
70. Nádudvar Agrochemical Ltd. Available online: www.nakft.hu (accessed on 6 November 2022).
71. Nzrt-Trade Ltd. Available online: www.genezispartner.com (accessed on 6 November 2022).

resources

Review

Application of New Hyperspectral Sensors in the Remote Sensing of Aquatic Ecosystem Health: Exploiting PRISMA and DESIS for Four Italian Lakes

Mariano Bresciani *, Claudia Giardino , Alice Fabbretto, Andrea Pellegrino, Salvatore Mangano , Gary Free and Monica Pinardi

Institute of Electromagnetic Sensing of the Environment, National Research Council of Italy (CNR-IREA), Via Bassini 15, 20133 Milan, Italy; giardino.c@irea.cnr.it (C.G.); fabbretto.a@irea.cnr.it (A.F.); pellegrino.a@irea.cnr.it (A.P.); mangano.s@irea.cnr.it (S.M.); free.g@irea.cnr.it (G.F.); pinardi.m@irea.cnr.it (M.P.)
* Correspondence: bresciani.m@irea.cnr.it

Abstract: The monitoring of water bio-physical parameters and the management of aquatic ecosystems are crucial to cope with the current state of inland water degradation. Not only does water quality monitoring support management decision making, it also provides vital insights to better understand changing structural and functional lake processes. Remote sensing has been widely recognized as an essential integrating technique for water quality monitoring, thanks to its capabilities to utilize both historical archive data for thousands of lakes as well as near-real time observations at multiple scales. To date, most of the applications developed for inland water have been based on multispectral and mid to coarse spatial resolution satellites, while a new generation of spaceborne imaging spectroscopy is now available, and future missions are under development. This review aims to present the exploitation of data gathered from two currently orbiting hyperspectral sensors (i.e., PRISMA and DESIS) to retrieve water quality parameters across different aquatic ecosystems, encompassing deep clear lakes and river dammed reservoirs.

Keywords: satellite images; hyperspectral data; water quality; PRISMA; DESIS; turbidity; chlorophyll-a; aquatic vegetation

Citation: Bresciani, M.; Giardino, C.; Fabbretto, A.; Pellegrino, A.; Mangano, S.; Free, G.; Pinardi, M. Application of New Hyperspectral Sensors in the Remote Sensing of Aquatic Ecosystem Health: Exploiting PRISMA and DESIS for Four Italian Lakes. *Resources* **2022**, *11*, 8. https://doi.org/10.3390/resources11020008

Academic Editor: Demetrio Antonio Zema

Received: 10 December 2021
Accepted: 16 January 2022
Published: 19 January 2022

1. Introduction

Inland water ecosystems are crucial environments as they provide multiple ecosystem services and are vital resources for drinking, irrigation, sanitation, industry and recreation [1,2]. Many regions of the world are now facing serious challenges to their freshwater resource including increasing water demand, water shortage, decreases in the level of groundwater, reduced water quality and increasing damages to dependent ecosystems. Overall, in the past decades and under the current global warming scenario the quality of inland waters has been severely endangered and often dramatically degraded [3–6]. An expected decline in the quantity of water available is likely to coincide with an increasing demand for freshwater globally. To understand and anticipate this process in order to mitigate the deterioration of inland water quality, the monitoring of bio-physical variables depicting water conditions is essential. Water quality monitoring actually supports water authorities to take decisions, while it provides useful insights to lakes scientists to understand processes.

In situ measurements have been used for water quality monitoring for many years allowing the gathering of information from the entire water column, including an extensive set of parameters (e.g., phytoplankton abundance, nutrient concentrations) many of which eventually became parameters subject to national and international standards (e.g., [7–9]). For more than three decades, remote sensing has been recognized as an essential integrating technique for water quality monitoring. Through acquiring timely, frequent synoptic water quality parameters, it also supports implementation of water protection directives [10], while it is a unique tool for a variety of scientific studies from local to global scale [11]. Satellite Earth Observation (EO) techniques are often used alongside in situ observations due to the possibility of having both historical archive data for thousands of lakes (since the 1970s) [12–14] and near-real time observational capabilities at multiple-scales.

Overall, the monitoring of water quality parameters in different types of waterbodies (lakes, rivers, etc.) can be optimally performed integrating traditional in-situ measurements with remote sensing and GIS techniques [15]. Water managers, in order to effectively deal with waterbodies ecological stressors, look to remote sensing as a useful tool to provide advanced spatial–temporal environmental information to support decision making [16–18]. An increasing demand for routine monitoring using remote sensing to support inland waters management to achieve water quality goals is now evident. Despite this recognized need and the numerous examples reported in the literature, water quality products derived by remote sensing are yet to be adopted on a routine framework, probably due also to the knowledge gap of water managers and policy makers to understand technical abilities and limitations of the different optical sensors [15]. For this reason, it is important that researchers and remote sensing experts communicate to water managers the appropriate available tools to address monitoring requirements [19–24]. To this aim international initiatives have been developed; among others GEO AquaWatch has several projects advancing in the coming years for promoting the use of scalable remote sensing and in situ water quality data. A key goal is to bridge the information gap between global decision-makers and water quality earth observation data providers.

To date, most of the applications developed for inland water quality monitoring and management have been based on multispectral and mid (e.g., Landsat constellation, Sentinel-2-MSI) to coarse (e.g., ENVISAT-MERIS, Sentinel-3-OLCI) spatial resolution satellites [25]. While research efforts are still ongoing to face the challenges typical of EO of inland waters, such as global chlorophyll-a concentration mapping or corrections for adjacency effects [26–30], it is also true that depending on an optical sensor's specifications, the EO-derived products vary from turbidity, transparency, and concentrations of chlorophyll, suspended particulate matter and colored dissolved organic matter, floating materials and, in the case of shallow waters, bottom depth and type. In the literature of the past and recent years, numerous articles and reviews [31] and reference therein addressed the optical water quality parameters that can be retrieved by remote sensing techniques (i.e., suspended sediments (turbidity), chlorophyll and other secondary pigments, color dissolved organic matter (CDOM), water clarity and temperature) (e.g., [13,32,33]), the different properties of sensors and platforms and their environmental applications (e.g., [34–36]) and the algorithms developed and implemented to retrieve water quality products (e.g., [37–39]). For the ecological analysis and correct management of lake ecosystems, it is of great importance to also have information on the benthic component of coastal areas and the presence of aquatic macrophytes. Most of the applications have been carried out with medium to high spatial resolution sensors (e.g., Landsat, Sentinel-2 MSI) supporting retrospective analysis, thematic mapping of species in valuable ecosystems such as wetlands.

In a recent review, Topp et al. [40] made a bibliometric analysis which highlighted the conspicuous increase of inland water quality remote sensing studies, particularly in the last 10–15 years studies were found to have moved beyond methods development towards water quality dynamics (e.g., trends, ecological and anthropogenic drivers, impacts of changing water quality on ecosystem function and water resources) thanks to an improved access to EO data and growing computational capabilities. In this review, the authors classified a subset of 236 articles into: Purely methodological, Methodological with pattern analysis, Trend/pattern analysis, and Water quality science research with a focus on impacts and drivers. In this latter more applied paper category a total of 40 articles were found, and the majority of them were related to lakes (85% of the applied articles). In addition, imaging spectroscopy has been also attracting a wide interest [28] and reference therein since its simultaneous collection of narrower and contiguous bands is improving aquatic ecosystem mapping for the retrieval of parameters describing water quality, aquatic vegetation (e.g., biomass [41], invasive species identification [42]) and benthic substrates that might be undetectable with broadband multispectral sensors [15,27,40,43–48]. In such a context, airborne data (e.g., APEX, AISA, MIVIS) have been providing unique data at high spectral and spatial resolution for performing advanced mapping as well to support satellite mission development and verification (e.g., [49–52]). With respect to spaceborne observations, following the launch of Hyperion which was followed a few years later by Chris-PROBA and HICO, a new generation of spaceborne hyperspectral sensors (e.g., PRISMA, GaoFen-5, DESIS, HISUI), is now available for improving water resources monitoring, while future missions are also under development (e.g., EnMap, CHIME, SBG). Accordingly, this study aims to present the exploitation of data gathered from two currently orbiting spaceborne imaging spectrometers, PRecursore IperSpettrale della Missione Applicativa (PRISMA) and the DLR Earth Sensing Imaging Spectrometer (DESIS), for water quality parameters retrieval, bottom type mapping and monitoring emergent macrophytes in four Italian water bodies, encompassing deep clear lakes, turbid shallow lakes and river dammed reservoirs.

2. PRISMA and DESIS Missions

PRISMA, a mission fully funded by the Italian Space Agency (ASI), is an EO system with innovative, electro-optical instrumentation that combines a hyperspectral sensor with a medium-resolution panchromatic camera [53]. The PRISMA orbit is characterized by a repeat cycle of approximately 29 days. The PRISMA payload is composed by an imaging spectrometer, able to take images at 30 m resolution in a continuum of spectral bands ranging from 400 to 2500 nm, and a 5 m resolution panchromatic camera. The PRISMA system provides the capability to acquire, downlink and archive images of all hyperspectral/panchromatic channels totaling 200,000 km^2 daily over the primary area of interest defined as: longitude in the range 180° W–180° E; Latitude: in the range 70° S–70° N. Although the revisit time in nadir-looking configuration is of 29 days, the system is capable of acquiring images distant 1000 km in a single pass (with a total rotation left to right side looking and vice versa) so that the temporal resolution can be significantly improved.

DESIS, the other imaging spectrometry system used in this study, is a hyperspectral instrument integrated in the Multi-User-System for Earth Sensing (MUSES) platform installed on the International Space Station (ISS). The mission is operated by Teledyne Brown Engineering, Alabama, USA, and the German Aerospace Center (DLR), Germany. DESIS is realized as a pushbroom imaging spectrometer spectrally sensitive over the VNIR range from 400 to 1000 nm with a spectral sampling distance of 2.55 nm. The main specifications of DESIS are shown in Table 1. The launch of DESIS to ISS was on 29 June, 2018. The commissioning and validation phases were finished in March 2019. DESIS is now working operationally and will continue until at least the end of 2023 [54].

Table 1. Main specifications of PRISMA (PRISMA Products Specification Document, 2019) and DESIS (from DESIS link https://www.dlr.de/eoc/desktopdefault.aspx/tabid-13614 (accessed on 20 October 2021)) instruments.

	PRISMA	DESIS
Launch	22 March 2019	29 June 2018
Coverage	70° N to 70° S	55° N to 52° S
Target lifetime	5 years	2018–2023
Orbit	SSO 615 km 10:30 LTDN	MUSES platform on ISS
Number of bands	VNIR: 66 (440–1010 nm), SWIR: 174 (920–2505 nm), PAN: 1 (400–700 nm)	235 (no binning), 60 (binning 4)
Spectral sampling interval	VNIR: 7.2–11 nm, SWIR: 6.5–11 nm	2.55 nm (no binning), 10.2 nm (binning 4)
Spectral coverage	440 nm to 2505 nm	402 nm to 1000 nm
Ground sampling distance	Hyperspectral: 30 m, PAN: 5 m	30 m
Signal-to-noise ratio	>160:1 VNIR, >100:1 SWIR, >240:1 PAN	195 (w/o binning), 386 (4 binning) (based on laboratory calibration) (albedo 0.3 @ 550 nm)
Radiometric resolution	12 bits	13 bits + 1 bit gain
Swath	30 km	30 km

3. Study Area, Imagery Data and Processing

For the purposes of this study, a set of PRISMA and DESIS images have been explored over four inland water systems located in Italy as shown in Figure 1. We have selected the case studies in order to include lakes of different trophic status, depth profiles and different hydrogeochemical and ecological processes.

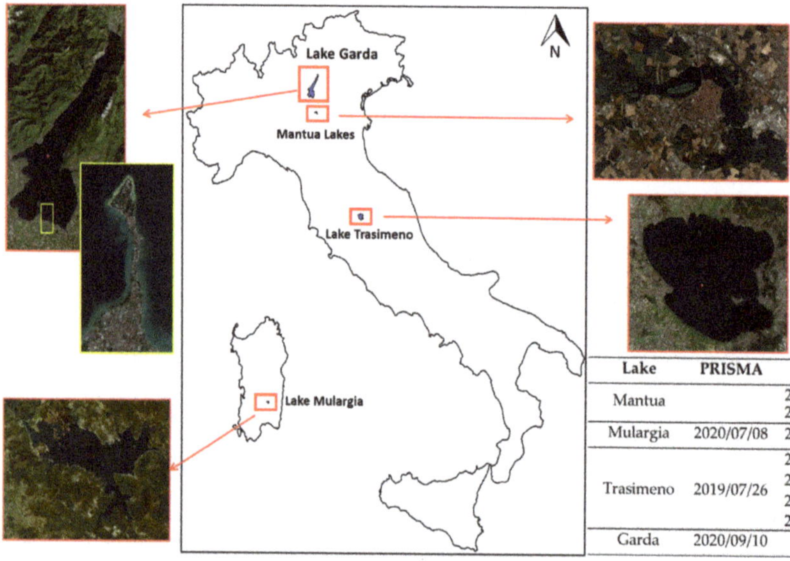

Lake	PRISMA	DESIS
Mantua		2019/08/07 2020/06/22
Mulargia	2020/07/08	2020/08/17
Trasimeno	2019/07/26	2019/06/18 2019/08/04 2019/09/05 2020/05/26
Garda	2020/09/10	

Figure 1. Maps of the lake case studies investigated in Italy. The table reports the dates of PRISMA and DESIS images acquired for Mantua, Mulargia, Trasimeno and Garda lakes.

RISMA and DESIS, the latter being the binned product, have been firstly corrected for atmospheric effects with the ATCOR tool (version 9.3.0) [55]. These instruments have already proved capable of providing valuable atmospherically corrected reflectance data for water quality applications [56,57], while further research efforts to improve the retrieval of water reflectance are still under development [58]. The result of atmospheric correction is presented in Figure 2; both PRISMA and DESIS show distinct features associated with different gradients of trophic status, water transparency, bottom properties, in addition to emerging aquatic vegetation.

PRISMA and DESIS data corrected for the atmospheric effects were then turned into aquatic ecosystems products by inverting a bio-optical modelling to both map water quality parameters and bottom types while, vegetation indexes, were used to estimate the macrophyte biomass. In particular, the bio-optical model used in this study is similar to previously published case-2 water models (e.g., [59,60]), where water reflectance is depending on concentrations of water quality parameters; then, in case of shallow waters, also on bottom depths and on a fractional cover of bottom reflectances. The bio-optical model parametrization and inversion were run with the BOMBER code, detailed described in Giardino et al. [61] while Table 2 and Figure 3 show the site-specific parameters which allowed the bio-optical model implemented in BOMBER to be calibrated. The parameters for the four sites, which are resulting from previous field measurements (e.g., [27,62,63]), are comparable to literature data for case-2 waters (e.g., [64–66]). For the specific application over shallow waters, the bottom reflectance used in this study can be found in Ghirardi et al. [67]. Finally, the inversion procedure was initialized with the average values of water quality parameters typical of each site, for the season for which PRISMA and DESIS imagery have been acquired.

Finally, for mapping the above water biomass of floating-leaved and emerging aquatic vegetation we used the algorithm proposed by Villa et al. [68] previously applied to airborne APEX hyperspectral data on Mantua lakes and valid for a biomass range of 0–0.5 kg m^{-2}. Briefly, the algorithm was built on a band ratio in the red-edge and near-infrared APEX wavelengths and, due to a similar hyperspectral band setting, it can be easily adapted to both PRISMA and DESIS data.

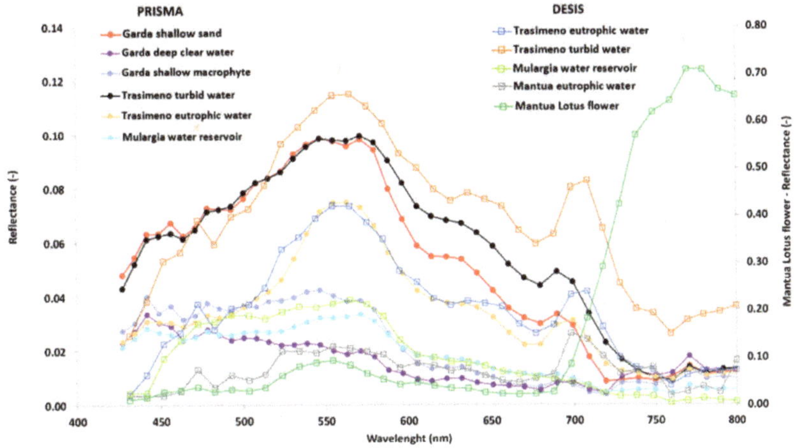

Figure 2. PRISMA and DESIS water reflectance. The spectra have been extracted to demonstrate the variability of hyperspectral reflectance across the water systems, from shallow to deep, and from clear to eutrophic or turbid and of aquatic vegetation (i.e., lotus flower) collected in the case study lakes investigated (Mantua, Mulargia, Trasimeno, and Garda lakes).

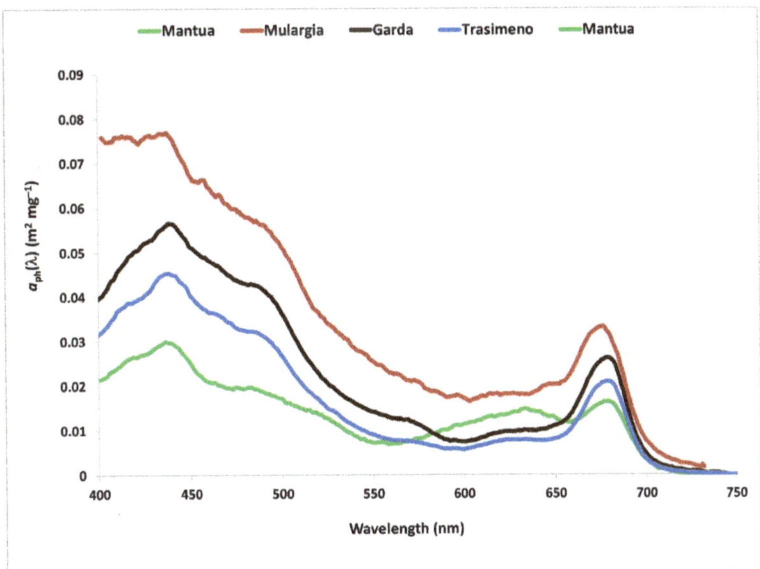

Figure 3. The specific absorption spectra of phytoplankton for the study areas used to calibrate the bio-optical model implemented in BOMBER.

Table 2. Calibration parameters of the bio-optical model used in BOMBER for estimating water quality parameters from PRISMA and DESIS imagery.

Parameter	Garda	Mantua	Mulargia	Trasimeno
Spectral slope coefficient of CDOM absorption	0.021	0.014	0.018	0.016
Specific absorption of NAP at 440 nm	0.05	0.30	0.11	0.20
Spectral slope coefficient of NAP absorption	0.012	0.008	0.012	0.013
Backscattering exponent of TSM	0.76	0.80	0.71	0.65
Specific back-scattering coefficient of TSM at 555 nm	0.0071	0.0111	0.0105	0.0119

4. Exploitation of PRISMA and DESIS Products for Water Quality Mapping

In this section we show how the PRISMA and DESIS hyperspectral images can generate aquatic ecosystems products for different types of water bodies. In particular, among the topics of greatest interest for lakes is the need to have spatial information on the distribution of primary producers (both phytoplankton and emerged and submerged aquatic macrophytes) and their temporal variability. Another key requirement for monitoring water quality is to have information relating to the transparency of the water, which in lakes with low humic substances, is directly related to the presence of total suspended solids in the water column. For the purposes of this review we limit the discussion of the products to the use of hyperspectral images for ecological studies, monitoring and management of aquatic environments.

4.1. Mantua Lakes

The fluvial-lake system of Mantua is located in Northern Italy (45°10′ N, 10°47′ E; surface area 6.2 km^2, mean depth 3.5 m; Figure 1) and was created in the 12th century by the damming of the Mincio River. The system is composed of three shallow lakes, Superior, Middle and Inferior lakes. In Superior Lake, the Vasarone sluice gate (dating back to 1190 A.C.) and the recently-opened Vasarina gate (2015) are regulated to produce electricity and maintain a constant water level in the three basins (17.5 m a.s.l. on the Superior Lake, and 14.5 m a.s.l. on the Middle and Inferior lakes) [69]. Elevated nutrient loads enter the fluvial-lake system from its main tributary, the Mincio River, that drains a watershed intensively exploited by agricultural and animal farming activities [70,71]. Therefore, this fluvial-lake system is highly productive with the coexistence of dense phytoplankton and macrophyte (emergent, submersed, floating-leaved and free-floating) communities [68,72]. Water is characterized by low transparency (Secchi disk depth < 1 m in summer) and high chlorophyll-a (Chl-a) concentration (up to 200 mg m^{-3}) [73,74]. The Mantua Lakes are part of the Mincio Regional Park, and since 2008 they have been part of the UNESCO World Cultural Heritage. Active management of the water and macrophytes is needed to ensure the conservation of the area, and enable the recreational and touristic activities developed on the lakes. Floating-leaved macrophytes (such as Nelumbo nucifera, Trapa natans and Ludwigia hexapetala) are kept under control by cutting and occasionally harvesting to avoid an excessive accumulation of organic matter on the bottom of the lakes that can lead to rapid infilling and the risk of hypoxia/anoxia events [68,75]. For a decade this study area has been investigated with different techniques which includes proximal and remote sensing techniques. For example, for the first application, Bresciani et al. [73] assessed short (intra-day) and medium (from weekly to monthly) term Chl-a dynamics by means of continuous spectroradiometric measurements, and Pinardi et al. [76] integrated the use of high frequency in situ measurements and multispectral satellite data to explore temporal (diurnal and seasonal) dynamics and spatial distribution of Chl-a concentration, together with physico-chemical water parameters. For the second application, various multispectral satellite (e.g., Landsat constellation, Sentinel-2, SPOT5) or hyperspectral sensors (e.g., APEX, MIVIS) were used to retrieve water quality [77], macrophyte products [68,78,79] or both [72,74,80] to assess the Mantua lakes system ecology at different spatial–temporal scales and to support management plans and actions. In this small fluvial-lake where there are multiple pressures and diverse uses of the water, hyperspectral data can guarantee spatial and temporal information suitable for the study of primary producers, including the identification of the different functional groups and their seasonal evolution.

In order to provide relative information of the different primary producers present in the Mantua Lakes we used two DESIS images (2019/06/21 and 2020/06/22), the products of which are shown in Figure 4. Chl-a and Phycocyanin (PC) concentrations were estimated by means of BOMBER as further described also in Giardino et al. [27], with specific optical properties of the waters of the Mantua Lakes (see Table 2 and Figure 3), and previously applied to airborne hyperspectral APEX images. The comparison of the products with field data indicated good agreement: for the 21 June 2019 the Chl-a measured in situ via fluorimetric probe in the Middle and Inferior lakes [76] had a value of 20.0 and 15.4 mg m^{-3} respectively versus satellite products of 21.9 and 16.0 mg m^{-3}; for the 22 June 2020 the in situ Chl-a was 23.5 and 17.7 mg m^{-3} in the Middle and Inferior lakes respectively, versus satellite products of 25.0 and 16.8 mg m^{-3} (Figure 4).

The possibility of having simultaneously spatial information related to the concentrations of the different phytoplankton photosynthetic pigments provides a high added value in the understanding of the quality state of aquatic environments. In the Chl-a maps a clear increasing gradient down the system from the Superior (median value of 13.65 and 16.08 mg m^{-3} for 2019 and 2020 images, respectively) to the Middle Lake (median value of 16.80 and 26.35 mg m^{-3} for 2019 and 2020 images, respectively) is shown. Notwithstanding the fact that portions of the Superior Lake had higher concentration due to water circulation driven by water discharge and by the presence of macrophyte meadows. Similarly, the lower average concentrations of PC in the Superior Lake on 22 June 2020 (median value of 10.98 mg m^{-3}) were due to its fluvial-lacustrine morphology which favors a shorter residence time compared to the Middle and Inferior lakes, in which the reduced current velocity of the waters facilitates the high growth of cyanobacteria that become dominant (for both lakes a median value ~26 mg m^{-3}).

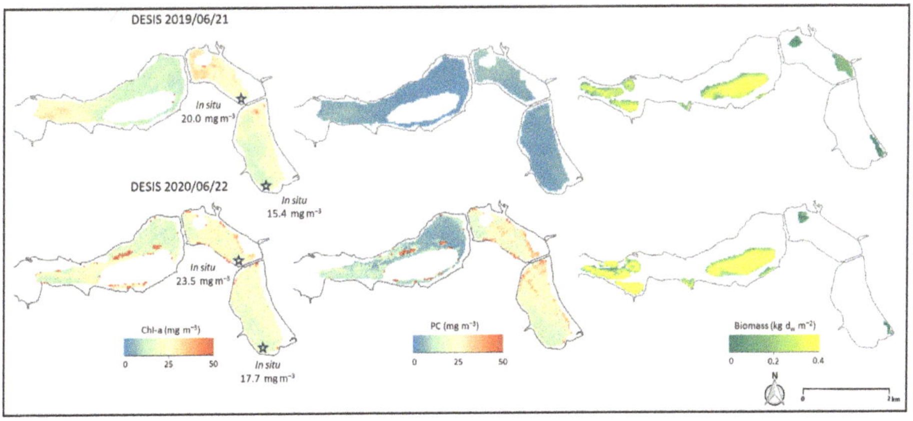

Figure 4. Maps of concentration of phytoplankton pigments (Chl-a and PC) and of floating-leaved and emergent macrophyte biomass obtained from DESIS images acquired on 21 June 2019 and on 22 June 2020. Blue stars (for Chl-a) are the locations of the in situ measurements with fluorimetric probes.

4.2. Lake Mulargia

On the Flumendosa river, located in Sardinia (Italy; 39°38′ N, 9°14′ E; Figure 1) a dam was built creating the Mulargia reservoir. Lake Mulargia has a surface area of 12 km^2 and a volume capacity of 347 hm^3, and it serves as a drinking water source (annually ~100 hm^3) for 700.000 inhabitants including the town of Cagliari and its hinterland as well as another

20 villages nearby. Soil erosion and nutrient loads (mainly phosphorous) from agricultural land can affect lake water quality leading to low transparency (elevated suspended solids) and high conductivity, with episodic algal bloom events resulting from the higher phosphorus loads. For this reservoir water temperature and water quality products, such as Chl-a, turbidity, transparency and a trophic status index, were mapped by means of multispectral satellite data (i.e., Landsat-7 and 8, Sentinel-2) for the period from 2013 to 2018 [17]. As for the previous case study, for these small waterbodies the use of hyperspectral imagery can provide added value as it allows the detection and quantification of algal blooms (such as cyanobacteria bloom) and the origin of suspended material (separation between the organic and inorganic) in the water column, supporting a warning system that is of crucial importance for a reservoir dedicated to drinking water supply.

The water quality maps retrieved from PRISMA (Figure 5, top) and DESIS (Figure 5, bottom), by means of the BOMBER code and related calibration (see Table 2 and Figure 3) describe the Chl-a, TSM and Inorganic Suspended Materials (Inorganic SM) concentrations (only produced with DESIS) with a range between 0 and 10 mg Chl-a m^{-3} or g TSM (Inorganic SM) m^{-3}, and CDOM concentrations within a range from 0 to 0.7 m^{-1}. All the parameters were higher in August compared to the July situation, and looking at the spatial distribution of water quality parameters, the highest concentrations (median value: Chl-a 4.51 and 6.29 mg m^{-3}, CDOM 0.16 and 0.47 m^{-1} and TSM 2.35 and 4.32 g m^{-3}, in July and August, respectively) were found in the eastern portion of the lake probably due to lower water circulation. This may result from the fact that a tributary is located in the western portion of the lake and the dam is in the south. Another area was the Chl-a concentration was high was the western portion of the reservoir close to the tributary inflow, probably due to nutrient enriched waters that favor phytoplankton blooms. The amount of yellow substances, indicated by CDOM, were significantly higher in August compared to the previous image in July (+70%). The inorganic component of the suspended material was about 66% of the total suspend matter in the map produced from the mid-August DESIS image. These findings are comparable to the results reported in Bresciani et al. [17].

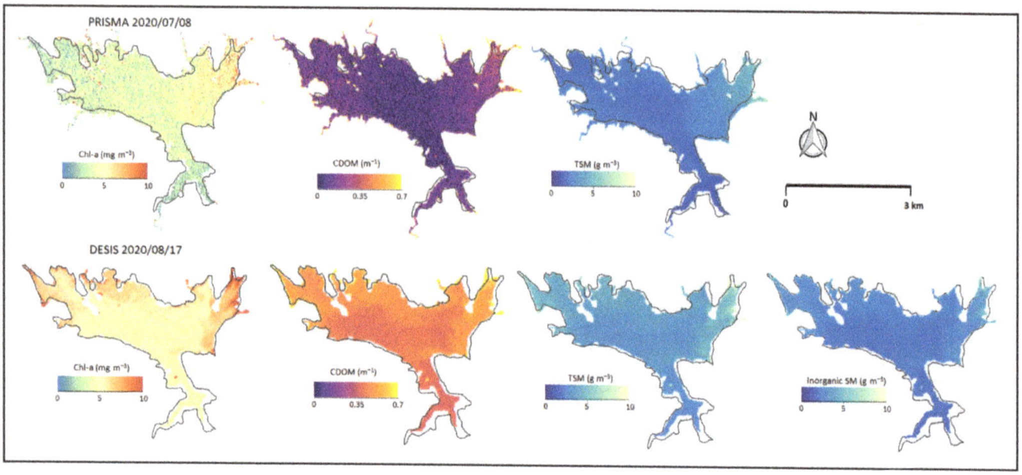

Figure 5. (**Top**) Maps of concentration of Chl-a, CDOM and TSM obtained from PRISMA image acquired on 8 July 2020. (**Bottom**) Maps of concentration of Chl-a, CDOM, TSM and Inorganic SM obtained from DESIS image acquired on 17 August 2020.

4.3. *Lake Trasimeno*

Lake Trasimeno is located in Central Italy (43°08′ N; 12°06′ E; surface area 120.5 km^2; Figure 1) and is characterized by shallow (maximum depth 6.3 m) and turbid (Secchi disk 1.1 m, and SPM 10 g m^{-3} for the period 2002–2008) waters. Precipitation is the main driver of lake water level as the lake is only fed by small watercourses and is essentially endorheic, having no natural outlet [81]. The water column is unstratified due to wind action which also favors sediment resuspension [81]. The lake's trophic status is eutrophic with Chl-a concentration up to 90 mg m^{-3} [82]. The primary producers are phytoplankton (e.g., chlorophytes and dinoflagellates) including cyanobacteria (e.g., *Cylindrospermopsis raciborskii*, *Planktothrix agardhii*), and emergent and submerged macrophytes located in littoral zones and in an open bay in the south-east portion of the lake (Oasi La Valle), respectively [83,84]. Agriculture and animal farming are the main pressures in the watershed, while fishing, tourism and recreational activities exert pressure directly on the lake. Lake Trasimeno is in fact an area of exceptional value for flora and fauna richness and for species biodiversity and it is part of a Natural Regional Park and of two Natura 2000 sites (IT5210018 and IT5210070). It is important in such kind of waterbodies to characterize the diurnal variation of phytoplankton abundance, by means of a proxy such as Chl-a concentration, because algal blooms can grow and dissipate within a few days [85]. In addition, the seasonal phytoplankton phenology and distribution are influenced by this short-term variability [86]. Hyperspectral sensors allow the retrieval of Chl-a and secondary pigments concentration together with other water quality parameters, and the use of continuous in situ spectroradiometers allow the capture of hourly variation of phytoplankton abundance (e.g., [87]). Merging high-frequency data with satellite observations allows one to go beyond the limited spatial coverage of in situ sampling [88], and provide spatial and temporal information for water monitoring and management [89]. Several studies using remote sensing techniques have been performed in this case study area. For example, the physical parameter surface water temperature and Chl-a concentration were both retrieved by satellite data (MODIS and MERIS) and they were positively correlated in the period from the spring to the end of summer (2005–2008) when biological activity was at its highest [90]. Recently, a study on Lake Trasimeno analyzed the integration of detailed spatial information from new generation remote sensing satellites (Sentinel 3-OLCI and Sentinel 2-MSI) with high frequency observations from in situ optical sensors (WISPstation) to improve water quality monitoring and identify drivers of Chl-a dynamics [84]. Giardino et al. [63] used a hyperspectral MIVIS image acquired from an airborne sensor to provide simultaneous maps of the concentration of the optically active parameters, water column heights and benthic substrate types. Recently Niroumand-Jadidi et al. [91] retrieved water quality maps from PRISMA images and the results suggested the high potential of PRISMA imagery in mapping water quality parameters in Lake Trasimeno.

For the estimation of the concentrations of Chl-a and TSM, the BOMBER code was still adopted with related site-specific parameters (see Table 2 and Figure 3), and as also previously applied to airborne hyperspectral MIVIS imagery. For this specific example the products were evaluated based on five match-ups of hyperspectral products (3 × 3 pixel boxes) and in situ data: for Chl-a and TSM with values of root mean square error of 11.1% and 15.8%, respectively, and mean absolute error of 0.94 mg m^{-3} and 1.58 g m^{-3}, respectively.

In Figure 6 is shown a series of maps retrieved in different seasons, which highlights the potential of using hyperspectral images for temporal analysis. In fact, the combination of different sensors can improve the temporal analysis useful to understand, for example, phytoplankton dynamics. Looking at the evolution of Chl-a concentration, there is a clear seasonal pattern with lower values in late spring (median Chl-a of 1.63 and 2.65 mg m^{-3} on 18 June 2019 and 26 May 2020, respectively) and progressively higher values from summer to late summer (median Chl-a value of 5.45, 8.77 and 31.54 mg m^{-3} in July, August and September, respectively). This result is in accordance with late summer phytoplankton blooms reported in previous work [84]. The same temporal pattern was found for the total suspended matter, which can reduce the penetration of light in the water column in some portions of the lake.

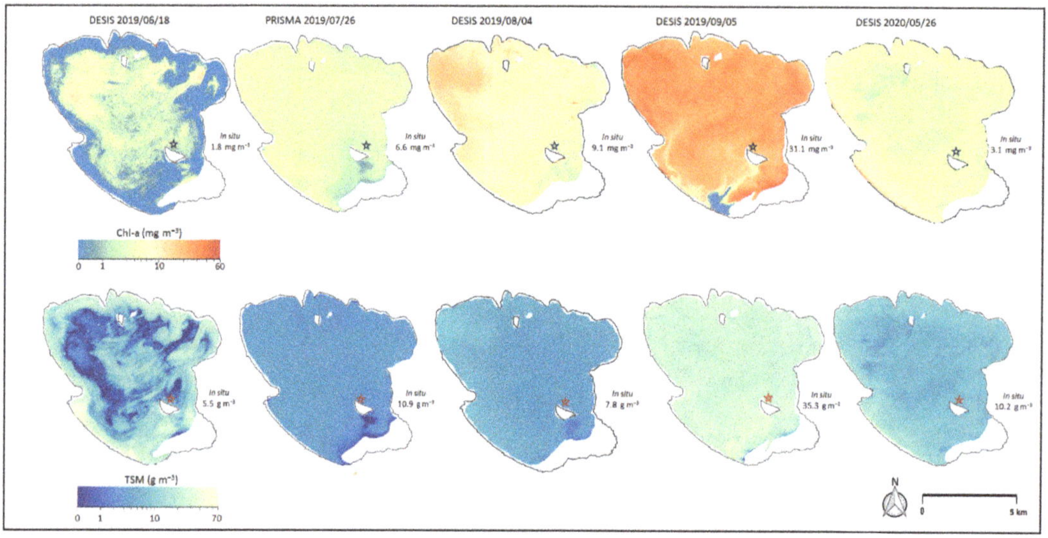

Figure 6. Maps of Chl-a (**top**) and TSM (**bottom**) concentration from DESIS and PRISMA images. Blue (for Chl-a) and red (for TSM) stars are the location of the in situ measurements.

5. Lake Garda

Lake Garda is the largest Italian lake and is located in the subalpine region (45°40′ N, 10°41′ E; surface area 368 km^2, water volume 49 km^3; Figure 1). In 1951 the Salionze dam was completed near the headwaters of the lake emissary (Mincio River), allowing the management of downstream water needs (e.g., crop irrigation) and prevent flooding through water level regulation [92]. As with the other deep subalpine lakes of the same region, the Lake Garda represents a strategic water resources for drinking, agriculture, industry, energy, fishing and tourism [62,93]. Lake Garda is classified as oligo-mesotrophic (mean Chl-a 3 mg m^{-3}; Secchi disk depth 5 m in summer and 15 m in late winter [94]) with two different morphological areas: the largest sub-basin extending from north to southwest area, characterized by the deepest depths, and the south-eastern shallower sub-basin [62]. In the south portion of the lake the littoral zone and the shallower water are characterized by the presence of emergent (e.g., *Phragmites australis*) and submerged macrophytes (e.g., *Chara globularis, Vallisneria spiralis, Lagarosiphon major*) [95]. Retrospective analysis of lake bottom and macrophyte meadows in the littoral zone around the Sirmione Peninsula (Figure 1) was performed for the period from 1997 to 2005 [96], and until 2010 [95] by means of MIVIS hyperspectral airborne images. Such studies allow the identification of the loss of well-structured submerged vegetation and its replacement by sparsely vegetated or unvegetated substrates, with repercussions on the sustainable use and conservation of natural lacustrine resources [95]. Remote sensing techniques have been used for a long time in Lake Garda to examine water parameters (from surface temperature to Chl-a and turbidity). To our knowledge the first study published on the retrieval of Chl-a concentration from multispectral data (Landsat TM images) was published by Brivio et al. [97]. Moreover, several methodological papers followed, with the use for example of the Hyperion image to retrieve Chl-a and other water quality parameters [98], or the use of MODIS and MERIS multispectral sensors to investigate relationships between lake surface temperature and Chl-a concentration in the shallower zone of the south portion of Lake Garda [90]. Applied studies were also performed to support water monitoring and management in accordance with the Water Framework Directive (e.g., [99]), to evaluate the influence of dust transport in phytoplankton growth [100], and to quantify the turbidity due to the discharge of the Adige-Garda spillway on Lake Garda [101].

As an example, from the PRISMA image of the 2020/09/10 different products of water quality and of lake bottom retrieved with BOMBER and related calibration (see Table 2 and Figure 3) are reported in Figure 7. The maps show the Chl-a and TSM concentrations of the south-west portion of the basin, which show also a good agreement with the in-situ data. The Chl-a and TSM concentrations were low on this date confirming the oligo-mesotrophic status of the waters, with higher values up to 3.05 mg Chl-a m^{-3} and 3.33 g TSM m^{-3} in the southern littoral portion of the study area. The map of the Sirmione Peninsula colonized by sparse and dense vegetation is also shown in Figure 7. This estimate of the macrophyte cover of the shallower portion of the Peninsula is in accordance with the results for September 2010 published by Bresciani et al. [95] that reported a reduction of the density of the macrophyte meadows compared to the previous decade. For the water quality products, the Chl-a concentration estimated from the PRISMA image was 1.30 mg m^{-3} with respect to 1.40 mg m^{-3} measured in situ, and the TSM concentration was 0.43 g m^{-3} compared to 0.50 g m^{-3} measured in situ (Figure 7).

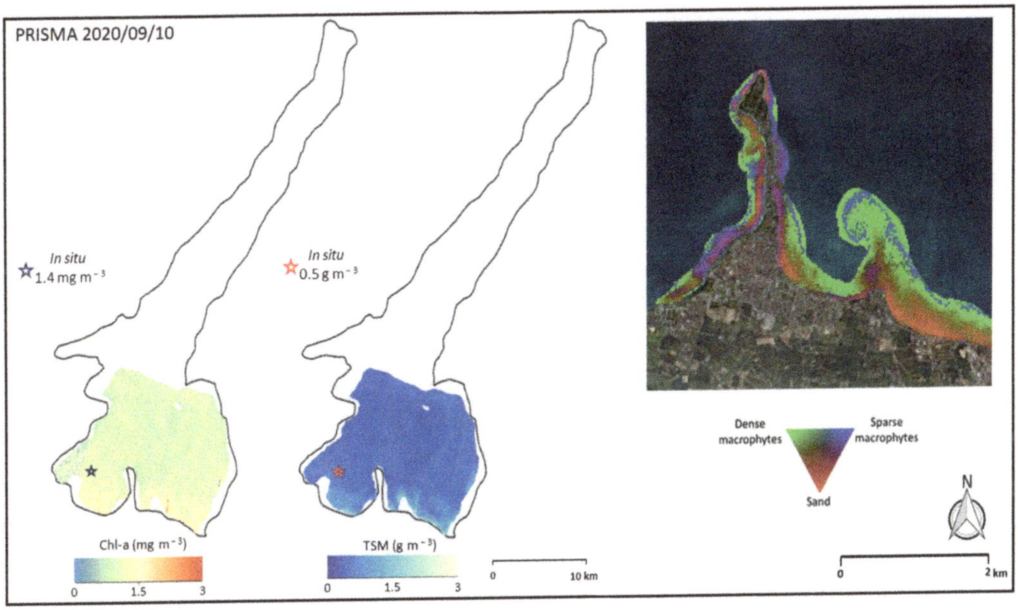

Figure 7. Maps of Chl-a (**left**) and TSM (**center**) concentration, and (**right**) of the bottom coverage of the Sirmione Peninsula in the south portion of Lake Garda (see also Figure 1 for the location of this area) retrieved by PRISMA image acquired on 10 September 2020.

6. Conclusions

Given the state-of-the-art algorithms and the improved EO technology, both multispectral sensors and ocean color radiometers are currently being used for a variety of purposes such as monitoring frameworks from local to regional scales (e.g., [10,39]), while the first set of satellite products for evaluating the effects of climate change on lakes has been also released [14]. Along with such developments, the field of aquatic hyperspectral remote sensing is also advancing rapidly and it is a fact that about 20 reviews have been published in the last four years on the subject [31] and references herein. Several spaceborne imaging sensors have been launched in recent years while in the near future new missions are planned. To focus on satellite sensors which provide a spatial resolution between ~17 and ~33 m, as recommended by Dekker [102], missions such as PRISMA [50], DESIS [103] or GaoFeng-5 [104,105] have to be mentioned as currently orbiting, while EnMAP [106], PRISMA-SG, CHIME and SBG [107] will follow in the next years.

The study presented in this review aimed to demonstrate some examples of water quality mapping based on PRISMA and DESIS imagery of four diverse Italian water systems supporting a variety of ecosystems services, from recreation to drinking water supply. The examples provided include the simultaneous retrieval of Chl-a, TSM, CDOM, the mapping of fractional cover of bottom types, the assessment of phytoplankton pigments as well as of emergent macrophytes biomass. For some of these parameters (e.g., TSM) multispectral sensors are already providing valuable products, while further studies [57] are also using data from the same missions for advancing the estimation of water quality parameters.

While a diversity of case studies was presented on the use of hyperspectral imagery indicating the enormous scientific potential of this approach, the benefits of its use are likely to be case-specific. A favorable cost–benefit analysis often results in situations where there are high numbers of recreational users, health implications or where it is important to manage reputation, for example to maintain confidence in a water supply [108]. Many of the examples presented here certainly represent such high value situations for tourism and recreation (Garda, Trasimeno, Mantua) or water supply (Mulargia). In particular Lake Garda was the second most visited tourist destination in Italy, with 24.6 million overnight stays in 2017 with environmental preservation and natural heritage ranking among six attributes in determining visitor satisfaction [109].

To conclude, the unique value of spaceborne imaging spectroscopy has to be considered in synergy with all classes of sensors as cooperation among missions remains the key approach to study aquatic ecosystems [110], and to monitor water quality by also addressing operational needs associated with policy obligations [111]. To this aim, some challenges in exploiting imaging spectroscopy are open and would require further research efforts. Among the others we would like to mention is the need of expanding hyperspectral validation sites and training the user community. This would enable the integration of datasets and amplify the value of information strands too often left separate.

Author Contributions: Conceptualization, M.B., C.G. and M.P.; data curation, M.B., M.P., A.F., A.P. and S.M.; methodology, M.B., C.G. and M.P.; validation, M.B., A.F., A.P., S.M. and C.G.; supervision, M.B. and C.G.; writing—original draft, M.B., C.G., G.F. and M.P.; writing—review and editing, M.B., C.G., G.F. and M.P. All authors have read and agreed to the published version of the manuscript.

Funding: Funding for satellite processing activities and data analysis was supported by the EU Horizon 2020 programme with the projects PrimeWater (GA No. 870497) and Water-ForCE (GA No. 101004186) and by the ASI-PRISCAV project (GA No. 2019-5-HH.0).

Institutional Review Board Statement: Not applicable.

Informed Consent Statement: Not applicable.

Data Availability Statement: The datasets generated during and/or analyzed during the current study are available from the corresponding author on reasonable request.

Acknowledgments: Part of this work was supported by EU Horizon 2020 programme with the project EOMORES (GA No. 730066) and of ESA-CHIME (4000125506/18/NL/IA). PRISMA and DESIS date were used upon agreement of respective licenses to use. We are very grateful to E. Lopinto from ASI and to N. Pinnel and U. Heiden from DLR for relevant discussion on PRISMA and DESIS missions. We would like to thank the three anonymous reviewers for comments that improved this manuscript.

Conflicts of Interest: The authors declare no conflict of interest.

References

1. Likens, G.E. Lake ecosystem ecology: A global perspective. In *Encyclopedia of Inland Waters*; Academic Press: Cambridge, MA, USA, 2010.
2. Hanjra, M.A.; Qureshi, M.E. Global water crisis and future food security in an era of climate change. *Food Policy* **2010**, *35*, 365–377. [CrossRef]
3. Ormerod, S.J.; Dobson, M.; Hildrew, A.G.; Townsend, C.R. Multiple stressors in freshwater ecosystems. *Freshw. Biol.* **2010**, *55*, 1–4. [CrossRef]

4. Carpenter, S.R.; Stanley, E.H.; Vander Zanden, M.J. State of the World's Freshwater Ecosystems: Physical, Chemical, and Biological Changes. *Annu. Rev. Environ. Resour.* **2011**, *36*, 75–99. [CrossRef]
5. Ho, J.C.; Michalak, A.M.; Pahlevan, N. Widespread global increase in intense lake phytoplankton blooms since the 1980s. *Nature* **2019**, *574*, 667–670. [CrossRef] [PubMed]
6. Visser, P.M.; Verspagen, J.M.; Sandrini, G.; Stal, L.J.; Matthijs, H.C.; Davis, T.W.; Paerl, H.W.; Huisman, J. How rising CO2 and global warming may stimulate harmful cyanobacterial blooms. *Harmful Algae* **2016**, *54*, 145–159. [CrossRef]
7. Carvalho, L.; Mackay, E.B.; Cardoso, A.C.; Baattrup-Pedersen, A.; Birk, S.; Blackstock, K.L.; Borics, G.; Borja, A.; Feld, C.K.; Ferreira, M.T.; et al. Protecting and restoring Europe's waters: An analysis of the future development needs of the water framework directive. *Sci. Total Environ.* **2019**, *658*, 1228–1238. [CrossRef]
8. Council of the European Communities. Directive 2000/60/EC of the European parliament and of the council of 23 October 2000 establishing a framework for community action in the field of water policy. *Off. J. Eur. Communities* **2000**, *327*, 1–72.
9. European Commission. Commission decision of 20 September 2013 establishing pursuant to directive 2000/60/EC of the European parliament and of the council, the values of the member state monitoring system classifications as a result of the intercalibration exercise and repealing decision 2008/915/EC. *Off. J. Eur. Union* **2013**, *266*, 1–47.
10. Papathanasopoulou, E.; Simis, S.; Alikas, K.; Ansper, A.; Anttila, J.; Barillé, A.; Barillé, L.; Brando, V.; Bresciani, M.; Bučas, M.; et al. Satellite-Assisted Monitoring of Water Quality to Support the Implementation of the Water Framework Directive. EOMORES White Paper; European Union's Horizon 2020 Project. Available online: https://zenodo.org/record/3903776#.YUDJ8J0zZPY (accessed on 1 November 2021).
11. Tyler, A.N.; Hunter, P.D.; Spyrakos, E.; Groom, S.; Constantinescu, A.M.; Kitchen, J. Developments in Earth observation for the assessment and monitoring of inland, transitional, coastal and shelf-sea waters. *Sci. Total Environ.* **2016**, *572*, 1307–1321. [CrossRef]
12. Kiefer, I.; Odermatt, D.; Anneville, O.; Wüest, A.; Bouffard, D. Application of remote sensing for the optimization of in-situ sampling for monitoring of phytoplankton abundance in a large lake. *Sci. Total Environ.* **2015**, *527*, 493–506. [CrossRef]
13. Ritchie, J.C.; Zimba, P.V.; Everitt, J.H. Remote sensing techniques to assess water quality. *Photogramm. Eng. Remote Sens.* **2003**, *69*, 695–704. [CrossRef]
14. Crétaux, J.-F.; Merchant, C.J.; Duguay, C.; Simis, S.; Calmettes, B.; Bergé-Nguyen, M.; Wu, Y.; Zhang, D.; Carrea, L.; Liu, X.; et al. ESA Lakes Climate Change Initiative (Lakes_cci): Lake Products, Version 1.1. 2020. Available online: https://catalogue.ceda.ac.uk/uuid/3c324bb4ee394d0d876fe2e1db217378 (accessed on 1 November 2021).
15. Gholizadeh, M.; Melesse, A.; Reddi, L.A. Comprehensive Review on Water Quality Parameters Estimation Using Remote Sensing Techniques. *Sensors* **2016**, *16*, 1298. [CrossRef] [PubMed]
16. Greb, S.; Dekker, A.G.; Binding, C.; Bernard, S.; Brockmann, C.; DiGiacomo, P.; Griffith, D.; Groom, S.; Hestir, E.; Hunter, P.; et al. *Earth Observations in Support of Global Water Quality Monitoring*; International Ocean-Colour Coordinating Group: Dartmouth, NS, Canada, 2018.
17. Bresciani, M.; Giardino, C.; Stroppiana, D.; Dessena, M.A.; Buscarinu, P.; Cabras, L.; Schenk, K.; Heege, T.; Bernet, H.; Bazdanis, G.; et al. Monitoring water quality in two dammed reservoirs from multispectral satellite data. *Eur. J. Remote Sens.* **2019**, *52*, 113–122. [CrossRef]
18. Seegers, B.N.; Werdell, P.J.; Vandermeulen, R.A.; Salls, W.; Stumpf, R.P.; Schaeffer, B.A.; Owens, T.J.; Bailey, S.W.; Scott, J.P.; Loftin, K.A. Satellites for long-term monitoring of inland US lakes: The MERIS time series and application for chlorophyll-a. *Remote Sens. Environ.* **2021**, *266*, 112685. [CrossRef]
19. Dekker, A.G.; Hestir, E.L. *Evaluating the Feasibility of Systematic Inland Water Quality Monitoring with Satellite Remote Sensing*; Commonwealth Scientific and Industrial Research Organization: Canberra, Australia, 2012.
20. Andres, L.; Boateng, K.; Borja-Vega, C.; Thomas, E. A review of in-situ and remote sensing technologies to monitor water and sanitation interventions. *Water* **2018**, *10*, 756. [CrossRef]
21. Sheffield, J.; Wood, E.F.; Pan, M.; Beck, H.; Coccia, G.; Serrat-Capdevila, A.; Verbist, K. Satellite remote sensing for water resources management: Potential for supporting sustainable development in data-poor regions. *Water Resour. Res.* **2018**, *54*, 9724–9758. [CrossRef]
22. Gómez, J.A.D.; Alonso, C.A.; García, A.A. Remote sensing as a tool for monitoring water quality parameters for Mediterranean Lakes of European Union water framework directive (WFD) and as a system of surveillance of cyanobacterial harmful algae blooms (SCyanoHABs). *Environ. Monit. Assess.* **2011**, *181*, 317–334. [CrossRef]
23. Alikas, K.; Kratzer, S. Improved retrieval of Secchi depth for optically-complex waters using remote sensing data. *Ecol. Indic.* **2017**, *77*, 218–227. [CrossRef]
24. Yigit Avdan, Z.; Kaplan, G.; Goncu, S.; Avdan, U. Monitoring the water quality of small water bodies using high-resolution remote sensing data. *ISPRS Int. J. Geo-Inf.* **2019**, *8*, 553. [CrossRef]
25. Claverie, M.; Ju, J.; Masek, J.G.; Dungan, J.L.; Vermote, E.F.; Roger, J.-C.; Skakun, S.V.; Justice, C. The Harmonized Landsat and Sentinel-2 surface reflectance data set. *Remote Sens. Environ.* **2018**, *219*, 145–161. [CrossRef]
26. Mouw, C.B.; Hardman-Mountford, N.J.; Alvain, S.; Bracher, A.; Brewin, R.; Bricaud, A.; Ciotti, A.M.; Devred, E.; Fujiwara, A.; Hirata, T.; et al. A consumer's guide to satellite remote sensing of multiple phytoplankton groups in the global ocean. *Front. Mar. Sci.* **2017**, *4*, 41. [CrossRef]

27. Giardino, C.; Brando, V.E.; Gege, P.; Pinnel, N.; Hochberg, E.; Knaeps, E.; Reusen, I.; Doerffer, R.; Bresciani, M.; Braga, F.; et al. Imaging Spectrometry of Inland and Coastal Waters: State of the Art, Achievements and Perspectives. *Surv. Geophys.* **2019**, *40*, 401–429. [CrossRef]
28. Dey, J.; Vijay, R. A critical and intensive review on assessment of water quality parameters through geospatial techniques. *Environ. Sci. Pollut. Res.* **2021**, *28*, 41612–41626. [CrossRef] [PubMed]
29. Pahlevan, N.; Mangin, A.; Balasubramanian, S.V.; Smith, B.; Alikas, K.; Arai, K.; Barbosa, C.; Bélanger, S.; Binding, C.; Bresciani, M.; et al. ACIX-Aqua: A global assessment of atmospheric correction methods for Landsat-8 and Sentinel-2 over lakes, rivers, and coastal waters. *Remote Sens. Environ.* **2021**, *258*, 112366. [CrossRef]
30. Warren, M.A.; Simis, S.G.; Selmes, N. Complementary water quality observations from high and medium resolution Sentinel sensors by aligning chlorophyll-a and turbidity algorithms. *Remote Sens. Environ.* **2021**, *265*, 112651. [CrossRef] [PubMed]
31. Dierssen, H.M.; Ackleson, S.G.; Joyce, K.; Hestir, E.; Castagna, A.; Lavender, S.J.; McManus, M.A. Living up to the Hype of Hyperspectral Aquatic Remote Sensing: Science, Resources and Outlook. *Front. Environ. Sci.* **2021**, *9*, 134. [CrossRef]
32. Palmer, S.C.; Kutser, T.; Hunter, P.D. Remote sensing of inland waters: Challenges, progress and future directions. *Remote Sens. Environ.* **2015**, *157*, 1–8. [CrossRef]
33. Ogashawara, I.; Mishra, D.R.; Gitelson, A.A. Remote sensing of inland waters: Background and current state-of-the-art. In *Bio-Optical Modeling and Remote Sensing of Inland Waters*; Elsevier: Amsterdam, The Netherlands, 2017; pp. 1–24.
34. Matthews, M.W. A current review of empirical procedures of remote sensing in Inland and near-coastal transitional waters. *Int. J. Remote Sens.* **2011**, *32*, 6855–6899. [CrossRef]
35. Giardino, C.; Bresciani, M.; Stroppiana, D.; Oggioni, A.; Morabito, G. Optical remote sensing of lakes: An overview on Lake Maggiore. *J. Limnol.* **2014**, *73*, 201–214. [CrossRef]
36. Pu, H.; Liu, D.; Qu, J.H.; Sun, D.W. Applications of imaging spectrometry in inland water quality monitoring—A review of recent developments. *Water Air Soil Pollut.* **2017**, *228*, 131. [CrossRef]
37. Odermatt, D.; Gitelson, A.; Brando, V.E.; Schaepman, M. Review of constituent retrieval in optically deep and complex waters from satellite imagery. *Remote Sens. Environ.* **2012**, *118*, 116–126. [CrossRef]
38. Malthus, T.J.; Lehmann, E.; Ho, X.; Botha, E.; Anstee, J. Implementation of a satellite based inland water algal bloom alerting system using analysis ready data. *Remote Sens.* **2019**, *11*, 2954. [CrossRef]
39. Sagan, V.; Peterson, K.T.; Maimaitijiang, M.; Sidike, P.; Sloan, J.; Greeling, B.A.; Maalouf, S.; Adams, C. Monitoring inland water quality using remote sensing: Potential and limitations of spectral indices, bio-optical simulations, machine learning, and cloud computing. *Earth-Sci. Rev.* **2020**, *205*, 103187. [CrossRef]
40. Topp, S.N.; Pavelsky, T.M.; Jensen, D.; Simard, M.; Ross, M.R. Research trends in the use of remote sensing for inland water quality science: Moving towards multidisciplinary applications. *Water* **2020**, *12*, 169. [CrossRef]
41. Turpie, K.R.; Klemas, V.V.; Byrd, K.; Kelly, M.; Jo, Y.H. Prospective HyspIRI global observations of tidal wetlands. *Remote Sens. Environ.* **2015**, *167*, 206–217. [CrossRef]
42. Chander, S.; Pompapathi, V.; Gujrati, A.; Singh, R.P.; Chaplot, N.; Patel, U.D. Growth of invasive aquatic macrophytes over Tapi river. *Int. Arch. Photogramm. Remote Sens. Spat. Inf. Sci.* **2018**, *5*, 829–833. [CrossRef]
43. Gupana, R.S.; Odermatt, D.; Cesana, I.; Giardino, C.; Nedbal, L.; Damm, A. Remote sensing of sun-induced chlorophyll-a fluorescence in inland and coastal waters: Current state and future prospects. *Remote Sens. Environ.* **2021**, *262*, 112482. [CrossRef]
44. Silva, T.S.; Costa, M.P.; Melack, J.M.; Novo, E.M. Remote sensing of aquatic vegetation: Theory and applications. *Environ. Monit. Assess.* **2008**, *140*, 131–145. [CrossRef]
45. Hestir, E.L.; Brando, V.E.; Bresciani, M.; Giardino, C.; Matta, E.; Villa, P.; Dekker, A.G. Measuring freshwater aquatic ecosystems: The need for a hyperspectral global mapping satellite mission. *Remote Sens. Environ.* **2015**, *167*, 181–195. [CrossRef]
46. Minghelli, A.; Vadakke-Chanat, S.; Chami, M.; Guillaume, M.; Migne, E.; Grillas, P.; Boutron, O. Estimation of Bathymetry and Benthic Habitat Composition from Hyperspectral Remote Sensing Data (BIODIVERSITY) Using a Semi-Analytical Approach. *Remote Sens.* **2021**, *13*, 1999. [CrossRef]
47. Jay, S.; Guillaume, M.; Minghelli, A.; Deville, Y.; Chami, M.; Lafrance, B.; Serfaty, V. Hyperspectral remote sensing of shallow waters: Considering environmental noise and bottom intra-class variability for modeling and inversion of water reflectance. *Remote Sens. Environ.* **2017**, *200*, 352–367. [CrossRef]
48. Sharma, L.K.; Naik, R.; Pandey, P.C. Efficacy of hyperspectral data for monitoring and assessment of wetland ecosystem. In *Hyperspectral Remote Sensing*; Elsevier: Amsterdam, The Netherlands, 2020; pp. 221–246.
49. Koponen, S.; Pulliainen, J.; Kallio, K.; Hallikainen, M. Lake water quality classification with airborne hyperspectral spectrometer and simulated MERIS data. *Remote Sens. Environ.* **2002**, *79*, 51–59. [CrossRef]
50. Cogliati, S.; Sarti, F.; Chiarantini, L.; Cosi, M.; Lorusso, R.; Lopinto, E.; Miglietta, F.; Genesio, L.; Guanter, L.; Damm, A.; et al. The PRISMA imaging spectroscopy mission: Overview and first performance analysis. *Remote Sens. Environ.* **2021**, *262*, 112499. [CrossRef]
51. Dekker, A.G.; Phinn, R.; Anstee, J.; Bissett, P.; Brando, V.E.; Casey, B.; Fearns, P.; Hedley, J.; Klonowski, W.; Lee, Z.P.; et al. Intercomparison of shallow water bathymetry, hydro-optics, and benthos mapping techniques in Australian and Caribbean coastal environments. *Limnol. Oceanogr. Methods* **2011**, *9*, 396–425. [CrossRef]
52. Giardino, C.; Candiani, G.; Zilioli, E. Detecting chlorophyll-a in Lake Garda using TOA MERIS radiances. *Photogramm. Eng. Remote Sens.* **2005**, *71*, 1045–1051. [CrossRef]

53. Coppo, P.; Brandani, F.; Faraci, M.; Sarti, F.; Dami, M.; Chiarantini, L.; Ponticelli, B.; Giunti, L.; Fossati, E.; Cosi, M. Leonardo spaceborne infrared payloads for Earth observation: SLSTRs for Copernicus Sentinel 3 and PRISMA hyperspectral camera for PRISMA satellite. *Appl. Opt.* **2020**, *59*, 6888–6901. [CrossRef]
54. Krutz, D.; Müller, R.; Knodt, U.; Günther, B.; Walter, I.; Sebastian, I.; Säuberlich, T.; Reulke, R.; Carmona, E.; Eckardt, A.; et al. The instrument design of the DLR earth sensing imaging spectrometer (DESIS). *Sensors* **2019**, *19*, 1622. [CrossRef] [PubMed]
55. Richter, R.; Schläpfer, D. Atmospheric/Topographic Correction for Satellite Imagery; ATCOR-2/3 User Guide, Version 9.4.0, July 2021. Available online: https://www.rese-apps.com/pdf/atcor3_manual.pdf (accessed on 15 September 2021).
56. Pahlevan, N.; Smith, B.; Binding, C.; Gurlin, D.; Li, L.; Bresciani, M.; Giardino, C. Hyperspectral retrievals of phytoplankton absorption and chlorophyll-a in inland and nearshore coastal waters. *Remote Sens. Environ.* **2021**, *253*, 112200. [CrossRef]
57. O'Shea, R.E.; Pahlevan, N.; Smith, B.; Bresciani, M.; Egerton, T.; Giardino, C.; Li, L.; Moore, T.; Ruiz-Verdu, A.; Ruberg, S.; et al. Advancing cyanobacteria biomass estimation from hyperspectral observations: Demonstrations with HICO and PRISMA imagery. *Remote Sens. Environ.* **2021**, *266*, 112693. [CrossRef]
58. Bracher, A.; Soppa, M.A.; Gege, P.; Losa, S.N.; Silva, B.; Steinmetz, F.; Dröscher, I. Extension of Atmospheric Correction Polymer to Hyperspectral Sensors: Application to HICO and First Results for DESIS Data. In Proceedings of the 2021 IEEE International Geoscience and Remote Sensing Symposium—IGARSS, Brussels, Belgium, 11–16 July 2021; pp. 1237–1240.
59. Lee, Z.; Carder, K.L.; Mobley, C.D.; Steward, R.G.; Patch, J.S. Hypespectral remote sensing for shallow waters: 1. A semianalytical model. *Appl. Opt.* **1998**, *37*, 6329–6338. [CrossRef]
60. Lee, Z.; Carder, K.L.; Chen, R.F.; Peacock, T.G. Properties of the water column and bottom derived from Airborne Visible Infrared Imaging Specrometer (AVIRIS) data. *J. Geophys. Res.* **2001**, *106*, 11639–11651. [CrossRef]
61. Giardino, C.; Candiani, G.; Bresciani, M.; Lee, Z.; Gagliano, S.; Pepe, M. BOMBER: A tool for estimating water quality and bottom properties from remote sensing images. *Comput. Geosci.* **2012**, *45*, 313–318. [CrossRef]
62. Giardino, C.; Bresciani, M.; Cazzaniga, I.; Schenk, K.; Rieger, P.; Braga, F.; Matta, E.; Brando, V.E. Evaluation of multi-resolution satellite sensors for assessing water quality and bottom depth of Lake Garda. *Sensors* **2014**, *14*, 24116–24131. [CrossRef] [PubMed]
63. Giardino, C.; Bresciani, M.; Valentini, E.; Gasperini, L.; Bolpagni, R.; Brando, V.E. Airborne hyperspectral data to assess suspended particulate matter and aquatic vegetation in a shallow and turbid lake. *Remote Sens. Environ.* **2015**, *157*, 48–57. [CrossRef]
64. Babin, M.; Stramski, D.; Ferrari, G.M.; Claustre, H.; Bricaud, A.; Obolensky, G.; Hoepffner, N. Variations in the light absorption coefficients of phytoplankton, nonalgal particles, and dissolved organic matter in coastal waters around Europe. *J. Geophys. Res. Ocean.* **2003**, *108*, 3211. [CrossRef]
65. Blondeau-Patissier, D.; Brando, V.E.; Oubelkheir, K.; Dekker, A.G.; Clementson, L.A.; Daniel, P. Bio-optical variability of the absorption and scattering properties of the Queensland inshore and reef waters, Australia. *J. Geophys. Res. Ocean.* **2009**, *114*, C05003. [CrossRef]
66. Kallio, K. Optical properties of Finnish lakes estimated with simple bio-optical models and water quality monitoring data. *Hydrol. Res.* **2006**, *37*, 183–204. [CrossRef]
67. Ghirardi, N.; Bolpagni, R.; Bresciani, M.; Valerio, G.; Pilotti, M.; Giardino, C. Spatiotemporal dynamics of submerged aquatic vegetation in a deep lake from Sentinel-2 data. *Water* **2019**, *11*, 563. [CrossRef]
68. Villa, P.; Pinardi, M.; Tóth, V.R.; Hunter, P.D.; Bolpagni, R.; Bresciani, M. Remote sensing of macrophyte morphological traits: Implications for the management of shallow lakes. *J. Limnol.* **2017**, *76*, 109–126. [CrossRef]
69. Pinardi, M.; Fenocchi, A.; Giardino, C.; Sibilla, S.; Bartoli, M.; Bresciani, M. Assessing potential algal blooms in shallow fluvial lake by combining hydrodynamic modelling and remote-sensed images. *Water* **2015**, *7*, 1921–1942. [CrossRef]
70. Pinardi, M.; Soana, E.; Laini, A.; Bresciani, M.; Bartoli, M. Soil system budgets of N, Si and P in an agricultural irrigated watershed: Surplus, differential export and underlying mechanisms. *Biogeochemistry* **2018**, *140*, 175–197. [CrossRef]
71. Pinardi, M.; Soana, E.; Bresciani, M.; Villa, P.; Bartoli, M. Upscaling nitrogen removal processes in fluvial wetlands and irrigation canals in a patchy agricultural watershed. *Wetl. Ecol. Manag.* **2020**, *28*, 297–313. [CrossRef]
72. Bolpagni, R.; Bresciani, M.; Laini, A.; Pinardi, M.; Matta, E.; Ampe, E.M.; Giardino, C.; Viaroli, P.; Bartoli, M. Remote sensing of phytoplankton-macrophyte coexistence in shallow hypereutrophic fluvial lakes. *Hydrobiologia* **2014**, *737*, 67–76. [CrossRef]
73. Bresciani, M.; Rossini, M.; Morabito, G.; Matta, E.; Pinardi, M.; Cogliati, S.; Julitta, T.; Colombo, R.; Braga, F.; Giardino, C. Analysis of within-and between-day chlorophyll-a dynamics in Mantua Superior Lake, with a continuous spectroradiometric measurement. *Mar. Freshw. Res.* **2013**, *64*, 303–316. [CrossRef]
74. Pinardi, M.; Bresciani, M.; Villa, P.; Cazzaniga, I.; Laini, A.; Tóth, V.; Fadel, A.; Austoni, M.; Lami, A.; Giardino, C. Spatial and temporal dynamics of primary producers in shallow lakes as seen from space: Intra-annual observations from Sentinel-2A. *Limnologica* **2018**, *72*, 32–43. [CrossRef]
75. Pinardi, M.; Bartoli, M.; Longhi, D.; Viaroli, P. Net autotrophy in a fluvial lake: The relative role of phytoplankton and floating-leaved macrophytes. *Aquat. Sci.* **2011**, *73*, 389–403. [CrossRef]
76. Pinardi, M.; Free, G.; Lotto, B.; Ghirardi, N.; Bartoli, M.; Bresciani, M. Exploiting high frequency monitoring and satellite imagery for assessing chlorophyll-a dynamics in a shallow eutrophic lake. *J. Limnol.* **2021**, *80*. [CrossRef]
77. Bresciani, M.; Giardino, C.; Lauceri, R.; Matta, E.; Cazzaniga, I.; Pinardi, M.; Lami, A.; Austoni, M.; Viaggiu, E.; Congestri, R.; et al. Earth observation for monitoring and mapping of cyanobacteria blooms. Case studies on five Italian lakes. *J. Limnol.* **2017**, *76*, 127–139. [CrossRef]

78. Villa, P.; Bresciani, M.; Bolpagni, R.; Pinardi, M.; Giardino, C. A rule-based approach for mapping macrophyte communities using multi-temporal aquatic vegetation indices. *Remote Sens. Environ.* **2015**, *171*, 218–233. [CrossRef]
79. Villa, P.; Pinardi, M.; Bolpagni, R.; Gillier, J.M.; Zinke, P.; Nedelcuţ, F.; Bresciani, M. Assessing macrophyte seasonal dynamics using dense time series of medium resolution satellite data. *Remote Sens. Environ.* **2018**, *216*, 230–244. [CrossRef]
80. Bresciani, M.; Giardino, C.; Longhi, D.; Pinardi, M.; Bartoli, M.; Vascellari, M. Imaging spectrometry of productive inland waters. Application to the lakes of Mantua. *Ital. J. Remote Sens.* **2009**, *41*, 147–156. [CrossRef]
81. Ludovisi, A.; Gaino, E. Meteorological and water quality changes in Lake Trasimeno (Umbria, Italy) during the last fifty years. *J. Limnol.* **2010**, *69*, 174–188. [CrossRef]
82. Giardino, C.; Bresciani, M.; Villa, P.; Martinelli, A. Application of Remote Sensing in Water Resource Management: The Case Study of Lake Trasimeno, Italy. *Water Resour. Manag.* **2010**, *24*, 3885–3899. [CrossRef]
83. Charavgis, F.; Cingolani, A.; Di Brizio, M.; Rinaldi, E.; Tozzi, G.; Stranieri, P. *Qualita' Delle Acque Di Balneazione Dei Laghi Umbri, Stagione Balneare 2019*; ARPA: Umbria, Italy, 2020.
84. Bresciani, M.; Pinardi, M.; Free, G.; Luciani, G.; Ghebrehiwot, S.; Laanen, M.; Peters, S.; Della Bella, V.; Padula, R.; Giardino, C. The Use of Multisource Optical Sensors to Study Phytoplankton Spatio-Temporal Variation in a Shallow Turbid Lake. *Water* **2020**, *12*, 284. [CrossRef]
85. Reynolds, C.S. *Ecology of Phytoplankton*; Cambridge University Press: Cambridge, UK, 2006.
86. Woods, T.; Kaz, A.; Giam, X. Phenology in freshwaters: A review and recommendations for future research. *Ecography* **2021**, *44*, 1–14. [CrossRef]
87. Free, G.; Bresciani, M.; Pinardi, M.; Giardino, C.; Alikas, K.; Kangro, K.; Rõõm, E.-I.; Vaičiūtė, D.; Bučas, M.; Tiškus, E.; et al. Detecting Climate Driven Changes in Chlorophyll-a Using High Frequency Monitoring: The Impact of the 2019 European Heatwave in Three Contrasting Aquatic Systems. *Sensors* **2021**, *21*, 6242. [CrossRef]
88. Dörnhöfer, K.; Klinger, P.; Heege, T.; Oppelt, N. Multi-sensor satellite and in situ monitoring of phytoplankton development in a eutrophic-mesotrophic lake. *Sci. Total Environ.* **2018**, *612*, 1200–1214. [CrossRef]
89. Arabi, B.; Salama, M.S.; Pitarch, J.; Verhoef, W. Integration of in-situ and multi-sensor satellite observations for long-term water quality monitoring in coastal areas. *Remote Sens. Environ.* **2020**, *239*, 111632. [CrossRef]
90. Bresciani, M.; Giardino, C.; Boschetti, L. Multi-temporal assessment of bio-physical parameters in lakes Garda and Trasimeno from MODIS and MERIS. *Ital. J. Remote Sens. Riv. Ital. Telerilevamento* **2011**, *43*, 49–62.
91. Niroumand-Jadidi, M.; Bovolo, F.; Bruzzone, L. Water Quality Retrieval from PRISMA Hyperspectral Images: First Experience in a Turbid Lake and Comparison with Sentinel-2. *Remote Sens.* **2020**, *12*, 3984. [CrossRef]
92. Hinegk, L.; Adami, L.; Zolezzi, G.; Tubino, M. Implications of water resources management on the long-term regime of Lake Garda (Italy). *J. Environ. Manag.* **2022**, *301*, 113893. [CrossRef]
93. Salmaso, N.; Mosello, R. Limnological research in the deep southern subalpine lakes: Synthesis, directions and perspectives. *Adv. Oceanogr. Limnol.* **2010**, *1*, 29–66. [CrossRef]
94. Salmaso, N. Long-term phytoplankton community changes in a deep subalpine lake: Responses to nutrient availability and climatic fluctuations. *Freshw. Biol.* **2010**, *55*, 825–846. [CrossRef]
95. Bresciani, M.; Bolpagni, R.; Braga, F.; Oggioni, A.; Giardino, C. Retrospective assessment of macrophytic communities in southern Lake Garda (Italy) from in situ and MIVIS (Multispectral Infrared and Visible Imaging Spectrometer) data. *J. Limnol.* **2012**, *71*, 180–190. [CrossRef]
96. Giardino, C.; Bartoli, M.; Candiani, G.; Bresciani, M.; Pellegrini, L. Recent changes in macrophyte colonisation patterns: An imaging spectrometry-based evaluation of southern Lake Garda (northern Italy). *J. Appl. Remote Sens.* **2007**, *1*, 011509.
97. Brivio, P.A.; Giardino, C.; Zilioli, E. Determination of chlorophyll concentration changes in Lake Garda using an image-based radiative transfer code for Landsat TM images. *Int. J. Remote Sens.* **2001**, *22*, 487–502. [CrossRef]
98. Giardino, C.; Brando, V.E.; Dekker, A.G.; Strömbeck, N.; Candiani, G. Assessment of water quality in Lake Garda (Italy) using Hyperion. *Remote Sens. Environ.* **2007**, *109*, 183–195. [CrossRef]
99. Bresciani, M.; Stroppiana, D.; Odermatt, D.; Morabito, G.; Giardino, C. Assessing remotely sensed chlorophyll-a for the implementation of the water framework directive in European perialpine lakes. *Sci. Total Environ.* **2011**, *409*, 3083–3091. [CrossRef] [PubMed]
100. Di Nicolantonio, W.; Cazzaniga, I.; Cacciari, A.; Bresciani, M.; Giardino, C. Synergy of multispectral and multisensors satellite observations to evaluate desert aerosol transport and impact of dust deposition on inland waters: Study case of Lake Garda. *J. Appl. Remote Sens.* **2015**, *9*, 095980. [CrossRef]
101. Ghirardi, N.; Amadori, M.; Free, G.; Giovannini, L.; Toffolon, M.; Giardino, C.; Bresciani, M. Using remote sensing and numerical modelling to quantify a turbidity discharge event in Lake Garda. *J. Limnol.* **2021**, *80*, 47–52.6p. [CrossRef]
102. Dekker, A.; Committee on Earth Observation Satellites (CEOS). Feasibility Study for an Aquatic Ecosystem Earth Observing System. 2021. Available online: https://ceos.org/ (accessed on 15 September 2021).
103. Müller, R.; Avbelj, J.; Carmona, E.; Gerasch, B.; Graham, L.; Günther, B.; Heiden, U.; Kerr, G.; Knodt, U.; Krutz, D.; et al. The new hyperspectral sensor DESIS on the multi-payload platform MUSES installed on the ISS. *Int. Arch. Photogramm. Remote Sens. Spat. Inf. Sci.* **2016**, *41*, 461–467. [CrossRef]
104. Liu, S.; Marinelli, D.; Bruzzone, L.; Bovolo, F. A review of change detection in multitemporal hyperspectral images: Current techniques, applications, and challenges. *IEEE Geosci. Remote Sens. Mag.* **2019**, *7*, 140–158. [CrossRef]

105. Ren, K.; Sun, W.; Meng, X.; Yang, G.; Du, Q. Fusing China GF-5 Hyperspectral Data with GF-1, GF-2 and Sentinel-2A Multispectral Data: Which Methods Should Be Used? *Remote Sens.* **2020**, *12*, 882. [CrossRef]
106. Bachmann, M.; Alonso, K.; Carmona, E.; Gerasch, B.; Habermeyer, M.; Holzwarth, S.; Krawczyk, H.; Langheinrich, M.; Marshall, D.; Pato, M.; et al. Analysis-Ready Data from Hyperspectral Sensors—The Design of the EnMAP CARD4L-SR Data Product. *Remote Sens.* **2021**, *13*, 4536. [CrossRef]
107. Rast, M.; Nieke, J.; Adams, J.; Isola, C.; Gascon, F. Copernicus Hyperspectral Imaging Mission for the Environment (Chime). In Proceedings of the 2021 IEEE International Geoscience and Remote Sensing Symposium—IGARSS, Brussels, Belgium, 12–16 July 2021; pp. 108–111.
108. Seifert-Dähnn, I.; Furuseth, I.S.; Vondolia, G.K.; Gal, G.; de Eyto, E.; Jennings, E.; Pierson, D. Costs and Benefits of Automated High-Frequency Environmental Monitoring—The Case of Lake Water Management. *J. Environ. Manag.* **2021**, *285*, 112108. [CrossRef]
109. Goffi, G.; Osti, L.; Nava, C.R.; Maurer, O.; Pencarelli, T. Is Preservation the Key to Quality and Tourists' Satisfaction? Evidence from Lake Garda. *Tour. Recreat. Res.* **2021**, *46*, 434–440. [CrossRef]
110. Guanter, L.; Brell, M.; Chan, J.C.W.; Giardino, C.; Gomez-Dans, J.; Mielke, C.; Morsdorf, F.; Segl, K.; Yokoya, N. Synergies of spaceborne imaging spectroscopy with other remote sensing approaches. *Surv. Geophys.* **2019**, *40*, 657–687. [CrossRef]
111. Taramelli, A.; Tornato, A.; Magliozzi, M.L.; Mariani, S.; Valentini, E.; Zavagli, M.; Costantini, M.; Nieke, J.; Adams, J.; Rast, M. An interaction methodology to collect and assess user-driven requirements to define potential opportunities of future hyperspectral imaging sentinel mission. *Remote Sens.* **2020**, *12*, 1286. [CrossRef]

MDPI
St. Alban-Anlage 66
4052 Basel
Switzerland
Tel. +41 61 683 77 34
Fax +41 61 302 89 18
www.mdpi.com

Resources Editorial Office
E-mail: resources@mdpi.com
www.mdpi.com/journal/resources

www.ingramcontent.com/pod-product-compliance
Lightning Source LLC
Chambersburg PA
CBHW040824250926
51850CB00007BA/1254
* 9 7 8 3 0 3 6 5 8 2 1 0 8 *